BTexact COMMUNICATIONS TECHNOLOGY SERIES 5

Broadband applications and the digital home

Other volumes in this series:

Volume 1 **Carrier-scale IP networks: designing and operating Internet networks** P. Willis (Editor)
Volume 2 **Future mobile networks: 3G and beyond** A. Clapton (Editor)
Volume 3 **Voice over IP: systems and solutions** R. Swale (Editor)
Volume 4 **Internet and wireless security** R. Temple and J. Regnault (Editors)

Broadband applications and the digital home

Edited by
John Turnbull
and
Simon Garrett

The Institution of Electrical Engineers

Published by: The Institution of Electrical Engineers, London,
United Kingdom

The Institution of Electrical Engineers,
Michael Faraday House,
Six Hills Way, Stevenage,
Herts. SG1 2AY, United Kingdom
www.iee.org.uk

British Library Cataloguing in Publication Data

Broadband applications and the digital home.–
(BTexact communications technology series; no. 5)
1. Broadband communication systems
I. Turnbull, J. II. Garrett, S. III. Institution of Electrical Engineers
621.3′821

ISBN 0 85296 428 5

Typeset in the UK by Bowne Global Solutions Ltd, Ipswich
Printed in the UK by T J International, Padstow, Cornwall

CONTENTS

PREFACE

It has been said that the move from narrowband to broadband access is the second revolution for the Internet — 'broadband is more bandwidth than you can use'.

Once users have experienced broadband access there is no turning back. A whole new world of applications and services becomes possible. No longer is it the 'world-wide wait'. The speed of response and visual quality enabled by broadband finally allows the Internet to reach its true potential.

In most homes there are several potential consumers/users of both information and content. Yet personal computers are still the prime access device to the broadband network, and only one person can use a personal computer at a time. The access device has become a bottle-neck.

Both the computing and entertainment industries have long been working on technologies for in-home distribution and communications.

But what makes a digital home? Is it wired or wireless? What standards will enable true interoperability? What terminal equipment can be used? What products are on the market? How easy is it to install and use?

And what will people use their broadband access for? What will be the 'killer' broadband application? Will there be an insatiable demand for more and more bandwidth?

This book tries to address these issues by providing an insight into the present and future of the digital home, the applications (and the technologies that enable them), and the users and their motivation to communicate. I hope that you enjoy its contents and that it will fuel both further debate and the increased development of those applications and services that will boost the use of broadband to and within the home.

Mike Reeve

Chief Technology Officer, BTexact Technologies

INTRODUCTION

Over the last few decades there have been many advances in technology in our homes. Electronic technology has been available for almost a century, and from an early stage made its impact in the home with the introduction of broadcast radio. This provided one-way communication into the home, and later the telephone brought two-way communication. Then came the television bringing entertainment and information to the home. These changes revolutionised home life in the second half of the twentieth century, but it was the invention of the transistor and the integrated circuit, which led to digital technology, that dramatically accelerated the change. All of a sudden previously undreamt-of processing power became available — all appliances became 'intelligent'. Moore's Law, defined by Charles Moore of Intel, says that the processing power of silicon integrated circuits doubles every two years (and usually the price halves). By way of comparison, the human brain evolved around ten million times more slowly than the silicon chip.

We now live in a digital age. Digital media such as CDs, DVDs, digital cameras, webcams and camcorders are now commonplace. Around 50% of us use PCs at home. Household appliances have digital processing to provide powerful capabilities and enhanced user control. Household devices such as set-top boxes, CD players and even cameras probably have more computing power than was used in NASA rockets to send men to the moon. Why on earth do we need it?

One of the principal uses to which digital technology is put is *communication*, and we have seen radical changes over the past decade. Cordless phones offer digital clarity, while mobile GSM digital networks provide voice and data. Data over the PSTN enabled a generation to get on-line to explore and benefit from the Internet. Broadband ADSL technology is poised to bring 1 million UK customers to high-speed 'always-on' digital services by the end of 2002, with a further 4 million by 2006. Digital radio, terrestrial and satellite technologies have all been deployed. All these technologies are made possible by powerful digital signal processing and computing.

It is *communication* enabled by digital technology that will change the way we use our homes. For example, the development of broadband to the home offers more than fast Internet; new types of application can now be developed and new services become practical and usable over this fast data link. A virtual connection is maintained continuously, so that incoming and outgoing data can be received and sent at will without the delay of setting up a connection, and this virtual connection is inherently multichannel. This 'always-on' multichannel capability of most broadband connections is almost as important as the high bandwidth it provides. Continuous services for surveillance, maintenance or monitoring can be offered and multiple services can be supported at the same time within the home. Currently, data packets may be lost or delayed on the Internet, limiting the use of the Internet for delivering real-time speech (including telephony), music and video. This too will change.

Broadband systems will, in future, allow a higher grade of service than today's Internet can offer for those applications that require a guaranteed quality. This will enable high-quality TV pictures to be delivered to the home over the Internet in real time. And for the first time, high-quality audio will be delivered in real time. Reliable telephony will be delivered via the Internet. With these capabilities new classes of application are possible. True integrated services are emerging where all communications services are sent over the same pipe and work together to create new synergies. New capabilities are leading to innovative applications.

With the increasing number of intelligent interconnected devices, past experience suggests that communication grows more than linearly with the number of devices; there is a virtuous circle as new application possibilities are created. It's as though, with more people to talk to, there is more reason to talk and communicate. The significance of new interconnected devices should not be underestimated. Originally, telecommunications networks were used mainly to connect telephones (and telex/telegraph machines). When other devices were connected — fax machines, computers, etc — the volume of telecommunications grew almost exponentially.

The Internet offers near-universal connectivity outside the home, but we also need standards for content to enable the applications and services that use this connectivity to operate successfully both outside and inside the home. It is not enough just to be connected to another device over the Internet, you have to speak the same language, and you have to have something useful to say. Basic connectivity standards are, in general, more advanced than the high layer standards needed to allow applications to talk to each other, but 'Web services' standards (principally XML, WSDL, SOAP and UDDI) are now well advanced. This should allow independently written application services to find each other and work together dynamically. This is analogous to the way modular programming allowed similar but static binding of independently developed software application elements.

While there are technology standards for connection to networks outside the home, there is a daunting array of technology choices and standards to achieve communication within the home. Communication within the home is, however, developing, but more slowly. We still need and use removable media such as cassettes or DVDs to view the same content on a TV, PC or games station. Where TVs are connected to devices such as VCRs or DVDs, the connection is usually analogue, even though the connected devices may well be digital. PCs connect to computer peripheral devices (such as printers and modems), but PCs rarely connect to home entertainment devices. Still, today, most household appliances are not connected to anything. This is changing.

We are starting to link our PCs and home entertainment systems. Simple 'Home Automation' systems to control and manage home applications, available for some years as expensive, hard-to-use, fringe market products, are becoming cheaper and more widely used. But it will probably be five years or more before all the devices that could communicate do communicate. There remain many unanswered questions about how this connectivity will develop. For example, how do we:

- share an Internet connection simultaneously among many different devices and services;
- share the output of devices such as DVD players and webcams with other computers and devices;
- access content and applications wherever we want in the home, and from the most convenient device;
- control and monitor other devices in the home — lights, fridges, cookers and other appliances?

The last question has intrigued people for many years; 'home automation' has been talked of by science fiction writers and visionaries, but until recently has been limited largely to a small enthusiast market[1]. Many imaginative ideas have been suggested:

- domestic appliances will call the repairer when they go wrong, saying what has gone wrong, so the repairer can bring the right parts;
- the freezer and larder will monitor food as it is used (by seeing bar codes of items as they are removed), allowing automatic grocery orders;
- the house can be remotely monitored for security, or the baby monitor can be checked while next door with your friends;
- the heating and lighting can be controlled remotely — for example, you can turn on the heating in time to warm the house for your arrival;
- even perhaps ... the home will monitor the use you make of it, and attempt to anticipate your requirements for heat, light, entertainment, food ...

Some suggestions are fanciful — remote control of most devices is pointless unless you are physically present, e.g. why run the bath unless you are there to take it. Other ideas sound good, but there may be no business case; remote fault reporting is probably not justified for devices that very rarely go wrong. And why call out a service engineer automatically if they need to talk to you anyway to arrange access? But there are good reasons to connect more devices in the home than at present, and

[1] In the USA though, there has been, for a decade or more, a sizeable market for simple products based on X-10 and similar technologies.

solutions to these connection problems are now emerging; these are described later in Part Two of this book.

There are other non-technical factors that will influence the growth of connectivity within homes. There are many different applications and appliances from many different industries, and taking a whole-home view is difficult. The different industrial players have different business priorities and motivations, different routes to market, and quite often very different product life cycles. The development of common connection and application standards has been a slow process. It is clear that there is a lack of understanding about what the digital home is or can be. We hope in this book to give the reader some ideas and pictures as to what the home will be in the future.

The following chapters focus on the impact of broadband on the home, the technologies available, and most importantly, the users of broadband applications and services. They are grouped into five parts.

Part One	Delivery to the home — connecting the home to external networks
Part Two	Networking the home — connecting devices and applications within the home
Part Three	Living in the digital home — how people use and relate to technology in the home, and how we should develop technology to meet their needs
Part Four	Applications — how applications are being developed to take advantage of digital technology
Part Five	The future digital home — two futurological views on what's to come

This book endeavours to provide you with an insight into what a digital home will be, what broadband applications and services will be made available, both today and in the future, and what the key technologies are to enable the future of such a digital home.

Simon Garrett
simon.garrett@bt.com

John Turnbull
john.g.turnbull@bt.com

BTexact Technologies

CONTRIBUTORS

C E Adams, Multimedia Terminals, BTexact Technologies, Adastral Park

B Anderson, Principal Researcher, Broadband, BTexact Technologies, Adastral Park

D Ballin, Creative Technologist, BTexact Technologies, Adastral Park

P R Benyon, Advanced Network Computing Concepts, BTexact Technologies, Adastral Park

N J Billington, Access Solution Designer, BTexact Technologies, Adastral Park

P M Bull, Network Computing Applications, BTexact Technologies, Adastral Park

D J Chatting, Future Content Researcher, BTexact Technologies, Adastral Park

J A Clark, Broadband Signalling and Design, BTexact Technologies, Adastral Park

A J H Fidler, Satellite System Technology, BTexact Technologies, Adastral Park

M A Fisher, Business Systems Researcher, BTexact Technologies, Adastral Park

E F France, Formerly Digital Living Researcher, BTexact Technologies, Adastral Park

R I Galbraith, Multimedia Terminals, BTexact Technologies, Adastral Park

C Gale, Principal Researcher, Broadband, BTexact Technologies, Adastral Park

M R Gardner, Principal Researcher, Broadband, BTexact Technologies, Adastral Park

S G E Garrett, Intellectual Property Exploitation, BTexact Technologies, Adastral Park

C Gibbs, Human Sciences Department, Loughborough University

A P Gower, Multimedia Designer, BTexact Technologies, Adastral Park

E Hendriks, Associate Professor, Delft University of Technology, The Netherlands

G Hernandez, Satellite System Technology, BTexact Technologies, Adastral Park

M L R Jones, Formerly Digital Living Researcher, BTexact Technologies, Adastral Park

H V Lacohee, Formerly Customer Behaviour Laboratory, BTexact Technologies, Adastral Park

M Lalovic, Satellite System Technology, BTexact Technologies, Adastral Park

M Lawson, Radical Multimedia Laboratory, BTexact Technologies, Adastral Park

B Lei, Researcher, Delft University of Technology, The Netherlands

R Limb, Network Computing Applications, BTexact Technologies, Adastral Park

M A Lumkin, Radical Multimedia Researcher, BTexact Technologies, Adastral Park

A McWilliam, Formerly Customer Behaviour Laboratory, BTexact Technologies, Adastral Park

K E Nolde, Multimedia Terminals, BTexact Technologies, Adastral Park

J Osborne, Virtual Humans Project, BTexact Technologies, Adastral Park

D Patel, Disruptive Technologies Researcher, BTexact Technologies, Adastral Park

I D Pearson, Futurologist, BTexact Technologies, Adastral Park

T Pell, Satellite Systems Technology, BTexact Technologies, Adastral Park

I G Rose, Satellite Systems Technology, BTexact Technologies, Adastral Park

P A Rout, eBusiness Customer Solutions, BTexact Technologies, Adastral Park

M Russ, Technology Analyst, BTexact Technologies, Adastral Park

J M Thorne, Future Technology Researcher, BTexact Technologies, Adastral Park

K Tracey, Prinicpal Researcher, Broadband, BTexact Technologies, Adastral Park

M Trimby, Concept Designer, BTexact Technologies, Adastral Park

A Tsiaparas, Broadband Network Engineer, BTexact Technologies, Adastral Park

J G Turnbull, Multimedia Terminals, BTexact Technologies, Adastral Park

T van Do, Application Research, Telenor, Norway

L-Q Xu, Principal Researcher, Visual Information, BTexact Technologies, Adastral Park

Part One

DELIVERY TO THE HOME

J G Turnbull

Access technologies are required to deliver the feature-rich multimedia content to the home. The two chapters in this first part describe, between them, the main access technologies of today and provide an interesting prologue to the digital home.

Chapter 1
Broadband Access Technologies

The first chapter is an overview of how to provide residential customers, and small businesses, with access to broadband services. It describes the different access technologies available today, and then goes on to give details of cable, wireless and satellite systems, with particular emphasis on the BT architecture for delivery of IP services over asymmetric digital subscriber line (ADSL). It explains the evolution of this two-wire delivery mechanism towards higher bandwidth hybrid fibre/copper access architectures using APON and VDSL.

The chapter closes by presenting a view of wireless broadband access networks, VDSL and the future. A number of broadband access technologies already exist and will continue to evolve, driven by consumer demand for greater bandwidth and speed at an acceptable cost.

Chapter 2
Satellite — a New Opportunity for Broadband Applications

The second chapter looks, in particular, at satellite delivery of broadband Internet and multimedia content to the edge of the network or end user. Looking at the different delivery topologies, this chapter goes on to appraise the underlying protocol aspects on which these are built.

Satellite delivery is particularly attractive in areas of low population density where it is uneconomic to provide ADSL over cable or the telephone network, or the connection to the local exchange is too long. This solution enables broadband access to be provided quickly to those premises that ADSL cannot reach.

What applications are appropriate to delivery by satellite? The chapter goes on to describe the unique characteristics that make satellite delivery attractive. The high latency makes it best suited for the transmission of broadcast information, and the types of application which do not require interaction are discussed. A satellite system, that by its nature broadcasts its signal over a large area, requires a higher level of security than delivery over a dedicated copper pair; a discussion of firewalls, conditional access and VPN security concludes this chapter.

1

BROADBAND ACCESS TECHNOLOGIES

N J Billington

1.1 Introduction

This chapter provides an overview of how to supply broadband access, i.e. high-speed Internet services, to the population at large. The chapter overviews digital subscriber line (DSL) technology, cable modems, wireless local multipoint distribution systems (LMDSs) and satellite technologies. It will allow readers to understand the terminology and help to explain how to use the technologies in real network configurations.

DSL and cable modems are currently being deployed across the globe. LMDS radio licences have been issued in several countries now and satellite technology has transitioned from a focus on the mobile, narrowband, backhaul and rural markets to being looked at as a viable broadband access system. In addition, an increasing amount of optical fibre is being deployed in access networks. Metropolitan fibre rings, from a range of different carriers, are now being constructed in many of the developed world's largest cities.

The chapter starts by providing a brief overview of some of the key broadband access technologies that are being used to address the residential and SME markets. The focus then switches to describe BT's approach to wired broadband access using ADSL and the potential evolution of the wired access network platform.

1.2 A Comparison of Broadband Access Alternatives

This section gives an overview of the asymmetric digital subscriber line (ADSL), cable modem, LMDS and satellite technologies for broadband access (see Fig 1.1). It highlights the salient features together with the strengths and weaknesses of each.

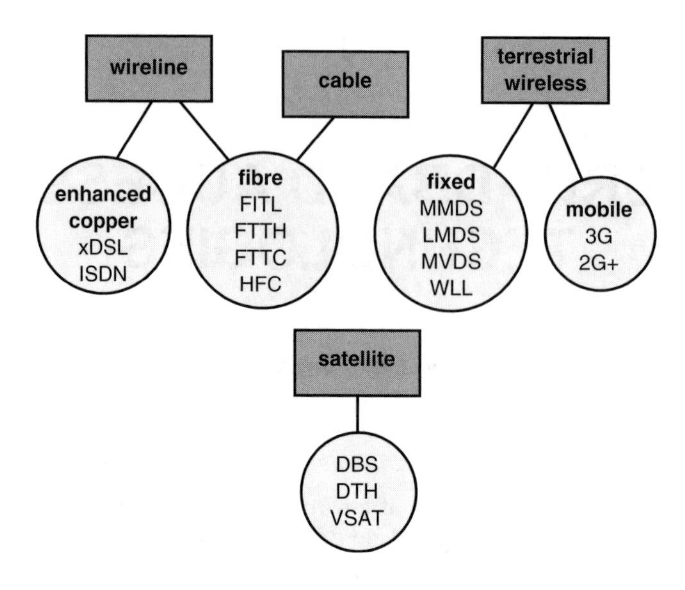

Fig 1.1 Broadband access technologies.

1.2.1 ADSL

ADSL is essentially a very high-speed modem enabling transmission of megabit rates over existing metallic telephone lines. The majority of access lines are copper, but in the past aluminium has been used as an expedient when copper prices made it economic to do so.

Although ADSL works perfectly well over aluminium cable, the extra signal loss of aluminium for a given cable gauge, can result in a lower range than would be expected over copper.

Where the term copper is used it can refer to the use of both types of access cable. Unlike voiceband modems, ADSL can provide the data transmission capability on the line at the same time as the existing analogue telephony service [1] as shown in Fig 1.2.

A major strength of this technology is that there are over 700 million copper telephone lines in the world. Most customers therefore already have the necessary bearer infrastructure connected to their premises. Capital expenditure to create an ADSL line is matched by a potential new revenue stream. Thus ADSL reduces up-front speculative expenditure by deferring customer equipment deployment until the customer requests service, i.e. just-in-time (JIT) provision of service with reduced sensitivity to service take-up.

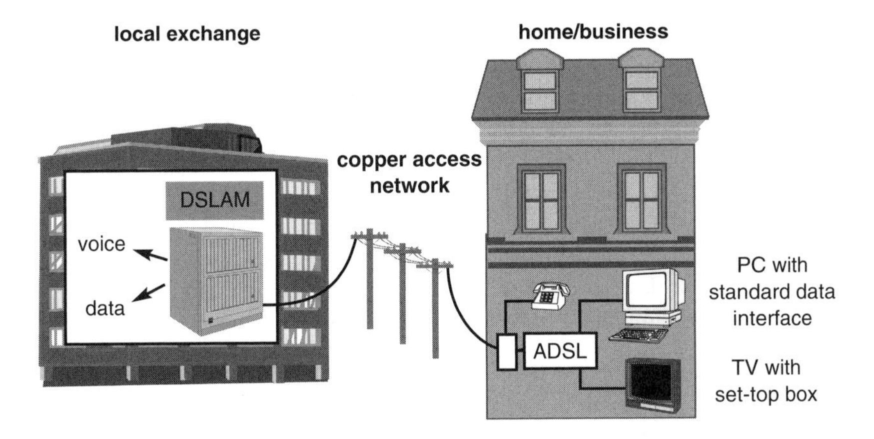

Fig 1.2 Broadband access over ADSL.

An individual copper pair is a point-to-point connection dedicated to a customer, thus making it possible to offer guaranteed performance bounds and an access connection that is more secure than wireless access technologies.

A weakness of DSL technology is that each individual customer's copper line is different (in terms of length). Hence for some service offerings a line qualification process may be required to determine exactly what bit rate the customer's line can support. Not all customers can receive the fastest rates that could vary from 8 Mbit/s, for customers close to an exchange, to 500 kbit/s for a customer on a longer line.

With the advent, in some countries, of local loop unbundling (LLU), operators other than the incumbent can connect their own DSL systems to the copper access network. Crosstalk between different DSL systems in the cable causes noise and interaction between adjacent systems. This has been well controlled in the hitherto single incumbent operator environment.

However, in the new era of LLU, agreement among operators, on which systems can be safely deployed in the network and at what locations, becomes necessary in order to avoid spectral pollution which could compromise network capacity for all users. Policing the deployment of equipment for mutual compatibility and identifying the source of service problems becomes a significant technical and administrative challenge.

1.2.2 Cable Modems over Hybrid Fibre Coax (HFC) Networks

The main advantage of cable networks over other broadband access networks such as ADSL, LMDS or satellite is the large amount of per-customer bandwidth available over the access transmission media. However, much of the downstream bandwidth is used for TV and, in general, the upstream bandwidth is limited both in terms of the spectrum available and by noise. The total bandwidth of such a system can be 860 MHz [2] or even approaching 1 GHz. This allows the use of frequency division multiplexing (FDM) to mix a large number of services and channels on a single broadband pipe into the home or business. This gives operators a great deal of flexibility in packaging service bundles for a 'one-stop shop'. Old cable networks used a large amount of coaxial cable (in a tree-and-branch topology) with the associated need for many amplifiers. Modern networks are HFC with an increasingly high fibre content, perhaps to within a few hundred metres of customers' homes (see Fig 1.3). The combination of deeper fibre penetration in the cable access network combined with modern digital modulation techniques has increased the bandwidth that can be delivered to cable customers.

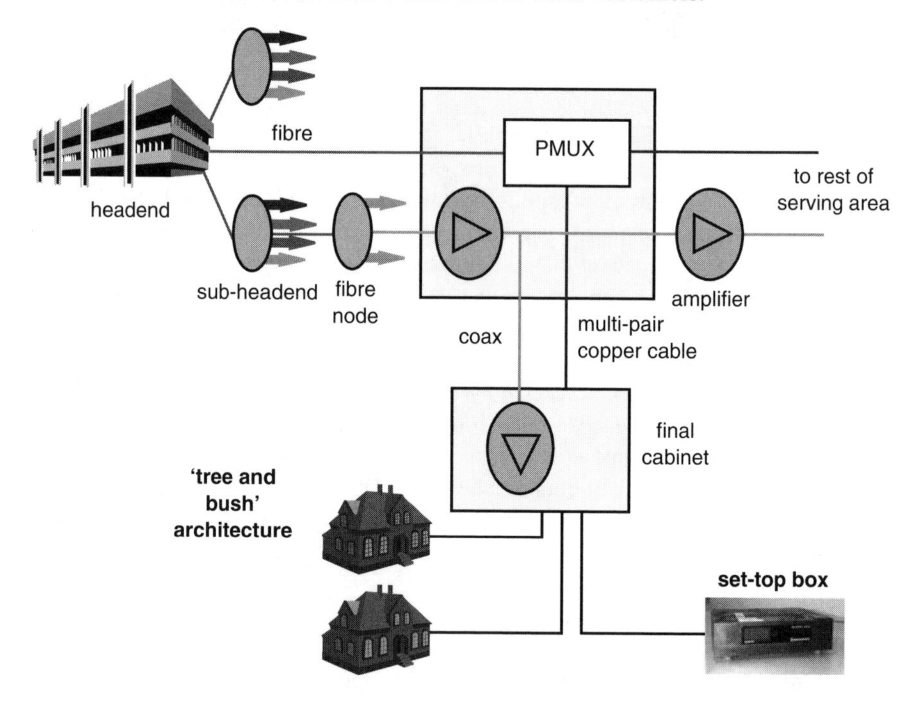

Fig 1.3 A typical modern cable (HFC) network.

Cable modems operate by using unused TV channels in the downstream direction to the customer and the lower frequency reverse path bandwidth for the upstream or return direction. A downstream bit rate of 30-40 Mbit/s per 8-MHz channel slot is possible. This capacity is shared across a number of customers[1]. The total aggregate upstream data rate is in the order of 10 Mbit/s shared across 50-2500 customers depending on the network topology. Most customers' cable modem units will not transmit at much more than 1 Mbit/s upstream. The downstream channel is

[1] Note that no single customer could receive this full rate since the 10Base-T Ethernet card in the PC would limit the burst speed to less than 10 Mbit/s.

continuous, but divided into cells or packets, with addresses in each packet determining who actually receives a particular packet. The upstream channel has a media access control that slots user packets or cells into a single channel.

The cable network is broadcast in nature with the head-end cable modem broadcasting to all customers on the cable network that can receive the 8-MHz channel containing the digitally encoded data. It is rather like a large Ethernet LAN spread across the locality. This has two major impacts on performance. On the positive side this means that there is efficient statistical multiplexing of capacity across active users. An individual customer's modem can burst up to very high rates. The downside is that during the 'busy hour' when many users are surfing the Internet, average throughput will be degraded. Hence performance is unpredictable and careful network dimensioning is needed to give a good experience to the end customer. Quality of service (QoS) will also degrade as Internet users on a network shift from text and simple graphics to high-quality graphics and multimedia, an inevitable trend if the Internet is in any way successful. The 'best-effort' nature of existing cable modems affects their ability to effectively support delay-sensitive services such as videoconferencing. It is possible to use the cable modem management system to impose limits on traffic sent downstream and to throttle back users trying to take more than a fair share of the capacity. Such management software may also be used to introduce a degree of QoS.

The broadcast nature of the cable modem means that each customer modem unit has to identify which IP packets are destined for itself and ignore the rest. This has led to security breaches where hackers have examined other people's traffic in the neighbourhood and also accessed other customers' PCs. These types of problem may be overcome using encryption and firewall techniques if the user is sophisticated enough to implement them and willing to pay for them. The improved DOCSIS 1.1 cable modem standard is designed to overcome such problems. Within Europe, there has not been widespread adoption of one single cable modem standard, giving rise to a potential 'battle' between DVB/DAVIC and EuroDOCSIS standards.

1.2.3 LMDS

LMDS is a radio technology that provides broadband network access to many customers from a single base-station [3], as shown in Fig 1.4. Although LMDS specifically refers to a frequency allocation in the United States, the term is generically used to refer to broadband multiservice radio access systems. These systems are also sometimes known by other terms, such as broadband wireless access (BWA), broadband point-to-multipoint (PMP), or broadband wireless local loop (B-WLL). A major benefit of broadband radio access is that once the base-station is in place, the remaining infrastructure required is only the customer units. Hence this enables provision of high-bandwidth access in a very expedient manner. LMDS can be used to extend the coverage of fibre rings without the need to

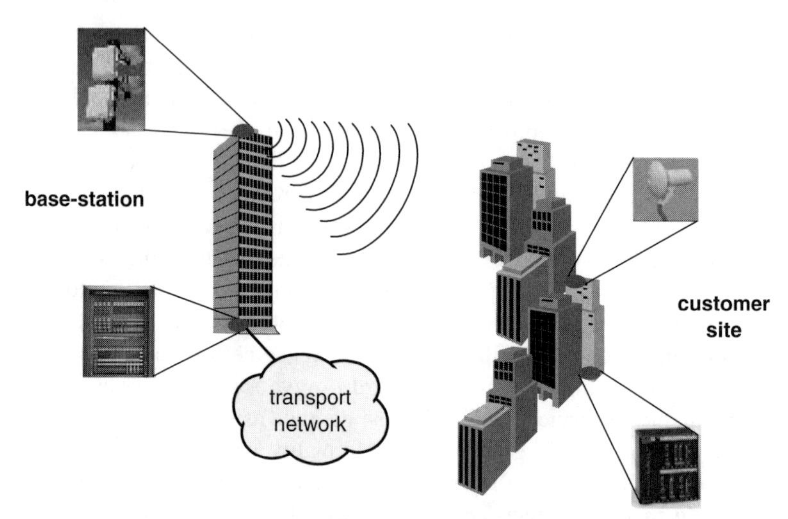

Fig 1.4 Broadband access using LMDS.

negotiate way-leaves and build infrastructure, such as cable ducts (although it is usually necessary to negotiate roof rights for both base-station equipment and customer equipment).

Customers are connected within a 'cell' that typically has a radius of around 2 km from the central base-station. The base-station is usually connected to the remainder of the network using fibre or point-to-point radio. The base-station acts as a hub for the network and provides service to customers who are in direct line of sight and within the cell radius. Generally, the cell area is split into a number of sectors which allows the cell radius, and the capacity offered within a cell, to be increased. LMDS allows flexibility in the way that capacity is allocated to customers, e.g. asymmetric circuits can be allocated in addition to symmetric circuits and quality-of-service options can be offered for data circuits. Many systems also allow bandwidth to be allocated on demand.

The point-to-multipoint topology of an LMDS system is not dissimilar to that of a cable modem in that the base-station sends information to all end customers within a radio sector on a single radio link and the customer premises equipment selects the information intended for it. Typically a base-station may provide a symmetric capacity of 500 Mbit/s to be shared between the users in a cell. The per-customer allocation may be around 7 Mbit/s symmetric, which can be further shared from the customer outstation (via multiple ports) between a number of offices or desks. LMDS can outperform ADSL in terms of upstream data rates. The core network infrastructure used to connect to LMDS base-stations is virtually the same as that used to connect ADSL exchange units (DSLAMs) and hence the technologies can be deployed in a complementary manner.

The radio spectrum allocated to access systems (usually around 10, 26, 28 or 38 GHz) is finite and therefore some degree of planning of the radio resource is required. The limited radio spectrum means that LMDS is better suited to bursty data than real-time high-quality video applications. That having been said, the support for real-time applications is probably better than on cable modems; however, there is not generally enough upstream bandwidth for this support to be comparable to that offered by ADSL. Another point to note is that, as LMDS systems tend to be based on ATM, their QoS support is not so immature.

One of the key limitations with LMDS is the requirement that there is a line-of-sight path between the customer and the base-station. The availability of this path is obviously dependent on the geography of any city where a network is being deployed. To ease the line-of-sight problem, it is usual to mount the base-station on a building with good visibility over the surrounding land (requiring negotiation with landlords). Millimetre wave systems are affected by rain and snow as these cause the signal to be attenuated leading to an increased bit error rate on a link. However, the impact of rain is well understood and can be accounted for by appropriate link margins in the network design. Typically, line-of-sight and interference considerations mean that only 50-70% of the potential customers in the nominal 2-km cell radius around a base-station can be provided with service. However, it should be noted that a few customers out to 4.5 km may still be reachable with good availability (the actual availability is highly system specific).

One of the factors that has the greatest impact on whether an operator can deploy LMDS is the availability of radio spectrum. Since radio spectrum is generally licensed by government agencies, an operator will have to bid for a licence before being able to use the technology. In Europe, the main band of interest is the 26 GHz allocation, although 28 GHz, which is used in the USA, is also being proposed in some countries. The UK is also considering 40 GHz. Due to this lack of global standardisation, the exact frequencies associated with a spectrum allocation may differ from country to country. In some countries, licensees may be obliged to provide coverage of a certain minimum area or certain number of base-stations within a time limit. The cost of the licence could affect the viability of using LMDS in certain geographies. In addition, the customer units are currently priced higher than equivalent DSL or cable modem units. Therefore use of LMDS has been more focused on the medium-size business market and not the residential market. Distribution of broadband to concentrations of customers in clusters (such as shared office buildings and business parks) has been used as a way of sharing the common costs among a number of users to reduce the cost per customer.

1.2.4 Satellite

Satellite is used to deliver broadcast entertainment TV to many consumers around the globe. More recently this ‘downstream’ broadband capability has been combined with narrowband terrestrial return paths (PSTN or ISDN) in order to add an interactive capability to services (see Fig 1.5) [4]. Two-way satellite systems, using very small aperture terminals (VSATs), have also been used by business customers. Assuming access to a satellite transponder is established, satellite systems can then be used to provision service rapidly across a wide geographic footprint to provide access to a large number of distributed customers. Since satellite was originally designed for broadcast, it has an inherent capability to support multicast services and 'push' technology. In an IP network, satellite can be used to update local caches and to deliver streaming IP video. A key strength is the ability to provide ubiquitous access at a consistent quality level across a large area. The cost of deployment is independent of geographic distance within the satellite footprint. Satellite access is less affected by local terrain than many alternative broadband technologies. This makes it an ideal technology to use in rural areas where it has the potential to be more cost effective than the alternative wired or terrestrial wireless systems. Satellite technology has been used for a while to provide business TV services to corporate sites. This can now be combined with IP in order to link this capability into intranets. For example, streaming IP video or software upgrades could be delivered to a site via satellite and then distributed to users' desktop PCs using the existing IP intranet over the site LAN.

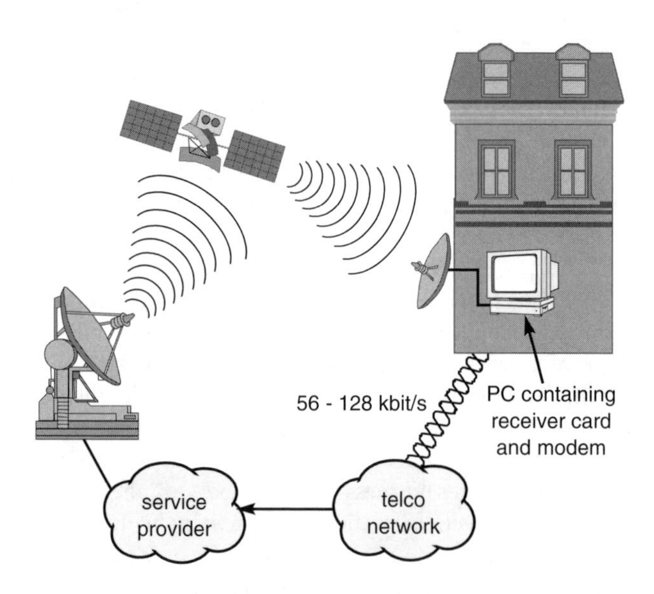

Fig 1.5 Broadband access using satellite.

1.3 Evolution of the Broadband Platform

In the longer term, the BT broadband platform is likely to evolve towards a feature-rich, multiservice delivery platform while at the same time supporting simple, low-cost, high-speed Internet delivery. ADSL network termination equipment (often referred to as the ADSL modem) will continue to be owned by the customer.

Innovation in the customer premises equipment (CPE) arena will place new requirements upon the network to which it is connected, while the desire for cheap, fast Internet connectivity will be addressed by simple ADSL modems and offer corresponding basic services.

1.3.1 Multiservice Delivery over ADSL

Innovation in the CPE arena is leading towards the ADSL modem becoming a portal into the home for a range of different services which is attractive to the customer by facilitating a one-stop-shop approach to communications into their home. However, bundling services to include a mix of video, voice and data communications requires an evolution in the capabilities of the access delivery mechanism. Differing services may place differing and possibly conflicting requirements on the access delivery mechanism. Streamed video is characterised by being tolerant of delay but very intolerant of transmission errors. Voice requires low latency, and is relatively tolerant of errors. Data is generically tolerant of both delay and transmission errors; however, in the context of TCP/IP the delay must be minimised in order to prevent a knock-on effect on data throughput.

ADSL was originally conceived as a delivery mechanism for streamed, compressed video over the telephone line access network. The high tolerance of video to delay meant that ADSL was able to employ physical layer forward error correction and interleaving techniques to mitigate the effects of impulse noise on the integrity of the bit stream. Interleaving, in particular, is critical in the defence against impulse noise events commonly encountered in the metallic access network. However, its downside is that it introduces significant delay. Hence, for ADSL delivery, low delay is synonymous with burst error events, while, if higher delay can be tolerated, virtually error-free transmission can be achieved. Typical round-trip delay figures for ITU-T compliant ADSL are up to 44 ms if interleaving is employed and 5 ms if it is disabled. When considering transmission of multiple, mixed services over an ADSL access system, the link between interleaving depth (i.e. delay) and error rate leads to the logical conclusion that different interleaving depths may be required for different services. ITU-T ADSL standards allow for two different interleave 'paths' through the ADSL link whereby the data stream can be split between a 'fast path' and an 'interleaved path', a concept referred to as 'dual latency'. However, dual latency is optional and is not commonly implemented. Furthermore it is not possible to vary dynamically the share of the available bandwidth between the two paths — hence the bandwidth allocation is usually fixed under management control. This leads to limitations for multiple service delivery where it may be desired to share ADSL line transmission capacity dynamically between more than one service with differing latency requirements. One possible way round this dual latency problem may be to carry all traffic over the same low-latency (non-interleaved) ADSL path and employ appropriate transport layer protocols (e.g. TCP) to ensure error-free delivery for error-sensitive applications.

However, this is not appropriate for real-time services that cannot tolerate any delay due to retransmission and buffering.

Using ADSL to carry multiple simultaneous services also requires some ability in the DSL access multiplexer (DSLAM) to allocate traffic resources according to the required QoS. Many implementations to date have relied on simple connection admission control rules at configuration time to allocate resources. However, where the mix involves some guaranteed (CBR) type traffic and some best-effort (UBR) type, then multiple QoS queues together with appropriate methods of dealing with excess traffic (e.g. EPD, PPD, weighted fair queuing, traffic shaping) will be required. DSLAM implementations are becoming more sophisticated in this area and such features will become a core requirement for reliable multiservice delivery.

An implication of using a single ADSL line to carry more than one service is that the transmission bandwidth requirement is likely to exceed that of the single service case. ADSL, like any other transmission system, is bound by Shannon's capacity theorem which defines the maximum transmission capacity for a given bandwidth and signal-to-noise ratio. In order to be able to guarantee ADSL service on a given line, a pessimistic assumption of crosstalk noise environment is made which holds true in all but the most extreme cases. From this assumption, for any given transmission bandwidth requirement, a corresponding maximum length of line can be defined. Figure 1.6 shows a typical relationship between ADSL transport capacity and line loss for 0.5 mm copper cables in the BT access environment. Thus as the transmission capacity requirement rises, the percentage of the customer base which can be served by ADSL reduces. ADSL is, by definition, an asymmetric transmission system (i.e. capacity in one direction is lower than in the other). The original requirements for ADSL, and the environment in which it has to operate, result in an asymmetry ratio of approximately 10:1 downstream to upstream, which

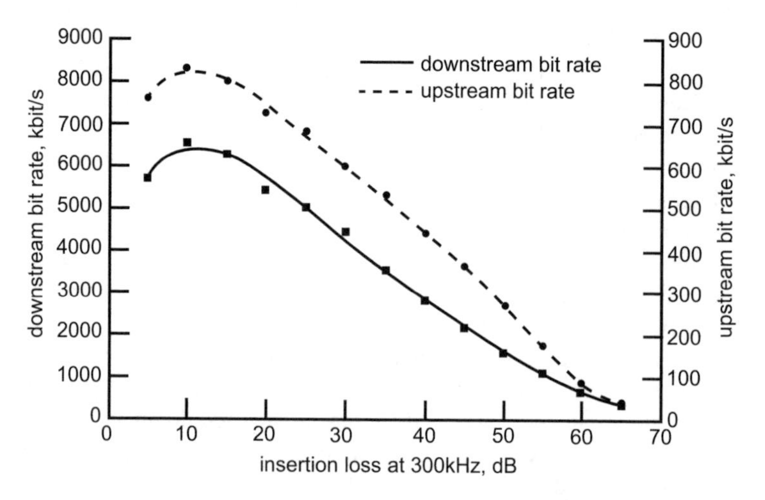

Fig 1.6 Data rate versus insertion loss at 300 kHz (0.5 mm Cu).

is fixed for standards-compliant ADSL equipment. However, such an asymmetry ratio is not ideal for all services.

ADSL is only one example of a family of xDSL transmission systems and others are able to exploit degrees of freedom not available for ADSL. For example SDSL (symmetric DSL) is optimised for symmetric operation over medium line lengths and IDSL (ISDN DSL) is optimised for low symmetric bit rates over long line lengths. However, neither of these alternatives deliver baseband PSTN on the same line. It is now standard practice for manufacturers to produce universal DSLAMs with the capability to support a variety of different line cards in the xDSL family. This gives operators the flexibility to optimise network penetration for all services based on optimal choice of line transmission technology. Trials are currently being undertaken by BT for symmetric variants of its IPStream product, delivered using SDSL technology.

1.3.2 Basic ADSL for Internet Access

At the opposite end of the scale, the requirements of the single PC Internet surfer are considerably simpler and are likely to account for the majority of the market in the near term. This type of user is not reliant (within reason) on a minimum access transmission rate, but delay is a key parameter. The interaction between the TCP window mechanism and the delay/bandwidth product of the communications link can limit the speed of data transmission if there is significant latency. This leads to operation of ADSL in 'fast' mode (i.e. non-interleaved) being popular for Internet connectivity. The resulting susceptibility to occasional error bursts can be taken care of by the TCP retransmission scheme.

The lack of a hard minimum bit rate requirement, means that the ADSL link can be operated on a best-effort basis in this case. Service follows the dial-up modem analogy where the bit rate is not guaranteed. Such an approach significantly increases the potential penetration of ADSL, because those installations on longer lines can continue to work, albeit at a lower bit rate.

The initial approach by BT was to offer three speed variants (576 kbit/s, 1.152 Mbit/s and 2.272 Mbit/s) all with a fixed upstream speed of 288 kbit/s. The fixed speed of the upstream for all products resulted in an equal range limit no matter which of the downstream speeds was selected. More recently the extra reach potential of the 576 kbit/s product has been realised by reconfiguring the upstream to be rate adaptive, with a minimum speed of 64 kbit/s and a maximum of 288 kbit/s. The downstream speed has remained fixed at 576 kbit/s. Apart from product positioning, and the fact that the downstream rate adaptive mode gives little or no advantage in this region, the main reason for this was that the IP RAS (remote access server) needs to shape its ATM traffic to equal or fall below the speed of the ADSL line. If the ADSL line were allowed to operate in rate adaptive mode downstream, then dynamic reconfiguration of the RAS shaper would be required in order that the shaper could be matched to the speed of the ADSL line.

Adoption of rate adaptation in the upstream direction has increased penetration of the IPStream 500 product from ~70% to well above 90% of customers on ADSL enabled exchanges. For most customers, the main flow of data to their PC is downstream from the network. As the upstream data flow is minimal, a reduced rate offered by the rate adaptive mode is an acceptable alternative to not being able to receive ADSL at all.

1.3.3 The ADSL Modem as CPE

There is a strong commercial case for moving away from the traditional view that the customer-end ADSL equipment is network terminating equipment (NTE) which is owned, installed, managed and maintained by the operator. However, from a technical viewpoint this raises a number of challenges:

- the baseband PSTN service must pass through a filter associated with the ADSL NTE in order that the ADSL and PSTN signals can be separated;
- there is still a variation in the quality of ADSL equipment from different manufacturers — the quality of the equipment in this context will determine the maximum length of line over which it will work and hence will affect the ability of the operator to predict if the service is suitable for a given line;
- the higher layer functionality of the ADSL NTE is closely linked to the requirements of the service — innovation in the NTE arena will require corresponding innovation in network services (functions above basic ATM transport are not covered by ITU ADSL standards).

The POTS splitter filter was traditionally included in the ADSL NTE or was closely associated with it. The specification of this filter is critical if good voice quality is to be maintained on the baseband PSTN service. BT's POTS filter requirements, in common with many other European operators, are far more stringent than those of the United States because a complex telephone line terminating impedance is used. There are two main candidate options for installation topologies which enable the customer to install their own ADSL modem — distributed micro-filters (Fig 1.7) and a BT-owned network terminating device containing the low-pass portion of the splitter (Fig 1.8).

BT's initial approach for launch of ADSL-delivered services has been to fit a line termination device containing the low-pass portion of the splitter (Fig 1.8). This is realised as a replacement plug-in front plate for the NTE5 PSTN master socket. There are two sockets on the front of the master socket — one phone socket and one for connection of the ADSL modem. The disadvantage of this approach is that BT has to install the new line termination device and that an extension cable is required from there to the desired location of the modem. On the other hand, the filter was designed to have minimal impact on voice quality while still enabling the customer to connect the ADSL modem themselves. This method is still used for the engineer install option on BT broadband products.

The distributed micro-filter topology has now been adopted as the default method for installation of the majority of ADSL connections today and is preferred by most customers for BT's self-install products (e.g. IPStream Home) (Fig 1.7).

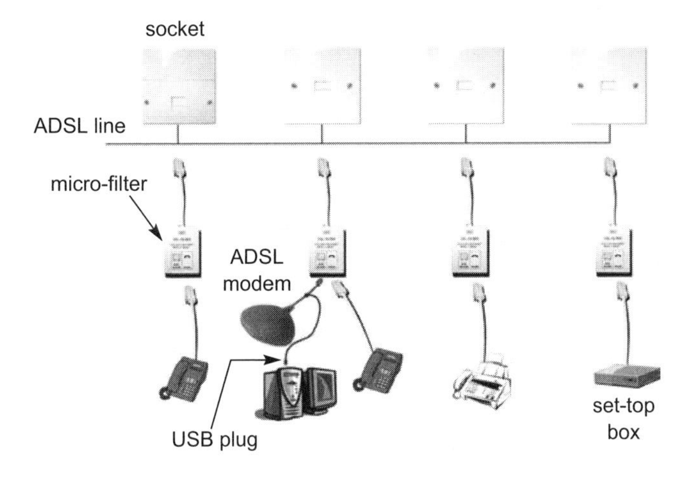

Fig 1.7 Distributed micro-filters.

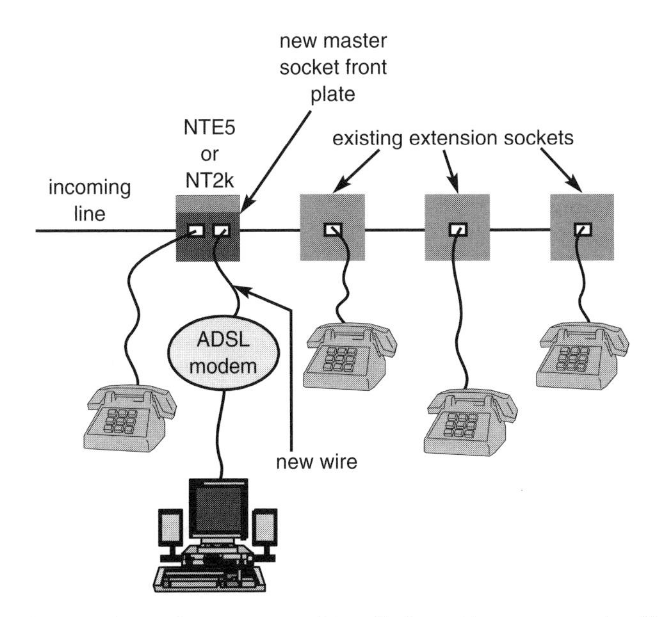

Fig 1.8 Topology of a customer-end installation with a master socket filter.

The major advantage of this approach is that the installation is entirely carried out by the customer at a time convenient to them, and that any telephone socket convenient for the location of the modem can be used to connect it to the network. Early micro-filters suffered from design constraints, when used on complex impedance networks such as the UK PSTN, that often resulted in significant degradation in the quality of the voice service. This issue has now been resolved and micro-filters suitable for the UK network are available today from several suppliers.

The issue of variability of quality in ADSL modem design is being addressed by industry and standards bodies. The DSL Forum has facilitated a series of industry 'plug fests' where vendors can meet and carry out interoperability experiments on neutral ground. It has also published a detailed ADSL interoperability test plan (TR-048) for use by operators and test houses to assess the performance of modems to a common standard. The ITU standards for ADSL include performance requirements for each major region of the world. Compliance with these standards should ensure that baseline performance targets are met.

Higher layer protocol requirements of services carried over ADSL are currently less well defined. Basic ATM and packet transport over ADSL recommendations have been produced by the DSL Forum, and ITU ADSL standards cover the transport of basic ATM cell streams. Innovation in CPE and services will require the definition of significantly more functionality than is covered by existing standards. In the near term it is likely to fall to the operator to publish interface specifications in order to define the requirements of specific services. In the UK, these interface specifications are referred to as supplier information notes (SINs).

1.4 The Evolution of Broadband Wired Access Networks

The lowest common denominator for the evolution of access networks is that they should provide ever 'fatter pipes'. This simplistic view is correct up to a point, but the demands that such networks must have the capability to support many IP and other services with varying characteristics mean that broadband access networks will have to support increased functionality, configurable features and intelligence. The FSAN (Full Service Access Network) initiative was set up in the mid-1990s with the vision of creating a world-wide common requirements specification for the next generation of fixed access networks [5]. Common themes to emerge from FSAN are the requirements for deeper fibre penetration and multiservice support.

1.4.1 Fibre in the Local Loop (FITL)

The FSAN approach to bringing fibre closer to the customer is based on a shared optical fibre feeder network called an APON (ATM passive optical network). An APON optical line termination (OLT) would feed a number of optical network terminations either on the customers' premises or at intermediate points in the access network. The main advantage of APONs over existing point-to-point optical fibre networks is that the interface electronics of an APON head-end are shared between many optical network terminations, thus reducing interface costs per customer. An APON network is broadcast in nature and requires a media access control (MAC) protocol to administer the ingress of upstream traffic on the PON. The FSAN requirements specification envisages the use of ATM to allow a range of traffic types, e.g. real-time, circuit-emulation, IP and quality-of-service levels. The perceived need to support different QoS guarantees and traffic types for many of the connected customers required layer-2 functionality. The challenge to FSAN network operators now is to support emerging end-to-end IP services.

The emergence of IP as the dominant service-carrying protocol has ramifications on the use of FSAN-type FITL networks. The bursty nature of much upstream IP traffic combined with the shared upstream access nature of the PON means that some existing 'static' MAC protocols (Fig 1.9) may inhibit the flow of upstream IP traffic on a contended or overbooked PON. The use of 'dynamic' MAC protocols

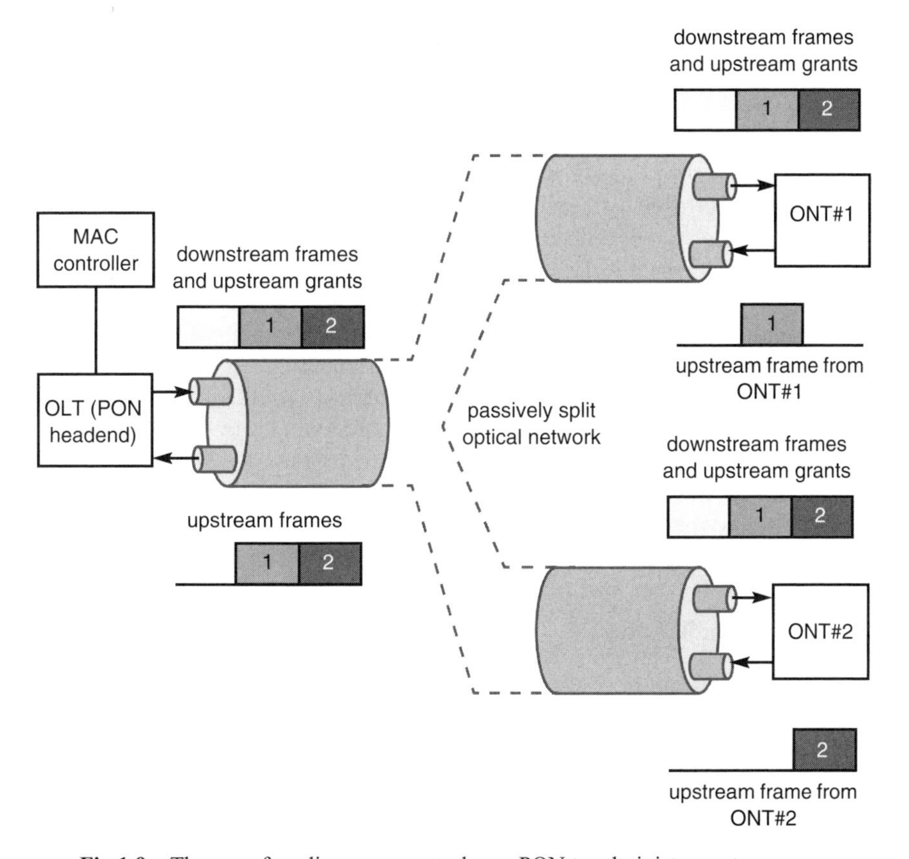

Fig 1.9 The use of media access control on a PON to administer upstream access.

may help to alleviate such problems on a highly contended PON [6]. The need to support more IP traffic, which is much more symmetric in nature than the highly asymmetric video-type applications, may also drive a requirement for PONs with larger upstream capacity. Studies have shown a modest component cost increase per ONU for a 622 Mbit/s symmetric PON.

A longer term approach may be to introduce wavelength division multiplexing (WDM) on a PON to provide additional capacity without having to build new fibre. Wavelengths could then be allocated to a particular ONU which would be equipped with a tunable receiver. A further likely development will be the integration of IP functionality, such as routing and IP QoS into the OLT and the ONU/T. The development of an IP-PON would require QoS issues to be resolved for all services and new MAC protocols to be developed. An optical overlay network based on WDM could provide a way to achieve an IP-PON type functionality upgrade.

1.4.2 VDSL

Fibre in the loop architectures may either connect fibre directly to customers via an optical network termination (ONT) or to an intermediate point in the network, thus 'shortening' the distance to the customer. The FSAN specification envisages the use of the 'fibre to the x' architecture where 'x' is an active node in the access network (Fig 1.10). Very high-speed digital subscriber line (VDSL) would then be used over the final copper drop to the customer. VDSL is fundamentally an extension of ADSL modem technology enabling higher (than ADSL) data rates over copper telephone lines. The main differences between ADSL and VDSL are that the latter uses wider band transmissions at lower power levels, can be operated in symmetric mode, and has a shorter effective range over copper telephone lines. However, both flavours of DSL can potentially be operated in the same cable set.

The large upstream capacity and the use of VDSL in symmetric mode would be particularly suitable to provide IP services that have symmetric bandwidth needs. This would require the bundling of multiple services over VDSL, such as multiple voice lines (VToA or VoIP), Internet access and Web hosting. The requirement to provide a circuit-type functionality for voice, for example, may facilitate the early introduction of IP functionality into an integrated VDSL NT.

1.4.3 Other Trends

The use of wireless interfaces on customer equipment and the next generation of mobile technology will have a profound impact on the functionality of emerging access networks. The need for seamless handover between the fixed and mobile networks, common applications, and multiple concurrent access per terminal for IP services, all have significant ramifications on the architecture of the broadband access platform. With the increasing bandwidth available from broadband wireline

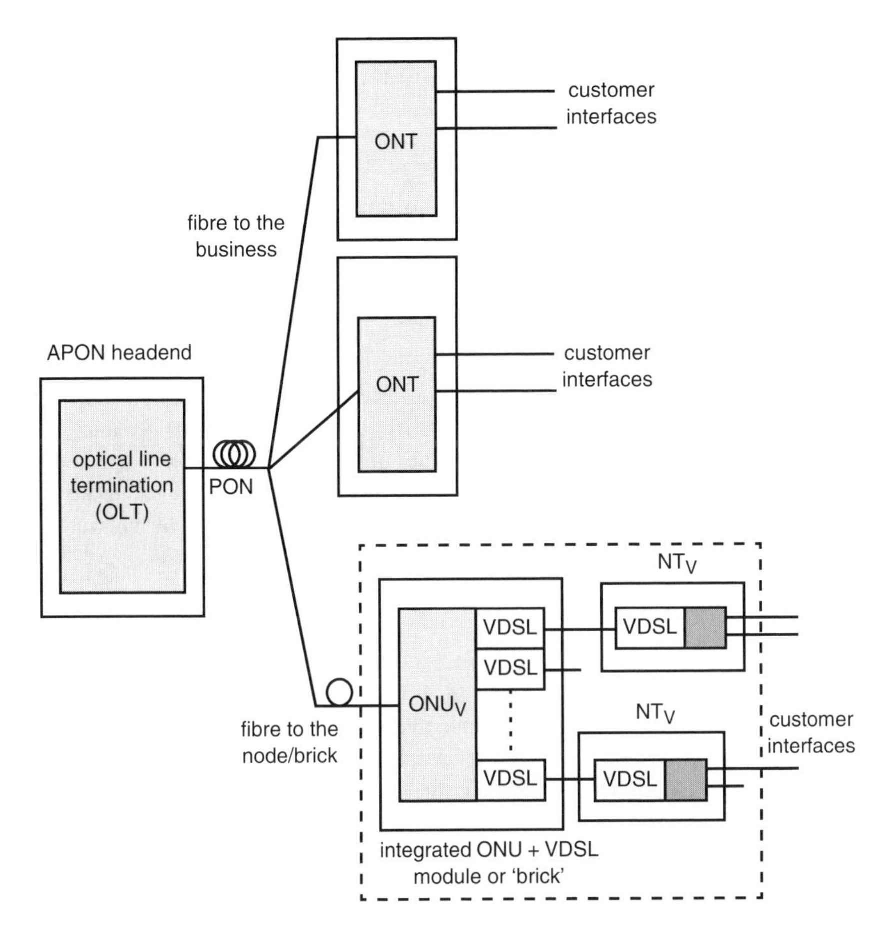

Fig 1.10 The use of a PON to feed compact VDSL modules ('bricks') and provide direct fibre connections to customers.

networks based on APON and VDSL, several services will be bundled. This means that any solution to managing the network and retaining QoS must be scalable.

The low customer density for many rural locations implies that investments in either DSL or optical fibre infrastructure may never earn a return. Investigations into alternative techniques have centred around fixed wireless (for which sufficient radio spectrum is not always available) or the use of satellite delivery with a low-speed return channel provided via the PSTN or ISDN. The other alternative is to make the provision of the exchange DSLAM cheaper. This has been recognised by the manufacturers and new, smaller and cheaper DSLAMs with fewer ports are now available. This will help make previously uneconomic exchanges viable for ADSL access provision.

Incumbent telco and cable company access networks have both begun to have an increasing amount of optical fibre infrastructure over the last few years. This enables homes and businesses to have improved broadband connectivity for future communications needs.

However, for years the 'bread-and-butter' business of telcos has been delivery of POTS using an access network that is largely copper based. The next evolutionary stages led to the use of ISDN and then ADSL to deliver increasingly broadband digital services over the existing copper access infrastructure. Investigations have shown that a fibre-to-the-cabinet architecture could potentially be a cost-effective broadband delivery solution [7] which further improves the access network capability. This high-speed access architecture could play a significant role in bringing multiple channels and higher quality channels, such as for HDTV, to those customers who require them. This will be required in the future as interactive multimedia services proliferate. Each member of a family may wish to access such services simultaneously, e.g. one member watching a movie, another shopping from home and children playing games or getting homework assistance over the network.

1.5 Summary

There are a number of broadband access technologies which are capable of effectively providing access to broadband IP services. These technologies and their associated network platforms will continue to evolve. Broadband access is no longer just a 'fat pipe' — many systems have increased functionality, configurable features and intelligence. Hence issues such as service provisioning, interoperability and CPE auto-configuration are increasingly important challenges that need to be overcome for mass-market viability of more complex services.

The various broadband access technologies have their relative merits and all will continue to prosper as the insatiable customer demand for bandwidth continues. Many operators and service providers will use a combination of these technologies to provide the best approach for their target markets. In the UK, some cable operators already use DSL on their 'siamese' cables which include coaxial cable (used by cable modems) and twisted copper pairs.

ISPs are using both DSL and broadband radio in the USA to increase customer coverage. Radio systems in conjunction with in-building distribution via DSL are being used to target office blocks. Satellite is increasingly being examined to expand broadband coverage to rural areas.

BT's current ADSL platform is well positioned to take advantage of the proliferation of broadband IP services. The platform has the capability to increase its functionality and evolve towards incorporation of APON and VDSL wireline technology.

References

1 Foster, K. T. et al: '*Realising the potential of access networks using DSL*', BT Technol J, **16**(4), pp 34-47 (October 1998).

2 Jewell, S., Patmore, J., Stalley, K. and Mudhur. R,: '*Cable TV technology for local access*', BT Technol J, **16**(4), pp 80-91 (October 1998).

3 Merrett, R. P., Beastall, P. V. E. and Buttery, S. J.: '*Wireless local loo*p', BT Technol J, **16**(4), pp 101-111 (October 1998).

4 Williamson, J.: '*High hopes*', Global Telephony (January 2000).

5 Quayle, J. A., et al: '*Achieving global consensus on the broadband access network — the Full Service Access Network initiative*', BT Technol J, **16**(4), pp 58-70 (October 1998).

6 Hoebeke, R., Ploumen, F. and Venken, K.: '*Performance improvements in ATM PONs in multi-service environments by means of dynamic MAC protocols*', Broadband Access and Network Management, NOC'98 (June 1998).

7 Olshansky, R. and Veeneman, D.: '*Broadband ADSL*', ADSL Forum contribution ADSLForum95-016 (March 1995).

2

SATELLITE — A NEW OPPORTUNITY FOR BROADBAND APPLICATIONS

A J H Fidler, G Hernandez, M Lalovic, T Pell and I G Rose

2.1 Introduction

The satellite communications industry is currently entering a new chapter by migrating from its traditional international private circuit (IPC), international direct dial (IDD) trunk connection, and broadcast TV/satellite news gathering (SNG) roles, into the new and exciting arena of delivering broadband Internet and multimedia content directly to end users [1—3]. Early examples of this migration are the successful operation of several satellite Internet services in North America and Europe. Furthermore, these satellite Internet operators are currently exploiting to some extent the unique ubiquitous coverage characteristics of satellite by being able to deliver broadband content to areas presently outside the reach of conventional terrestrial xDSL networks (see Chapter 1 for more details).

2.2 Satellite Delivery Systems

Current satellite delivery systems can be summarised into three basic architectures:

- one-way satellite systems;
- hybrid satellite systems;
- two-way satellite systems.

All of these different topologies utilise three fundamental mechanisms — unicast, where content is delivered to individual users on a point-to-point basis, broadcast, where content is indiscriminately delivered to all users, and multicast, whereby content is delivered to selected user groups. Content is typically delivered to the end user either by using the internationally recognised digital video broadcasting (DVB) transmission standard or by using proprietary implementations [1].

2.2.1 One-Way Satellite Systems

Looking into the different network topologies in more depth, one-way satellite systems typically comprise just a one-way link through the satellite, which delivers the selected content directly to the end users, utilising unicast, broadcast or multicast connections. This topology is used to support applications or services that do not require any end-user interaction, hence the content delivery is merely scheduled by software hosted either at the content provider premises or at the hub satellite earth station. Examples of broadband applications which use this topology are business TV channels and store-and-forward video content distribution systems. Figure 2.1 shows the typical network topology for a one-way satellite solution.

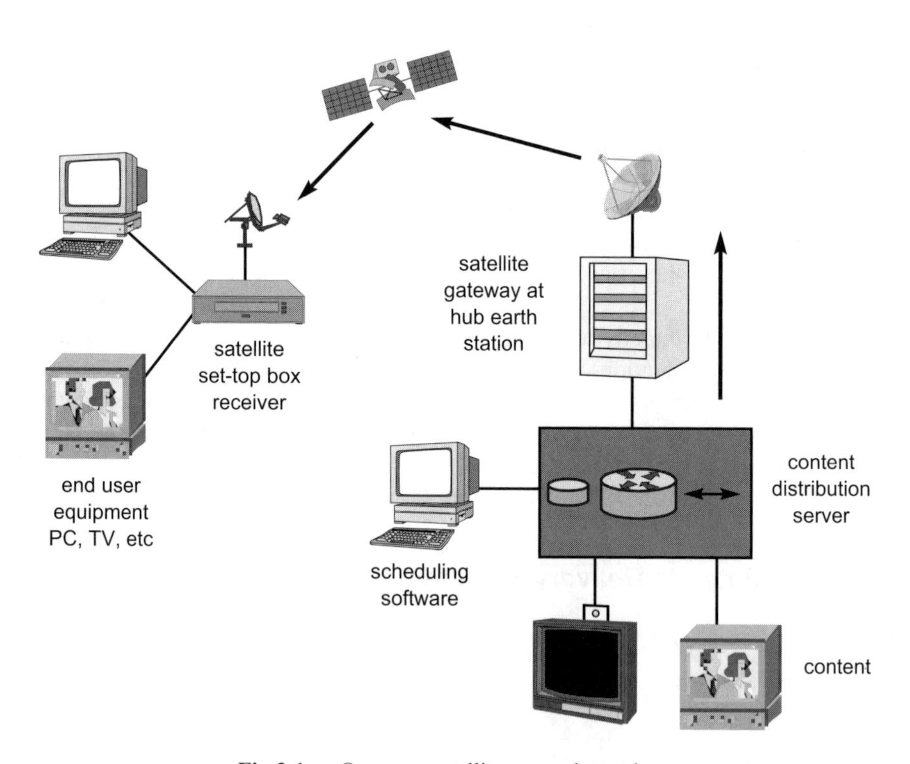

Fig 2.1 One-way satellite network topology.

2.2.2 Hybrid Satellite Systems

Hybrid satellite systems differ from the one-way systems in that they have a terrestrial return path to allow real-time end-user interaction, e.g. to request specific content from the Internet. Typically the terrestrial return path is achieved with either a PSTN dial-up [4], ISDN dial-up or leased-line connection. This topology is inherently asymmetric since it typically consists of a broadband satellite connection in the downstream (Internet-to-user) direction and a terrestrial narrowband connection in the upstream (user-to-Internet) direction.

Hence all downstream content (e.g. Web page text) is delivered via satellite, while all upstream information (e.g. Web page requests and TCP acknowledgements for received data) is sent terrestrially. For optimal TCP/IP performance, care should be taken to correctly dimension the terrestrial return path for the particular application service set required. Unless the TCP stack is optimised, TCP acknowledgements typically require about 20% of received bandwidth, and therefore the bandwidth of the return channel is very important. Examples of applications that use this topology are satellite Internet services and corporate virtual private networks (VPNs). It should be noted that, in addition to the delivery of the usual unicast Internet and data services, this topology can support the bundled delivery of multicast and broadcast services, e.g. electronic newspaper delivery and local Web caching services. Figure 2.2 shows the typical network topology for a

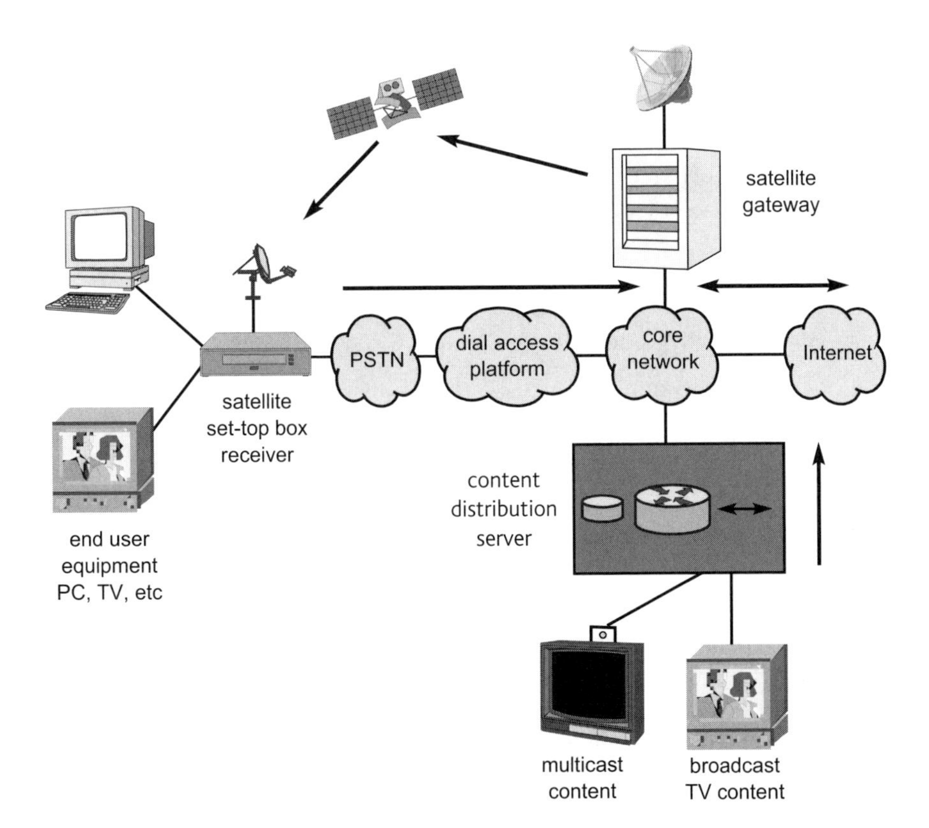

Fig 2.2 Hybrid satellite system.

hybrid satellite solution. Of the several design options available for integrating the terrestrial return path and associated dial access platforms into a hybrid satellite access network, two are described here.

- Direct-connection solution

  This is a solution whereby the Internet service provider (ISP) undertakes the responsibility of routing downstream traffic to the satellite hub earth station via either a direct leased-line or fibre connection, or across the Internet using tunnelling protocols such as generic routing encapsulation (GRE) and point-to-point tunnelling protocol (PPTP). This principle is shown schematically in Fig 2.3.

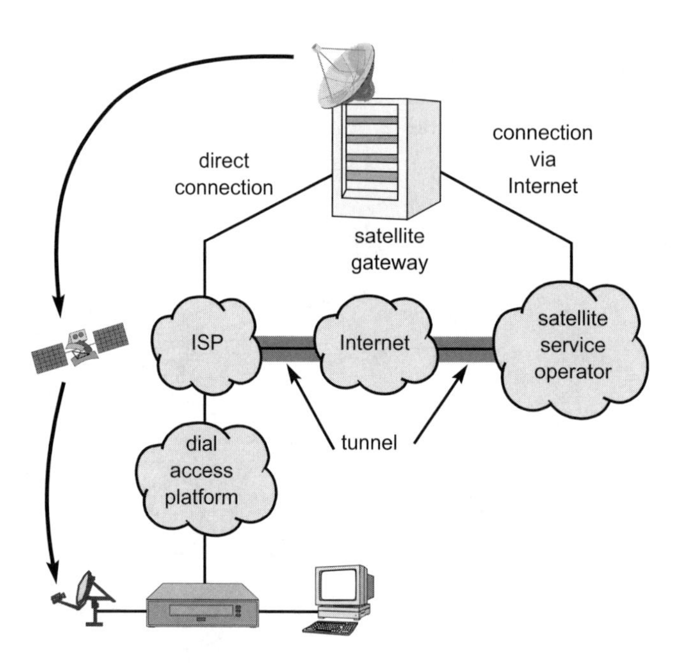

Fig 2.3 Direct connection solution.

- Proxy solution

  A proxy solution is one whereby Internet requests (e.g. HTTP and HTTPS) from end-user PCs are routed to a designated proxy server from the Internet. The network is then configured to route all downstream traffic from this designated proxy server via the satellite hub earth station. With this solution the proxy option within the end user's Internet browser needs to be enabled. Some solutions also require the installation of customised local proxy software that resides on the user's PC. With this approach the end user should be able to gain access to the service using any terrestrial ISP. This principle is shown in Fig 2.4.

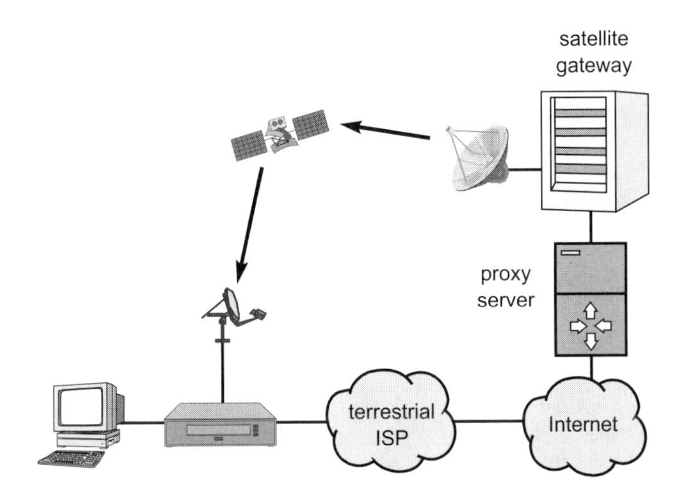

Fig 2.4 Proxy solution.

For simplicity Fig 2.4 assumes collocation of the satellite gateway and proxy server and therefore shows a direct connection between the two. In reality, if these units are not physically collocated, they could be connected via leased line, fibre link or across the Internet.

2.2.3 Two-Way Satellite Systems

In contrast, two-way satellite systems consist of asymmetric satellite paths — a broadband downstream path for the delivery of the actual content and a smaller upstream path for the carriage of the user-to-Internet requests. As with hybrid satellite systems, the upstream path needs to be carefully dimensioned and contended to ensure that it does not limit the overall performance of the service.

The majority of solutions available today use the internationally recognised DVB standard for the downstream satellite carrier. However, there are presently two alternative options for the upstream carrier — the use of the new DVB return channel via satellite (DVB-RCS) open standard or the use of proprietary solutions. Figure 2.5 shows the typical network topology for a two-way satellite solution.

The main applications for two-way satellite solutions are broadband satellite Internet solutions and corporate VPN solutions. In addition to supporting point-to-point Internet applications, two-way solutions can also support the delivery of broadcast and multicast content. There is considerable interest in the delivery of broadband content to a remote single office/home office (SOHO) and small to medium enterprises (SMEs) via two-way satellite systems. For example, several ISPs now offer two-way-based satellite Internet services in the UK.

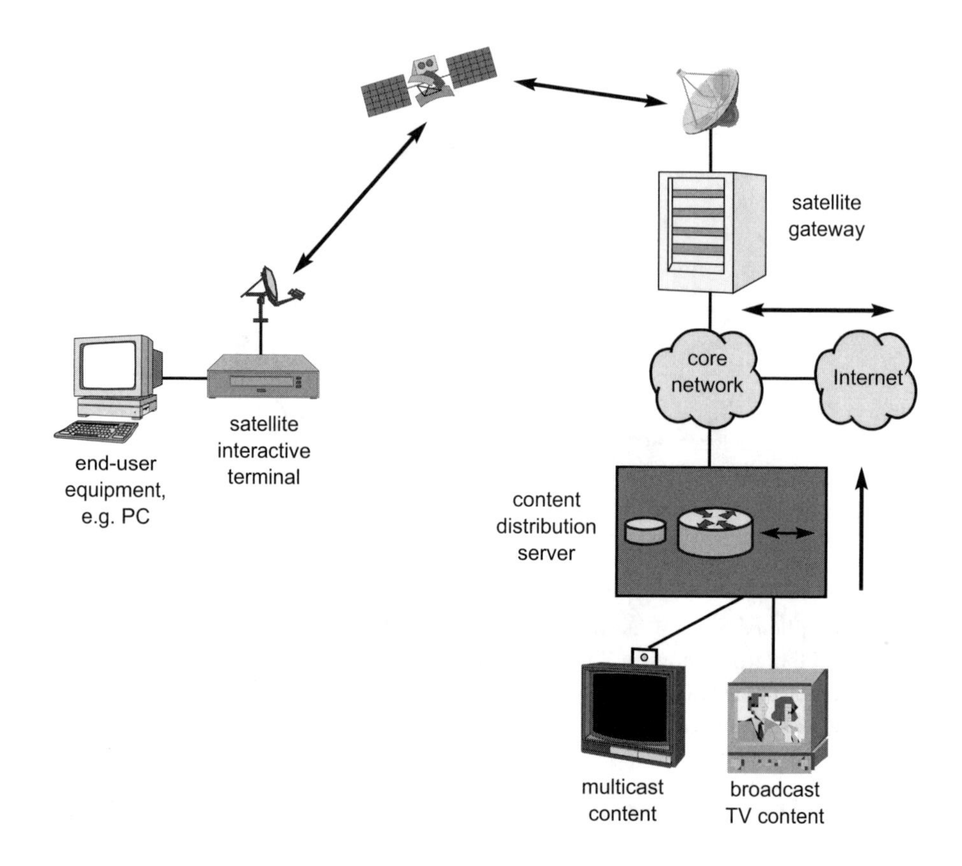

Fig 2.5 Two-way satellite network.

2.2.4 Hybrid Versus Two-Way Satellite Internet Solutions

The main advantages of hybrid satellite Internet solutions are:

- low-cost customer premises equipment (CPE) since these are largely the same as existing mass-market TV receive-only (TVRO) components;
- the ability to self-install satellite antenna and associated indoor CPE.

The main disadvantages with hybrid satellite Internet solutions are often focused around the use of narrowband return paths, for example:

- not having an 'always-on upstream connection' and hence having to dial up for Internet access;
- the call blocking and contention nature of narrowband dial-access platforms;
- the optional cost to ISPs of bundling 'anytime' ports with a satellite Internet offering to emulate an always-on service.

Although hybrid satellite Internet solutions cannot provide an always-on pull Internet service analogous with xDSL offerings, they can provide 'pseudo always-on' in the downstream path in that they can always receive off-line content such as e-mail, pre-fetch files, multicast/broadcast feeds and news ticker feeds.

In contrast, the main advantages of two-way satellite Internet solutions are:

- their ability to offer a true always-on solution, similar to that of terrestrial xDSL solutions;
- their self-contained nature and non-reliance on other telco platforms such as dial access — hence their ability to be rolled out quickly to target areas once the hub earth station is installed and operational.

The main disadvantages of two-way satellite Internet solutions are:

- current high-cost CPE compared to hybrid solutions;
- requirement for the terminal to be licensed by the Radiocommunications Agency to transmit in the UK;
- requirement for all satellite antennas and associated indoor CPE to be professionally installed;
- requirement for larger CPE antennas;
- requirement for additional satellite space segment in the upstream direction and careful dimensioning of this to ensure that, when heavily contended, the overall service performance is not limited.

2.3 Protocol Attributes

Broadband applications rely on a variety of protocols at each layer of the communications stack. This section outlines a number of issues that may have an impact on the performance over satellite networks of protocols in the TCP/IP family.

2.3.1 Transport Layer

The two most widely used transport layer protocols are UDP and TCP. UDP is an unreliable transport mechanism that does not provide any error checking, retransmission or congestion avoidance facilities. UDP therefore works seamlessly over satellite and the only characteristic of satellite networks that may have an influence on the performance of UDP-based applications is the latency of the satellite link and the very rare occurrence of packet loss — with the typical end-to-end network latency of a one-way satellite system being approximately 300 ms[1] and a two-way satellite system approximately 700 ms[1].

[1] These are typical values experienced when pinging IP addresses or Web sites over a satellite access network.

On the other hand, TCP is a reliable transport protocol that provides facilities like error checking, retransmission and congestion avoidance. While TCP works well over satellite links, its performance may in some cases be limited by the latency and bandwidth characteristics of the link. However, there are a number of well-understood mechanisms that can be used to improve the performance of the protocol over satellite networks. RFC 2488 [5] describes a number of performance-enhancing techniques that have been standardised by the IETF.

The techniques recommended in RFC 2488 include:

- the use of large TCP windows to increase the throughput of individual TCP connections;
- the use of fast retransmit/fast recovery algorithms for better congestion handling;
- the use of selective acknowledgements to avoid unnecessary retransmissions in case of lost segments.

In addition to the standard techniques proposed in RFC 2488, there are a large number of alternative TCP enhancements that are currently being investigated. The items being researched have been recently compiled and published as the informational RFC 2760 [6]. These include additional techniques to mitigate the effect of the slow-start algorithm, techniques for speedier recovery from loss situations, TCP/IP header compression techniques, and the use of multiple TCP connections.

Another research area outlined in RFC 2760 aims at improving TCP performance over highly asymmetric networks. This is of particular interest because the topology of most satellite networks used to deliver broadband applications presents a bandwidth asymmetry — a very large data rate in the downstream direction together with a low-speed upstream link.

TCP can tolerate a certain amount of bandwidth asymmetry. However, if the network asymmetry is too large, TCP performance may be reduced, for example through there being insufficient return channel capacity for the TCP acknowledgements (unless the TCP stack is optimised, TCP acknowledgements typically require about 20% of received bandwidth). Although many of the techniques already mentioned can be effectively used to improve this situation, there are several mechanisms (like ACK congestion control and ACK filtering) specifically aimed at reducing the limiting effect that a low-speed upstream link can have on the data flow over the faster downstream link.

An alternative approach to improving TCP performance is the use of performance enhancing proxies (PEPs), a generic term for a device in the path intended to improve performance. PEPs are used to overcome certain link characteristics that have an adverse effect on protocol performance. In theory, they can operate at any layer in the protocol stack, but they are usually found either at the transport or at the application layer. Although several transport and application layer PEPs are currently available, there is no standard implementation.

A survey of PEPs (not only for satellite networks but also for wireless WAN and LAN environments) is available as informational RFC 3135 [7]. For example, one form of PEP is a TCP spoofer that sends timely acknowledgements back to hosts while packets are in transit over the satellite link.

2.3.2 Application Layer

Even though a lot of emphasis is placed on improving the performance of transport layer protocols over satellite links, it is important not to overlook the improvements that can be made to the performance of application layer protocols.

Some application layer protocols are particularly 'chatty' or use inefficient headers. In this case, application-layer PEPs can be effectively used to improve protocol performance over satellite links.

In other cases, performance improvement can be achieved by making an effective use of features already present in the protocol. An example of this approach can be found with HTTP, where the use of persistent connections and request pipelining (both available with HTTP/1.1, but not with HTTP/1.0) can lead to a significantly improved Web-browsing experience.

2.4 Applications

2.4.1 Satellite Application Delivery Characteristics

Satellite provides several unique characteristics that make it very attractive for the delivery of broadband content directly to end users and to the edge of networks:

- wide area coverage — current satellites have the capability to cover areas the size of sub-continents from a single-hub earth station, e.g. Fig 2.6 shows the European footprint of a Eutelsat satellite;
- broadband bandwidth capabilities — satellite transponder payloads can typically support downstream data rates of up to 55 Mbit/s based on 36-MHz broadcast satellites, dependent on choice of modulation scheme used, e.g. QPSK, 8-PSK, 16 QAM;
- distance-independent costs — the wide-area coverage from a satellite means that it costs the same to receive the signal from anywhere within the footprint area;
- fast access — once the hub-satellite earth station and network connection are in place, which is necessary for the first user, more users can be added in the time it takes simply to install the equipment, pending regulatory requirements.

These attributes combine to make satellite a flexible and fast way to connect users. For example, they can be used:

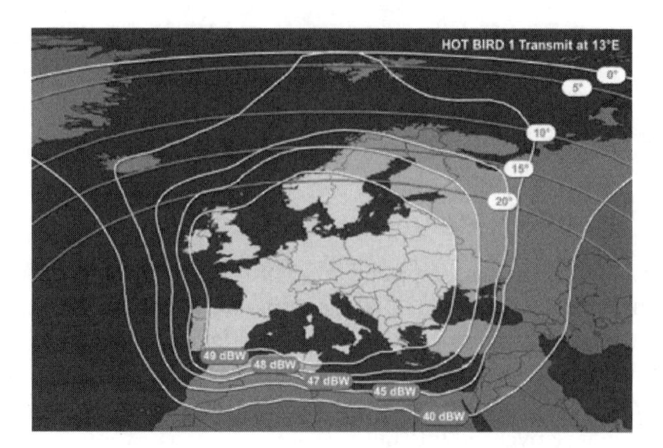

Fig 2.6 Satellite footprint example.
[Courtesy Eutelsat]

- as a strategic tool to bring services to market quickly and to bring in early revenue;
- to reach outlying company and residential premises in rural areas currently not covered by terrestrial xDSL solutions.

Due to the higher latency times associated with satellite access networks, there are naturally some types of applications which are very suitable to satellite and others which are less suitable. Generally the best-suited are those that broadcast information to a large number of dispersed users, since satellite is a natural broadcast medium.

In contrast, applications that are less suited to satellite are those that require quick interaction and response times, such as gaming and voice without adequate echo cancellation. However, these limitations can be mitigated to some extent by implementing some of the performance-enhancing features already discussed in section 2.3.

Another concern people often associate with satellite services is that of outages due to heavy rain, snow and sun transit. The truth is that these aspects have been thoroughly analysed since the beginning of the satellite era and are now well understood and hence are taken into consideration when dimensioning the link parameters of new satellite services. Furthermore, today's satellite services routinely implement fade countermeasures (e.g. automatic uplink power control) and forward error correction (FEC) schemes in order to enhance transmission reliability. With these features in place and careful link budget design, satellite network operators guarantee reliability figures of typically 99.95% over the satellite link.

2.4.2 Satellite Application Service Set

The typical application service set that can be supported by satellite can be subdivided into two categories — those utilising broadcast/multicast streams and those utilising point-to-point unicast streams, as shown in Table 2.1.

Table 2.1 Satellite application service set.

Broadcast and multicast	Unicast
Digital TV Digital radio Business TV Video streaming News tickers Content delivery Software updates Caching Stagger-cast movies, or Near movies-on-demand	Web browsing FTP e-mail VoIP Gaming True video-on-demand Caching

Traditionally people associate satellite broadcast with the delivery of direct-to-home TV and business TV applications using broadcast streams. However, more recently satellite has been used to deliver unicast Internet and multimedia services such as Web browsing, e-mail and FTP downloads. While there is nothing technically wrong with delivering these services via unicast streams, they do not make the most efficient use of expensive satellite transponder capacity, the reason being that this unicast content is broadcast to all users in the downstream direction, irrespective of the fact that it is only required by one particular end user. In the future, the solution to this may be to migrate services, where possible, to a multicast scenario. For example, with the advent of large hard disks integrated into set-top boxes, it may be advantageous to multicast common Web-page content, such as electronic newspapers and news sites, directly to the end-user's terminal for local caching and access. However, this in itself introduces new aspects that need further investigation, such as how to handle personalised content in a multicast environment, and the copyright and regulatory issues associated with caching content locally on customer equipment [3]. Multicast is an area where satellite could lead the way, since:

- it offers an efficient delivery mechanism due to its natural broadcasting nature;
- it offers 'multicast' solutions that are available now, while, in contrast, terrestrial network operators are still considering the merits of rolling out relatively immature true IP multicast protocols (presently only a few terrestrial ISPs offer multicast services);

- it provides a reliable method for delivering multicast streams, e.g. through the use of additional coding in the forward path combined with acknowledgements sent back via a return path.

2.5 Security Design Considerations

Security of user information is essential if the use of satellite communications is to become widespread in the delivery of broadband content and applications to residential, SOHO and SME communities.

In line with terrestrial networks, security within satellite networks can be applied at various levels within the OSI seven-layer model, primarily at either the application layer (e.g. SSL), the network layer (e.g. IPsec), or at the physical layer (e.g. smartcards and conditional access).

Although the majority of terrestrial-based Internet traffic is sent unencrypted, on the whole it is perceived by the general public as being secure, mainly due to the fact that terrestrial solutions involve the end users having their own dedicated line between their premises and the local exchange.

In contrast, with a radio delivery mechanism (such as satellite or wireless LANs), eavesdropping is seen as a major concern since the user information is broadcast over an open medium, and, with the right equipment, it can be relatively easy to eavesdrop on unencrypted data.

Looking closer at the satellite Internet scenarios discussed within this paper, the downstream satellite carrier can generally be considered as the least secure component in the network due to its broadcast nature, e.g. all users (both authorised and rogue) are capable of receiving unencrypted downstream content. With hybrid satellite Internet topologies, the upstream path can be deemed relatively secure due to its terrestrial nature. With two-way satellite Internet topologies, the upstream path can also be deemed relatively secure due to the frequency hopping nature of both proprietary and DVB-RCS access schemes used.

Although radio is inherently insecure as an access medium, secure services can be delivered effectively via satellite by employing the following precautionary measures.

2.5.1 Firewalls

As the satellite-based Internet topologies can be considered as 'always open to attack' on downstream satellite carriers, even when a terrestrial link is used, a firewall is recommended at the customer site to prevent unauthorised access into the customer's network. This recommendation aligns with the advice given to terrestrial-based Internet customers (both residential and business) that use access networks such as ADSL.

2.5.2 Conditional Access

Smartcard technology, similar to that employed by satellite broadcast TV operators, could be used to encrypt sensitive streams and content. A number of these systems are on the market that encrypt user data (either at the physical layer or by IP payload scrambling) and employ public key/digital certificate mechanisms to allow decryption. However, the key distribution and management of these systems is not currently perceived to be scalable to the size required for mass-market penetration, but could be applied to small or medium-sized VPNs.

2.5.3 VPN Security

For corporate and SME environments, a secure hybrid or two-way satellite service can be offered by implementing a VPN solution to enable remote access to a company intranet. This process involves setting up a secure tunnel between the VPN server, located at the corporate or SME hub, and the remote worker's PC. Standard terrestrial VPN client and server software can be used, since the VPN software will just see the satellite link as a very long leased line and thus should operate in its normal way.

VPN solutions are typically based on tunnelling protocols such as L2TP, PPTP and GRE, which are then encrypted using various security and encryption protocols. It is also becoming popular to instigate secure VPNs using just IPsec without the extra tunnelling protocol, the main driver being to reduce the size of the overhead introduced when tunnelling.

IPsec is a suite of open standards, defined by the IETF to provide network layer security to IP for both private and public network infrastructures. These services include confidentiality, authentication, integrity and compression. Tests carried out by BTexact Technologies at Adastral Park have revealed that IPsec can operate fully over satellite access networks, but that this is heavily dependent on where exactly IPsec is implemented within the network. For example, the use of performance-enhancing proxies that spoof TCP can cause problems for IPsec-enabled networks, dependent on the order that IPsec and the proxy enhancement occur. The reason for this is that performance-enhancing proxies generally cause IPsec to think that packets have been intercepted or tampered with, which results in the IPsec client software discarding them. However, this problem can be overcome easily by performing the PEP function before the IPsec operation. Furthermore, IPsec can work either in a transport mode, where the IP payload is encrypted but the IP header information is left untouched, or in a tunnel mode, where the IP address and payload are hidden by the IPsec header.

Presently it is perceived that the IPsec transport mode may be a more attractive solution for satellite networks where multicast operation is preferable to unicast operation.

2.5.4 Application Security

Application security within the Internet is primarily implemented using secure servers, where such techniques as secure socket layer (SSL) WWW sites and secure e-mail are utilised. Tests carried out by BTexact Technologies have revealed that these services work perfectly well over hybrid and two-way satellite networks, as the security is implemented above the transport layer and is therefore considered part of the user data as far as the satellite networks are concerned. However, these secure features require the service provider to invest time and possibly technology in setting up these capabilities.

2.6 Summary

In summation, satellite networks provide a unique local access delivery opportunity to the broadband applications area:

- through the efficient delivery of on-line content such as distance learning, news or near-VoD;
- through delivery of off-line download services such as virus and OS updates or e-mail;
- by virtue of their natural support for the future generation of multicast-based applications.

References

1 Fitch, M., and Fidler, A.: '*An overview of satellite access networks*', in Willis, P. J. (Ed): '*Carrier-scale IP networks: designing and operating Internet networks*', The Institution of Electrical Engineers, London, pp 157-170 (2001).

2 Wakeling, J. F. and Dobbie, W. H.: '*Satellite access services*', BT Technol J, **16**(4), pp 112-121 (October 1998).

3 Wakeling, J. F.: '*Can satellites move into the access space?*', Future Comms (2001).

4 Chuter, J.: '*Dial access platform*', in Willis, P. J. (Ed): '*Carrier-scale IP networks: designing and operating Internet networks*', The Institution of Electrical Engineers, London, pp 143-155 (2001).

5 Allman, M., Glover, D. and Sanchez, L.: '*Enhancing TCP over satellite channels using standard mechanisms*', BCP 28, IETF RFC 2488 (January 1999).

6 Allman, M. et al: '*Ongoing TCP research related to satellites*', IETF RFC 2760 (February 2000).

7 Border, J. et al: '*Performance enhancing proxies intended to mitigate link-related degradations*', IETF RFC 3135 (June 2001).

Part Two

NETWORKING THE HOME

J G Turnbull

The second part of the book looks at what happens when broadband information arrives at the home. How is it delivered around the home? This part comprises five chapters which introduce the home area network, talk about the practical issues of installing the home network, describe the residential gateway, and outline the network technologies, with the final chapter discussing the standards required for home networking.

Chapter 3
Introducing the Home Area Network

This chapter assesses commercial developments, against the backdrop of a difficult time for the telecommunications industry, in order to give a view of the future — when, where and how the digital home will emerge.

Chapter 4
Broadband in the Home

What actually happens when broadband reaches the home? The author of this chapter reviews current broadband technology, examing how broadband has been introduced, the practical difficulties of installing broadband in the home, the move to self-installation, and the need for a home gateway.

Chapter 5
Residential Gateways

This chapter describes a view of the residential gateway ('... a key infrastructure component in the future home network ...') as the connection point for networks, computers, telecommunications, entertainment systems and management schemes within the digital home. It discusses its evolution and the role it could play. Is there a future in home network management? Do we need devices to communicate? Should we strive for plug-and-play or is it too far out and expensive? These and other issues are identified and discussed.

Chapter 6
Home Area Network Technologies

Once connected to the broadband network, what are the options for extending broadband delivery within the home? This chapter takes a deeper look at the possibilities for networking the home, describing those technologies that are available now. The author assesses their relative strengths and weaknesses and identifies possible eventual winners. One view of the future is of a networked home with many types of device attached — some already exist, and these are mentioned in the chapter, some are in the future, and some are not so far away but too expensive for the average domestic user.

Chapter 7
Standards for Broadband Customer Premises Equipment

In the future there will be many service providers offering broadband applications and content. Standards for devices in the home will be essential to allow connection to any service provider, to support many different applications and services, and to give the user a rich multimedia experience. The final chapter in this part introduces the major standards bodies and describes the activities in the key forums, highlighting the need for compatibility between devices and multi-service networks.

3

INTRODUCING HOME AREA NETWORKS

J G Turnbull

3.1 Introduction

The transformation from an analogue telephone network to a digital network in the 1980s has made the transmission of digital information possible across the network to the domestic customer. In the early 1980s, with the development of ISDN, it was possible to transmit data at rates of 64 kbit/s and 128 kbit/s to and from the home, enabling faster and improved communication both for data, speech and facsimile. The personal computer (PC) continued to make major technical advances through the 1990s, increasing in both speed and memory, and also reducing in cost to such a degree that the PC became part of the domestic scene.

The years of growth in the domestic market-place have presented the end user with a terminal capable of receiving and transmitting digital information, but it was the development of the Internet and the user-friendly browser in the 1990s that finally opened the digital information channel to the domestic user. Personal computers now arrive ready equipped with a modem. The data rates are still slow (56 kbit/s) and connections are often time-consuming to set up, typically taking two to three minutes. Using the 64 kbit/s or 128 kbit/s ISDN service improves access times but connecting to the Internet can only be improved by moving to the always-on service.

Developments in ADSL technology in the 1990s led to the video-on-demand trials in Colchester and Ipswich. The trials showed that, by using this emerging technology, it was technically possible for higher data rates to be sent to the customer over the telephone line at the sacrifice of a reduced data rate on the return path from the customer, but still sufficient for the signalling and necessary return data. Tests showed that customers within a 3.5 km reach of the ADSL equipped exchanges had an opportunity to leap from data rates of 56 kbit/s to 2 Mbit/s. The shorter the line to the exchange, the more the data rate could be increased. This opened up a new and exciting entertainment, information and communication channel, using the copper network of telephone lines, to a potential 22 million end users.

The product currently available from BTopenworld offers rates of 512 kbit/s and 2 Mbit/s. This allows data rates ten times greater than those previously available and, at the 2 Mbit/s data rate, it enables real-time video to be streamed, albeit at 'VHS' quality and not DVD or broadcast quality. Over 1000 exchanges are now equipped with the technology to deliver these data rates to customers on the new broadband service. It is the start of 'broadband to the home' and with it the start of home networking. End users will be able to realise the potential and the benefits of having access to a range of content and applications from anywhere in the home.

3.1.1 Service Take-up Forecasts

How quickly will users adopt the new broadband technology? It is predicted that two thirds of the PC users in North America will have broadband service by 2005, compared to just over a third in Europe [1]. The process of local loop unbundling may increase access by other broadband operators, but the initial cost of deploying broadband across the country has been high, slowing down the take-up by the other broadband operators and customers in the UK. This will change. On the other hand, in Germany, one of the leaders in broadband technology, Deutsche Telekom, has divested its cable interests to Liberty Media, thus freeing itself to deliver video content over the network using DSL technology. Adoption of the broadband service in Germany is rapid and perhaps could be attributable to the cost —$22 per month compared to $57 in the UK.

Enders Analysis predicts [2] that there will be 2 million broadband homes in the UK by 2005. This is a reduction from the 4.5 million predicted by Forrester [3] but they believe this reflects the current climate. They also predicted that the market would have six major Internet service providers (ISPs) by the end of 2001. This was based firstly on the economics of unmetered access and secondly on the lack of development in on-line advertising and eCommerce through portals that have in the past justified loss-making ISPs. This will lead to consolidation in the industry, re-evaluation of the cost model and only the strongest players will survive.

Research from Morgan Stanley Dean Witter (MSDW) [4] shows the penetration (see Fig 3.1) of the Internet with 16 million homes in the UK having either narrowband or broadband Internet access by 2005. Out of this 16 million, MSDW predicts that 4.9 million will access the Internet using a broadband connection (DSL or cable). The split across the technologies for the UK is shown in Fig 3.2 with dial-up modems continuing to dominate until 2004/2005 when the broadband access is provided either by using DSL technology over telephone lines or by cable. Broadband access to the home heralds the development and growth of the home network.

It is predicted that DSL technology will grow at a greater rate than the cable modem market, but that dial-up modems over the PSTN will fall by the end of 2010 as broadband access becomes available to all homes.

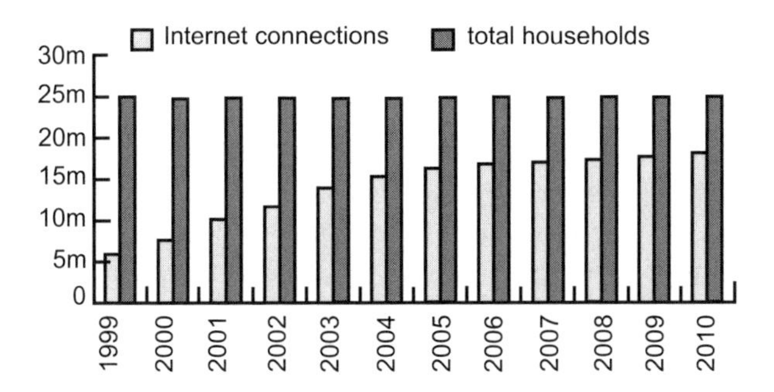

Fig 3.1 The predicted Internet growth over the next ten years compared to a near static population. The figures are for both narrowband and broadband Internet access.

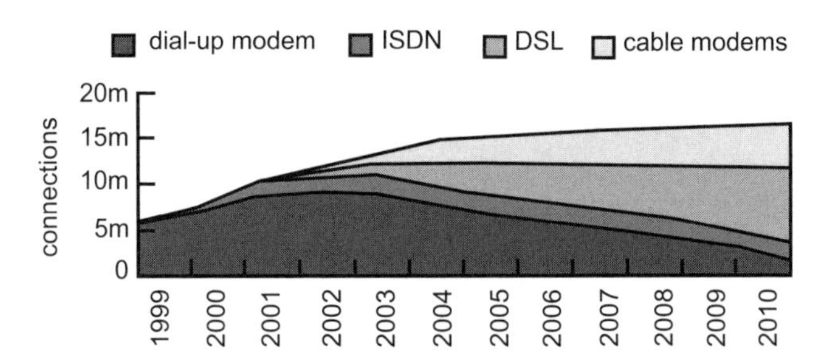

Fig 3.2 The predicted change of four access technologies over the next ten years. It is forecast that DSL technology will grow at a greater rate than the cable modem market, but that dial-up modems over the PSTN will fall by the end of 2010 as broadband access becomes available to all homes.

There is a diversity of opinion between the analysts on the predicted number of homes connected by 2005, varying from 2 to almost 5 million. There are many factors affecting the growth of broadband within Europe and North America — these include cost, coverage and content. The downturn in the economy in 2001 has also played a part in slowing down the broadband growth. Deutsche Telekom is cashing in on the problems that are being faced by the cable operators in Germany. The large losses and debts carried by the cable operators are halting the upgrades required for broadband Internet access. Deutsche Telekom is exploiting a three-year lead on the cable competitors and were aiming for 90% coverage by the end of 2001.

In a recent survey carried out by Mori for Oftel in August 2001, 39% of the surveyed adults claimed to have Internet at home: "... almost half of the UK residential consumers are connected to the Internet at home and/or via an alternative access point ..." [5]. The survey also found that 40% of the Internet-connected homes have fully or partially unmetered packages (based on a monthly fixed charge).

The way the Internet is used has changed with the introduction of an 'always-on' broadband service and the unmetered narrowband service (e.g. the BTopenworld 'Anytime' package). These services enable instant connection in the case of the broadband service or connection via a dial-up modem at no extra cost for the narrowband service (in the future there may be an always-on narrowband ISDN service available). These consumers were found to spend twice as long on-line as those using call-based packages. "Growth in connections to the Internet is mainly in the fully unmetered packages ..." [5] and this in turn will increase the Internet usage.

This may stimulate data download and promote services such as audio streaming. In a recent report [6], Enders Analysis states that there are three on-line activities favoured by the home Internet user — communication, searching for information, and home shopping. Based on data from the British Market Research Bureau (BMRB), Enders Analysis reports that 93% of the home Internet users draw on the service for e-mail and 39% for listening to or downloading music. The narrowband service is inadequate for watching or downloading video, but the new broadband services should be capable of providing this facility as the quality of service (QoS) issues are tackled.

Almost 100% of households now have a telephone connection to the local exchange [5]. The growth in the mobile cellular market (predicted to have grown in the UK from 33 to 41 million users between March 2001 and March 2002 [7]) will have an impact on the usage of the telephone line into the home for person-to-person calls. The increased number of Web sites and the vast amount of information which can be accessed over the Internet has changed the usage of the telephone connection from that of speech to that of being a data connection. The report by Oftel [5] states that 55% of UK homes with Internet access spend up to 5 hours on-line per week and 20% spend between 6 and 10 hours per week.

This increase in usage has caused a trend towards an attractive unmetered (subscription) service. It is predicted that there will be a continual move to unmetered access as the cost of a monthly subscription, compared to the cost of the calls, becomes attractive. Cost remains the key to the choice of service and ISP for the home user [8].

3.1.2 Alternative Technologies

Currently connection to the Internet is made from a PC, but there is no reason why the connection into the home cannot be shared between more than one terminal and the terminals need not be a PC. The PC is not suitable for the living room as it can be noisy and not aesthetically pleasing, although this could change. Developments of a set-top box (STB) suitable to carry out all the functions required and sited close to the entertainment centre, using the television as the screen, could open up new areas and applications for the home Internet user.

There were 24.5 million households in the UK in 2001 with a television, representing 99% of all households [4] compared to 10.5 million households with a

PC. A good strategy may be to target those households who have a television but no PC, and to offer these customers access to the Internet and to other sources of digital entertainment such as video-on-demand service providers. With increased data rates there will be an increase in downloading data such as MP3 audio files and video content. Quality of service for streaming real-time video and audio is both a bandwidth issue and a network issue; there is little scope for end-to-end error correction and bandwidth must be maintained for the length of the audio or video. In addition, the content owners will want to ensure that their material is not illegally copied, although, unlike satellite delivery, content via ADSL over the telephone network has one point of entry to the home and the home is identified by a unique number. The set-top box could also offer content security whereas the PC cannot.

Within the home itself there is a growth in the technologies available for entertainment. DVD players, video recorders, audio CD players, video cameras (both digital and tape), etc, are all available to the end user. A number of these can play different media, for example the music CD can be played on the CD player or PC, the DVD can be played on DVD player or PC, the digital camera can display on the television or PC and be downloaded to a hard disk drive (HDD) on the PC for storage. Music can be stored on a PC's HDD, and HDD storage can also be used to record and keep television programme material for viewing later (TiVo[9]) via an STB. Moxi and Pioneer presented digital entertainment centres at the Consumer Electronics Show, Las Vegas, in January 2002, which play DVDs, and store video, audio and on-line content. SONICblue's Rio Riot allows consumers to take their CD collection on a hard-disk-based, handheld music player. Personal video recorders are available which store digital recordings of favourite television shows and enable them to be shared over a broadband connection (SONICblue ReplayTV 4000). To add to the complexity, different people in the household want to view or listen to different media.

3.1.3 Content Sharing

The need for home networking is growing. There is a developing need to share the Internet between two or more terminals and to share peripherals such as scanners and printers. There is a desire to share material delivered over the broadband network. There is a need to store information and media and to make it accessible anywhere in the home. The development of media (MP3 music files, video, data, etc) within the home — storing content, retrieving content and moving content to different rooms — raises issues of security (both of content and from external attack) and management of the device technology. The networked devices will need to be simple to install, inexpensive, backed up by a very good help desk, and impervious to virus attack. This is a major challenge for the device manufacturers.

But what is the situation today and how will it develop? MSDW predicts that, by 2005, 65% of all UK households will be connected to the Internet [4], and of these 47% will use a broadband connection, with 23% of all connections using DSL over the telephone line, 19% using cable modems and the remainder using other broadband access technologies.

Current access to the narrowband or broadband service requires customer premises equipment (CPE) in the form of a modem and terminal equipment (e.g. PC, STB, WAP telephone). Currently most connections stop at a personal computer (PC) and go no further. Forrester states that almost half (46%) of the broadband homes in the USA claim more than one PC [10]. Only half of the 46% have networked the PCs and the other half have either no Internet access on the second PC or have PSTN Internet access. The device manufacturers aim to exploit this market.

The figures for the UK are more difficult to find with only 1% of households connected to the Internet using broadband and 43% of the households with dial-up Internet access [3]. The trend to multiple PCs and home networking will undoubtedly follow the USA lead.

The challenge is to enable the development of networking in the home. The service provider has the challenge to provide applications, facilities and premium rate services for the home user in order to extract more revenue, given that the broadband service is delivered at a flat rate and there is no additional return in generating more traffic. The terminal providers have the challenge to provide devices and technology that are wanted by the customer and work across the home network.

This growth in access to the Internet using broadband connectivity leads on to the need for, and development of, the home area network.

3.2 Why Do We Need a Home Area Network?

In 2000 there were 10 million (42%) of homes in the UK with at least one PC and 24.5 million homes (99%) with a television [4]. Streaming data need not be just to a PC. The STB could output directly to the TV and lend itself to displaying real-time streamed-video pictures. Data will be received and stored locally to be shared between STBs or other terminals. The growth of Internet traffic has been steady over the last seven years (see Fig 3.1) and the forecast is for 65% of the population in the UK to have at least one PC by the year 2005. If we follow the American trend, half will have two or more PCs. The development of both the PSTN and ADSL delivery systems has and will encourage shared access to the Internet from a PC, an STB or a second PC. The greater bandwidth overcomes the slow access, allows greater data rates and file downloads, assists in the development of more video-rich Web sites, overcomes the need for a second telephone line, and enables sharing of the Internet. This will lead to an increasing demand for home networking.

There are a number of industry groups (the Digital Video Broadcasting Group is developing the standards in Europe) that have an interest in developing home networking [11] — but there is currently a split between the PC industry and the entertainment industry. Two services were modelled by Forrester in a recent report [3].

- Access plus value-added PC services

  The telco delivers to the PC over broadband, adding services, e.g. gaming, music, remote office, and security systems such as antivirus and firewall.

- The triple play — access, value-add services and television over DSL

  This retains the value-added services of the first option but adds what chello and NTL can already offer over cable — television, video-on-demand, time-shifting personal video recorder (as TiVo offers). This delivers to the television via an STB and not the PC.

There will be a need to share the content around the home, to the PC and to a home storage server. These technologies will eventually merge as home networking develops.

Others who have an interest in home networking include PC manufacturers (e.g. Compaq), IC manufacturers (such as Intel), the networking and communications industry (e.g. 3Com) and consumer electronics manufacturers (e.g. Sony), software and application developers (such as Microsoft), content and eCommerce providers, and broadband service providers.

It is the consumer or end user, however, who will have the final say, based on the perceived benefit and cost. Home networks may first develop in the entertainment space driven by players such as Sony.

Sony is currently looking at IEEE1394 as a means to manage the multimedia data around the home for entertainment purposes. Source material could be stored on hard disk and, using the standard IEEE1394 to format and send the data, could then be sent to any location within the home for playback or viewing. DVD-quality video requires between 3.5 Mbit/s to 6 Mbit/s, HDTV requires 19 Mbit/s, TV requires 4 Mbit/s, and high-quality fast-moving video requires 15 Mbit/s. These high data rates can be managed within the home with current technology, but, with today's access technology, it is not possible to stream these data rates over the network to be viewed in real time. The content could, however, be streamed at a lower data rate and downloaded on to a local HDD as a background activity to be played at a later time.

The home network can be thought of as a way to interconnect your home — to connect the hi-fi to the home server, to connect the PC to the STB, to enable media to be shared across the premises and be accessible by anyone, anywhere in the home. This sharing will require networked computers, networked devices and, as the complexity grows, a home gateway and hub to manage the interface between devices.

Telecommunications can be considered as covering three main areas.

- Communication

  We communicate using speech over the telephone network, by the mobile wireless network or over the Internet.

- Information

  Information has traditionally been gathered from books and records. In future we will no doubt make more use of on-line libraries and CD-ROMs. Access to a global source of both written and visual text is available electronically over the Internet.

- Entertainment

  Entertainment comes in many forms in the home — videotape, CD, MP3 disk or from data stored on the HDD of a computer. In the future, content will be in digital form, delivered by media such as satellite dish, digital terrestrial aerials or the copper telephone connection from the local exchange. It will be stored either locally on a computer or in the network. Transmitting this multimedia-rich content around the home, to be viewed on a number of different terminals, will require a home area network.

3.3 What is a Home Area Network?

A home area network is the connection of a number of devices and terminals in the home on to one or more networks which are themselves connected in such a way that digital information and content can be passed between devices and any access 'pipe' to the home. Simply, it is technology that connects terminals.

In its simplest form it would be a connection between two PCs perhaps using a USB link, but, as different network technologies develop, a hub or residential gateway (see Chapter 5) is required to manage the data flow, set up the devices and add protection and security.

Instances of possible terminals in the home that may require networking are shown in Fig 3.3. The home gateway will act as the hub and manage the devices and the content flow between devices. Starting as a low-cost home hub device it would evolve to become an intelligent gateway that manages the terminals and provides security and firewall protection.

Figure 3.4 shows the home entertainment gateway that may be integrated into the STB. It will focus on entertainment within the home and not be associated with the PC. The STB will be aimed at households without a PC and who have no desire to own or operate a PC. It will enable these homes to have access to richer entertainment media, such as video-on-demand, interactive television and music download. Communication and e-mail via the television and STB is available, as well as networking around the home to allow more than one television to have access to the media and content. Cost and benefit are the keys to entering, and having a major impact on, this market.

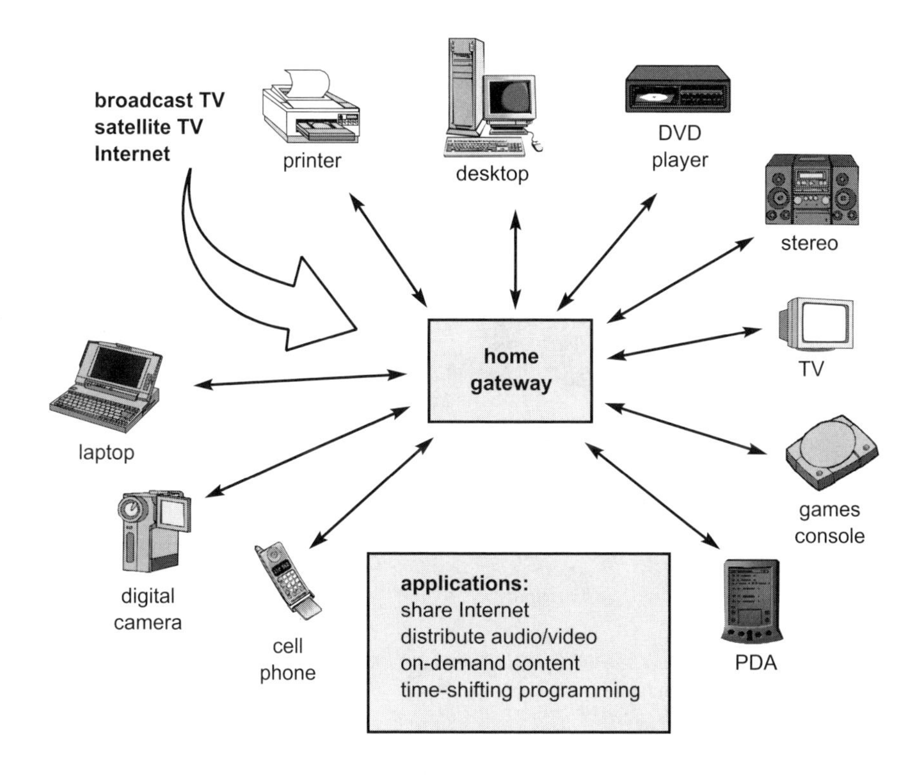

Fig 3.3 The home gateway, to which are connected a number of terminals or devices. It needs to manage the devices, the data flowing between these devices, and the data flowing into the home.

Figures 3.3 and 3.4 show two totally separate home networks, each adding benefit to the customer for two different reasons. The analysts are divided on the future of home networking. One Forrester analyst [12] does not see the convergence of the broadband Internet with the home entertainment market, advising that PC manufacturers should concentrate on providing very good access to the Internet, while carriers and content providers should focus on the entertainment gateway (STB). The home networking market debate between analysts has yet to reach a conclusion.

The home area network provides the means to connect more than one terminal with another or to another terminal device, e.g. a printer, within the home. It allows shared resources, and enables information to be sent between devices and the narrowband or broadband access into the home to be shared. The home area network may be in the form of a wired network, using for example Ethernet as the data carrier and Cat5 cable as the physical layer, or a wireless network using, for example, the IEEE802.11 protocol. The development of broadband modems, which allow greater data rates to the home, has enabled both higher speed and shared

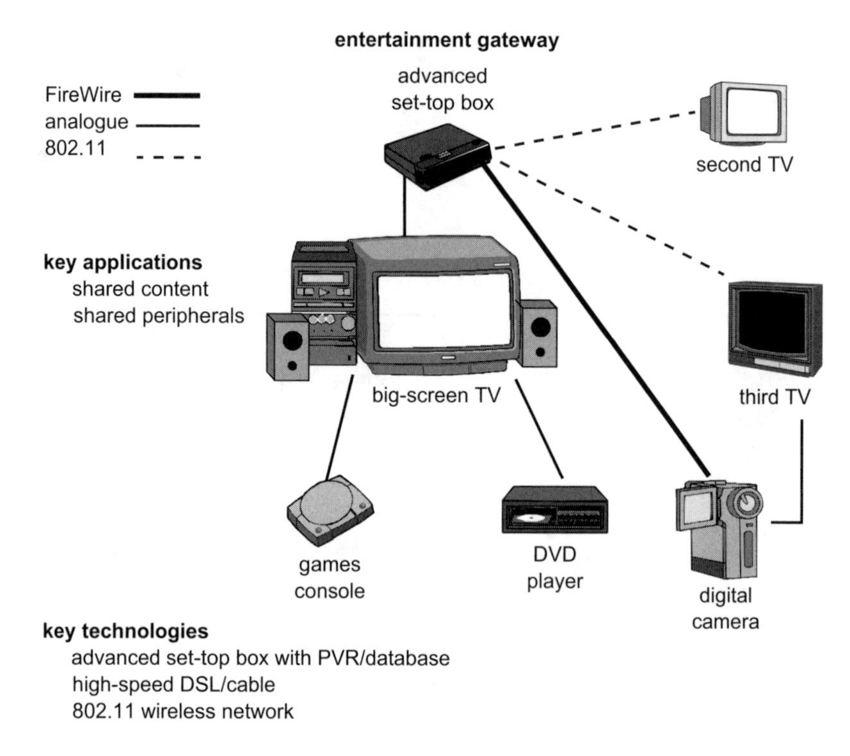

Fig 3.4 The entertainment gateway manages a number of entertainment terminals and the media flow between devices.

Internet access, and the downloading of files or the streaming of video, to become viable.

The home pictured in Fig 3.5 shows devices connected by both wired and wireless methods. Several technologies could be used and the home gateway becomes the interface to these technologies. The wired home area network may use existing telephone extension wiring or use new cables fitted around the home. The wireless home network, a more expensive but less obtrusive technology, is easier to fit and removes the need for new wires and is an attractive option.

Homes are being built today with new wiring to carry the multimedia data around the home in the future [13]; but what of those homes already built, are they going to install new wiring? The development of wireless technologies, the development of data over the existing telephone wiring (HomePNA2), or the use of the mains power wiring around the home to carry high data rates (Powerline) could help to stimulate the home networking market.

The home networking market is immature and fragmented. There are currently numerous competing technologies and alliances. HomePNA, Ethernet, IEEE802.11, HomeRF, Home Plug, and Bluetooth are all available today and all have their good

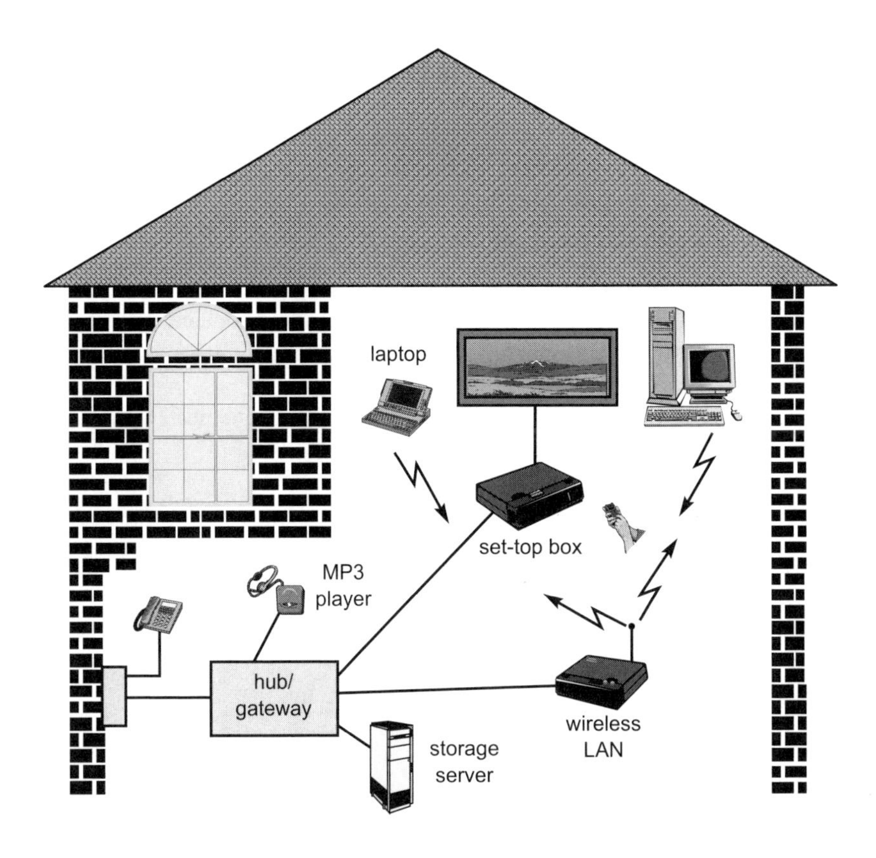

Fig 3.5 The home area network connects CPE within the home using one of several technologies.

and bad points. It looks likely that IEEE802.11 and Ethernet will become the *de facto* standards.

The key requirements of such a network are:

- low cost;
- easy installation;
- easy network management;
- good data rates and QoS;
- sufficiently high data rate around the home to carry video;
- security.

There are five interconnect architectures to consider. These are:

- Universal Plug and Play (UPnP);
- Jini;
- Open Standards Gateway Initiative (OSGi);
- Home Application Programming Interface (HomeAPI);
- Home Audio Video Interoperability (HAVi).

UPnP and Jini were initiatives of Microsoft and Sun Microsystems, the others being initiatives of the industry forums. HomeAPI has subsequently merged with UPnP and OSGi merged with Jini. The three remaining groups will become commercially available during 2002.

At the physical layer there are several types of interconnect technology to consider:

- Phoneline;
- Powerline;
- wireless;
- Ethernet;
- IEEE1394 (FireWire);

and there are also several variations of the wireless technology:

- HomeRF (Compaq, Ericsson, Hewlett-Packard, IBM, Intel, Microsoft and Motorola);
- IEEE802.11 (a, b and e); HiperLAN (1 and 2);
- Bluetooth;
- DECT;
- ShareWave.

There are many variants from which to choose and some will invariably fall by the wayside.

It is the wireless standards IEEE802.11 and HiperLAN2 which will dominate the market-place in Europe according to Datamonitor [9], replacing Ethernet which is dominating the fixed-line interconnect technology. Phoneline is not suitable in Europe due to the low number of telephone socket outlets around the home, Powerline has problems meeting the European regulations, and Ethernet requires new wiring but is relatively low cost. Wireless technology requires no change to the infrastructure and is more flexible, although power levels must be kept low to be safe, limiting the range of the RF signal. This may be an issue and succeed in keeping Ethernet as a preferred network solution.

3.4 Issues with Home Networking

As both 'no new wires' and 'low cost' are of major importance to the success of home networking, Powerline, HomePNA and wireless technologies may be seen as the main technologies in this area. Developments in the 5-GHz frequency spectrum (data rates up to 54 Mbit/s), will overcome the interference issues around the 2.4-GHz frequency spectrum shared by microwave ovens and cordless telephones and which provide sources of interference.

The main stumbling block is producing the chipsets at a realistic price for the consumer market. The standards for 5-MHz wireless were ratified by the IEEE in March 2000, but today there is no sign of a product within the price range for a domestic user. Distance limitations and high power consumption make it currently unattractive.

There is a lack of global agreement, with a single standard in the USA (IEEE802.11a approved in September1999) which is different to Europe. IEEE and ETSI are addressing this issue (vendors supporting this are Atheros, Radiata and ShareWave). This standard does not address the quality-of-service problems (the standard does not have the extensions required), but both 802.11a and 802.11b aim to have QoS extensions by late 2002. This will form part of the 802.11e standard (including media access control (MAC) enhancements that add QoS and security to 802.11a and 802.11b based equipment) and hopefully be accepted globally. However, the use of WEP (Wired Equivalent Privacy) in 802.11b has been shown to be insecure and open to hacking, and work using IPsec as the security technology is currently under investigation within the industry. The security issues rumble on!

Other home networking technologies exist but it is the number and the lack of interworking that causes problems. No one vendor will 'go it alone' as Sony did over the Betamax VCR system. Customers will find it confusing, but, if they see a benefit at a price, then certain technologies will evolve before others. Home gateway devices have been produced to enable different network types to co-exist but currently such a device is not available at an affordable price for the home user.

Vendors supporting wireless home networking, which is based on the IEEE802.11b standard, are Aironet/Cisco, Apple and Lucent. Vendors supporting HomeRF include Motorola, Proxim and Siemens. The contender to IEEE802.11 is HiperLAN2 and vendors are in the early stages of developing a unified protocol to allow these two wireless technologies to interoperate.

At the time of writing, wireless home networks based on IEEE802.11 were reported in the USA as being the future for the home networking market [14, 15]. Prices are becoming affordable and the 'no new wires' adds great benefit for the home user.

The challenge is to make such systems easy to self-install and needing little maintenance.

3.5 Who are the Key Players?

Finally, who are the key players in the area of home networking? The industry suffered in 2001 due to the recession in industry in general, and also the bursting of the dot.com bubble as investors realised that there would be little return in the short term and in some cases the business model could not be supported. Growth in Internet use has also slowed recently, adding to this problem.

The main players in the home networking market are listed below.

- Cisco and the millennium home — the Web site gives you access to the home and shows what happens in each room of the house. Intel was also producing Internet appliances. Nokia offers a wireless LAN card to enable the laptop to access the home networks. IEEE802.11b standards were adopted based on the 2.4-GHz direct sequence radio technology. Security of the wireless data was accomplished using WEP protocol with 128-bit encryption.
- Motorola has a family of home networking products based on wireless technology. The solution is based on 'no new wires' philosophy. Mainly targeted at the cable market, the company offers a residential gateway device and a range of wireless networking devices. It supports Ethernet, USB and HomeRF networks around the home, has the ability to share printers, scanners and cameras, and enables the distribution of files around the home using the in-house Netmanager. Firewall software to protect the in-home network is also built in.
- SerCoNet has a number of solutions. The data solution uses existing telephone wiring to transport the data around the home. The wireless solution uses Bluetooth (future), IEEE802.11b for wireless and IEEE1394 for multimedia around the home.
- Maxgate is a cable/xDSL wireless-based gateway, using TCP/IP and IEEE802.11b. Network address translation (NAT) runs DHCP and allows PPTP or IPsec VPN connection to corporate servers. It also runs a dynamic domain name server allowing easy domain name setting, and an Ethernet-based solution running 10/100baseT, supporting the same range of options as for wireless.
- Intellon — this company has developed a high-speed home powerline networking technology with data rates up to 14 Mbit/s.
- Coactive Connector is also in the powerline area for home networking — it is low bit rate and aimed at control and not multimedia data.
- CoNet — its key product is the residential gateway which includes the G.DMT ADSL modem. Features include TCP/IP, NAT, and DHCP server. Also available are Ethernet gateways, switches and wireless Ethernet home networking.
- Actiontec uses wireless technology to provide simultaneous on-line access around the home.

- 3COM — home wireless gateway — supports TCP/IP, IPX/SPX, NetBEUI, DHCP, IEEE802.11b, PPoE and NAT. Also in the range is the dual-speed hub — connecting up to five devices using Ethernet. They have developed a proprietary network technology called HomeConnect.
- Intel — the AnyPoint product range is a wireless technology using IEEE802.11b technology in the 2.4-GHz frequency range. It supports TCP/IP, IPX and 128-bit WEP encryption.

  The gateway product has wireless and wired (Ethernet) technology, and supports NAT, DHCP server and client, PPoE, VPN, TCP/IP, IXP, and NetBEUI. The range also covers networking over existing telephone lines, and Phoneline Home Network products, but these are not compatible with the wireless products.

3.6 Summary

This chapter has examined the analysts' reports in an attempt to shed light on how and when home area networks may become a reality. It is clear that there are a number of different networking technologies and, with the current state of the telecommunications market, the predictions for in-home networks are in years rather than months. Benefit, price and the access technology (broadband to the home) are all contributing factors.

Connecting PCs on a local network to access the Internet for information will eventually merge with networked entertainment systems. This will generate a need for the residential gateway. Initially this is seen as part of the CPE (PC or set-top box) until costs fall and benefit grows. In reality, it is a confusing picture with many players and variables.

It is the task of those in the technology areas to simplify, clarify and show the home users a clear path to networking their digital homes.

References

1 Strategy Analytics: '*The Interactive Home*', (June 2001).

2 Enders Analysis: '*Broadband Update*', (January 2001).

3 Forrester: '*Making ADSL Broadband Profitable*', (December 2001).

4 Morgan Stanley Dean Witter: '*Telecommunication and Internet services: industry overview*', (June 2001).

5 Franck, S. and Hayes, L.: '*Consumers' use of Internet*', Oftel Residential Survey (August 2001).

6 Enders Analysis: '*UK Internet Trends*', (December 2001)

7 Enders Analysis: '*Mobile Market Trends*', (May 2001).

8 Enders Analysis: '*The Past, Present and Future of Internet Access in the UK*', (June 2001).

9 TiVo — http://www.uk.tivo.com/

10 Forrester: '*Data overview — devices and access North America: consumer technographics data overview*', (July 2000).

11 Datamonitor: '*Home Networking Markets to 2005*', (December 2000).

12 Forrester: '*Broken Home Gateways*', (September 2001).

13 Homenet — http://www.business-systems.bt.com/cabling/homenet/index.htm

14 International Consumer Electronics Show, Las Vegas —http://www.cesweb.org/

15 Consumer Electronics Association (CEA) — http://www.ce.org/

4

BROADBAND IN THE HOME

K E Nolde

4.1 Introduction

To drive down costs and ensure as competitive a product offering as possible, each area of the broadband delivery needs to be examined carefully to see if there are ways in which costs can be reduced. A good example of this is in the home installation where the recent launch of self-install ('Plug & Go'), which allows the customer to install the modem and drivers, has led to the reduction in the price of the broadband service. The main areas where savings could be made are in the registration process, installation of the customer premises equipment (CPE), and in the regulation of the broadband service which affects the delivery process.

4.1.1 Regulation

For regulatory reasons, and other operational factors, a telco, in this case BT, has a number of different roles and responsibilities in the delivery process. For example, the customer registers for service and makes a contract with an Internet service provider (ISP) (in this case BTopenworld); their request for service is in fact dealt with by BT Wholesale, the supplier of the ADSL service. This wholesale/retail split ensures that other ISPs also have access to the ADSL service on the same terms and conditions as BT's own ISP (BTopenworld). While this is good from a regulatory point of view, it has, to date, been difficult to ensure a 'seamless' customer experience as the customer has to deal with different parts of the telco at different times during the installation cycle. This adds unnecessary delays and costs.

4.1.2 Engineer Installation Versus Self-Installation

It has always been the aim for ISPs to be able to deal directly with the customer throughout the installation and on-going life cycle of the product. Until recently, installation of the modem was never considered appropriate for the end user and not feasible for an ISP to consider undertaking this task itself. All this is set to change now that the self-installation option is available to the customer. This is not without potential problems (dealt with in more detail later in this chapter), which with careful management can be overcome and so enable the real broadband revolution in the home to take place.

4.2 Installation Issues

Prior to the introduction of self-install, installation of the ADSL broadband connection within the home was carried out, on behalf of the ISP and end-customer, by a BT engineer. This added both delay and cost to the process, which, if removed, could potentially stimulate broadband take-up further by allowing lower entry costs for those who wish to carry out the installation themselves. But the question 'How feasible is this?' is one that has concerned network and service providers alike.

At present two main tasks are carried out by the appointed engineer:

- installation of a line filter;
- installation of the CPE.

4.2.1 Line Filter Installation

It is essential that the line filter or 'splitter' is fitted correctly as the purpose of the splitter is to separate the ADSL higher frequencies from the lower frequency baseband telephony. Without this splitter, the service will not work properly, as the telephone tends to attenuate the ADSL signals (due to its low impedance at higher frequencies). A high level of background noise can also be heard in the telephone as a result of the ADSL high frequency signals. Installation of the filter appears straightforward. It should be a matter of simply removing the current frontplate or 'customer connection unit' (CCU) and replacing it with the new frontplate containing the low-pass filter. In practice, the CCU is the connection point for the extension sockets around the home and these will require disconnection and reconnection on the new frontplate. This task alone can be problematical as the end-customer's 'method' and choice of wiring for the extensions may not be standard and require some re-work, especially if the changes introduce new faults. Figure 4.1 shows a standard NTE5 with the replacement ADSL frontplate.

4.2.2 CPE Supply and Installation

The engineer currently supplies the CPE in the form of a modem along with the required software drivers. Although network providers such as BT Wholesale do not offer assistance at this stage, it is sometimes difficult for a well-intentioned engineer not to offer some support to the customer if they encounter problems when installing the software on their PC. ADSL installers carry laptop computers so that connectivity can be checked before leaving the site without having to rely on the customer's PC. While demonstrating the service working on the engineer's laptop can help in any disputes about where a fault may lie (and indeed it has been shown

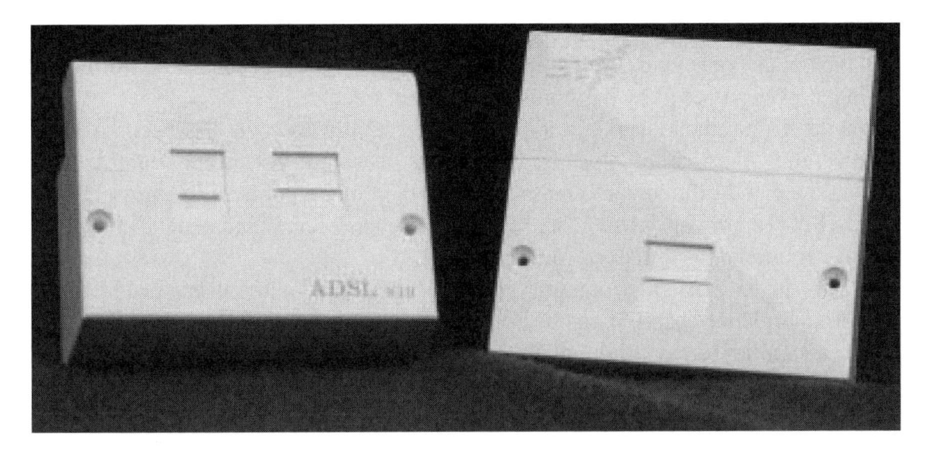

Fig 4.1 Standard NTE5 'master socket' and replacement ADSL frontplate as fitted by the BT engineer.

during the trials of ADSL that a number of faults have been on the customer's PC), it is clearly not ideal when trying to create the best end-customer user experience.

4.2.3 Self-Installation — Learning From Experience

A quick look back in history reveals some interesting parallels on the potential benefits/pitfalls of moving to a self-installation approach:

- prior to liberalisation in the telephone CPE market (c1981), the policy was that virtually all CPE provided for connection to the UK network had to be supplied by the Post Office;
- while other non-telephony terminal equipment was supplied and sold by third parties, the process of obtaining connection approval was lengthy and complicated and involved compliance to Technical Guides published by the Post Office;
- these documents were forerunners to the British Standards and documented the requirements for various types of equipment;
- after liberalisation, the Post Office monopoly was withdrawn for the supply of telephones and almost overnight there began a whole industry supplying telephones in every shape, size and colour.

To ensure that telephony products from any supplier would maintain the standard traditionally expected, the government appointed the 'British Approvals Board for Telecommunications' (BABT) to oversee both BT's own and other approved test houses. Further, to ensure consistent interpretation of the UK connection requirements and to align UK standards more closely with other product standards, the British Standards Institute (BSI) were given the task of creating new standards to define fully and unambiguously the requirements for connection approvals. This task took many years and resulted in a whole set of standards used for conformance testing, now superseded by harmonised European standards.

While at first glance it might appear that the approvals route has become more bureaucratic over time, this approach has ensured consistent quality and performance and offered some reassurance to the buying public in the guise of the 'Green Spot' approvals label. More recently there has been a fundamental change in the way equipment is approved for connection.

Again, in an attempt to simplify the procedure (and ultimately speed up the process and reduce costs), it has been decided that suppliers can 'self-certify'. This allows a supplier to essentially claim compliance with the necessary standards and so avoid potentially lengthy laboratory testing. Indeed the only formal evaluation that may be required is safety testing. Ultimately it is left to market forces to determine which products will survive competition and which will not. The reasoning behind 'self-certify' is that it places the incentive on manufacturers and suppliers to ensure good-quality products that function as expected or otherwise lose sales and money as the product fails to sell in sufficient numbers.

It is almost certain that the supply of ADSL terminal equipment will follow a similar model. The development of an accepted standard for ADSL has made this feasible, allowing equipment interoperability, i.e. the ITU-T G.DMT standard [1] has enabled self-install to take place.

4.2.4 Standards for ADSL Self-Installation

Now that the physical ADSL line interface standard (often referred to as G.DMT standard) and associated protocols for the broadband service are available in the UK, it is possible for potential equipment suppliers to design equipment for this service. In addition, BT has published Supplier's Information Notes (SINs) that complement the ADSL specifications and give additional information specific to the implementation in the UK, i.e. which protocols are supported on the ADSL data stream and the specifics of ATM VCI/VPI numbering, etc. These SINs are available for downloading [2].

4.2.5 Interoperability Testing

It is unlikely that we will ever see formal conformance testing again in the way that has been carried out in the past for telephony equipment. However, within the industry, there are already moves by technology companies (including BTexact) to provide testing services to support potential manufacturers and provide interoperability testing between items from different vendors of modem and exchange equipment (i.e. digital subscriber line access multiplexer — DSLAM). Such interoperability testing is now regarded as being as important as traditional parametric and conformance testing.

The customer experience is equally poor if the modem, home gateway, etc, cannot be made to operate with the exchange equipment used by the customer's service provider because of incorrect device configuration. In the past this has not been a great problem because of the limited market penetration of ADSL and the low number of suppliers, but now, as many Far-Eastern companies are targeting Europe as a potential growth area for ADSL equipment, this is likely to change quite rapidly.

Self-installation has the potential to stimulate this market even more, in just the way it did for telephones in the last two decades.

4.3 Self-Installation

There are clearly good reasons for seriously considering how self-installation of ADSL CPE can work effectively. Some of the potential issues which have prevented this happening are discussed below.

4.3.1 Limited Equipment Choice

Until recently, there has been little choice in the ADSL modem market. This has been particularly true for products intended for mainly consumer or home use. Figure 4.2 shows the Alcatel SpeedTouch USB modem widely used among the ADSL subscribers in the UK. This product was initially supplied to network providers and not intended for sale direct to the end user. Again history shows how PSTN analogue modems were once considered complicated 'data terminal equipment' (DTE) and as such required a trained engineer to install. Now we think nothing of buying a modem in any computer/electrical retailer and fitting it ourselves. The same is about to happen with ADSL modems now that G.DMT and related protocol support is well understood. Such a change will almost certainly alter the relationship between end users and service provider, as the end user has a choice over which ADSL modem to use and the service provider has to support this choice.

4.3.2 Frontplate Filter Installation

Figure 4.1 showed the modern network terminating equipment (NTE5) or master socket used in the UK. Also shown was the replacement frontplate that contains the low-pass filter which prevents the telephone impedance at ADSL frequencies from affecting the broadband signal. On the initial deployments of ADSL in the UK, during the video-on-demand (VoD) trial in 1996/97, an active filter was used to ensure the best possible performance was achieved for the ADSL system. This solution required three separate units within the home — an active splitter, an ADSL modem and a set-top box (STB), each of which required their own power

Fig 4.2 Alcatel SpeedTouch USB modem as currently supplied by BT Wholesale.

supply. This was clearly not feasible for a large-scale roll-out. Studies were carried out to see what type of low-cost passive filter could be used in place of the active splitter and eventually a design was selected. The main purpose of the filter was to ensure that the telephone impedance at frequencies from about 25 kHz to over 1 MHz would not affect the ADSL signal.

It is ironic that, until fairly recently, it was considered good design practice to ensure that the telephone set impedance was relatively low at these frequencies so as to reduce their susceptibility to radio frequency interference (RFI immunity) and other electromagnetic emissions (EMC). Indeed many designs include a simple LC filter just for this purpose as part of the line interface.

Thus the frontplate filter acts as a buffer between the line where the ADSL signals are 'tapped-off' and the path for the baseband POTS (plain old telephone service), as shown in Fig 4.3 which depicts the standard UK 3-wire extension plan.

One major advantage of an engineer fitting the frontplate is that they are able to ensure that no extensions are improperly connected 'behind' the frontplate as shown by the diagonal (dashed line) connection in Fig 4.3. Clearly any standard telephony equipment (telephone, fax, modem, etc) connected at this point will again cause potential interference to the ADSL signals, as they are not isolated by the frontplate filter. Also an engineer may move the master socket to collocate it nearer the required termination point for the broadband, e.g. a back bedroom used as the 'office' in the home. This is, of course, possible, again providing the engineer takes the necessary precautions to ensure that there are no additional extensions connected of which they are unaware. Thus having moved the master socket/frontplate, they can again ensure all extensions are connected so as to ensure the filter is correctly isolating all the baseband CPE.

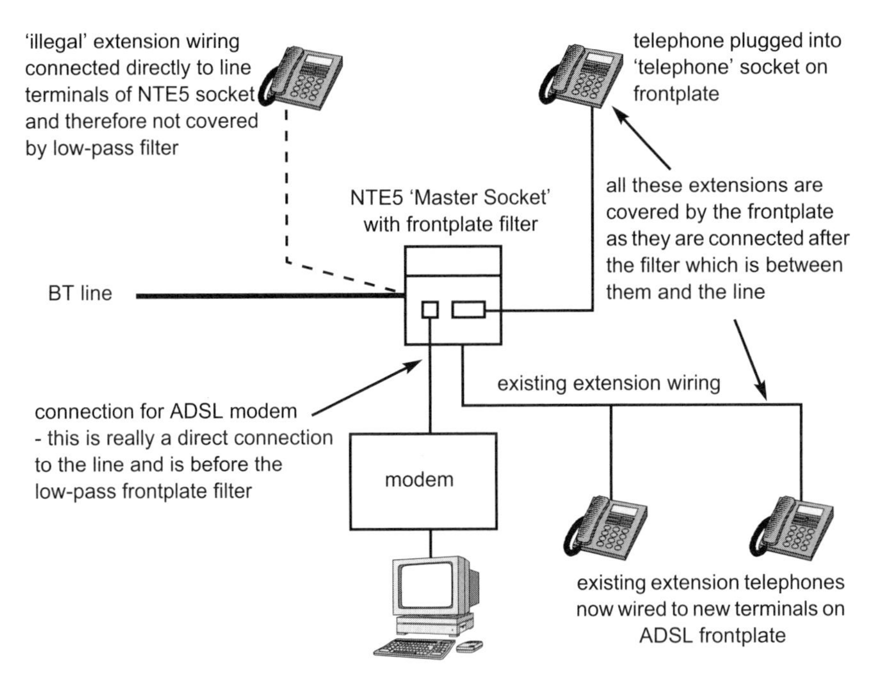

Fig 4.3 Standard wiring arrangement for UK extension telephones (illegal extension shown by diagonal connection (dashed line).

While it is possible to produce detailed end-user instructions that cover the installation of the frontplate filter, in practice it is easy to imagine some of the potential problems end users could face. Even the basic assumption that everyone now has the correct type of master socket (i.e. NTE5) cannot be made and problems will arise as the frontplate is only designed to work with this type of NTE. Therefore it seems likely that, when self-installation is rolled out, alternative micro-filters as shown in Fig 4.4 will be used instead.

Fig 4.4 Typical micro-filter as supplied with ADSL equipment to be fitted by the customer.

Micro-filters provide the same isolation as is provided by the frontplate filter but are required to be used with every baseband CPE connected via one or more extension sockets. It is essential that the end-user takes the time to ensure that all their equipment has a filter and that they remain fitted all the time the line is used for ADSL services. Such an approach has the advantage that the customer no longer has to worry about fitting a new frontplate and they can exploit their house extension wiring by picking up the ADSL signals on any extension nearest the desired point of termination for the modem.

There are, of course, several potential pitfalls with this approach:

- numerous micro-filters may be required (how many does the ISP supply — 3, 4 or 5, etc?);
- the end user must fit them correctly and on all extensions;
- one may be added by mistake to the extension used by the ADSL modem and prevent the modem working;
- multiple filters may affect telephone performance and even ADSL throughput; the wiring in the home must be suitable for ADSL signals.

Current thinking is that in spite of these issues, there is still benefit in using this approach. Limited trials have taken place which suggest that the initial concerns are far outweighed by the benefits of ease, speed and flexibility of installation which directly benefit the customer.

4.3.3 Modem Installation and Connection to the PC

For self-installation to really work, the modem must be simple and quick to install for both the physical network connections and software drivers. The frontplate or micro-filter approach are both intended to make the physical connections to the network quick and straightforward — but what about to the PC itself? For baseband modems this has not been a problem, since irrespective of whether an internal or external modem was used, the required interface was ubiquitous and well defined, i.e. serial port or ISA/PCI card.

When the first trials of high-speed Internet were offered using 2-Mbit/s ADSL in the late 1990s, the only practical solution was to terminate the service using an Ethernet LAN card in the PC. This upgrade had to be either undertaken by the engineer who installed the modem, or carried out by a third party such as a computer reseller. At the time this was extremely problematic. An Ethernet presentation version of the service still exists which is aimed mainly at businesses. Fortunately, with newer PCs using the latest versions of the Windows^{TM} operating system and with the wider use of PCI cards, this is no longer a major problem.

In spite of this, however, the lessons learned from the early BT Interactive Trials (trials in the late 1990s to test all aspects of broadband delivery to the home using ADSL transmission technology) led to a desire to achieve a 'plug-n-play' type of installation for both hardware and software. To this end, a conscious decision was made that, for the consumer (home-oriented) product offering, the modem installation should not necessitate opening the PC, in order to avoid potential pitfalls and problems following the fitting of internal PC cards. Since the traditional PC serial interface was not fast enough for the planned speeds of ADSL, an alternative solution had to be found. By coincidence certain equipment suppliers were themselves addressing this same issue and the obvious solution was to exploit the new universal serial bus (USB) interface developed by a consortium of several PC vendors and component suppliers.

The original USB specification [3] was designed to have a throughput of 12 Mbit/s —certainly fast enough for an ADSL modem whose maximum speed was typically 4 or 8 Mbit/s. The other desirable feature of USB was that it has been designed to be a 'hot-pluggable' interface so that peripheral devices could be plugged and un-plugged without power cycling the PC. Also such an operation would prompt the user to install the necessary drivers if they had not already been installed and so was considered quite user friendly. Finally, the USB specification defined a power budget of up to 500 mA that can be used by a peripheral. This suggested that a modem could be solely powered via the USB port and so eliminated the need for a plug-top power supply, making the modem even cheaper and not country specific. For these reasons, the USB SpeedTouch ADSL modem made by Alcatel was selected as the initial consumer modem supplied by BT for its new ADSL service. Thus it seemed that USB offered the 'perfect' interface for this type of application.

The drawback was that only PCs manufactured from around 1997 had basic USB support (i.e. not always enabled in a PC's BIOS and not always provided on the outside case) and only those from about 1998 had support by default. An innovative PC-checking utility was developed by BTexact Technologies and made available for download which checked to ensure the PC specification (USB support, CPU speed, memory, hard disk space, etc) was sufficient to support the modem. Using this utility, the customer is able to check their PC prior to ordering the service and so determine whether they would first have to upgrade their PC.

4.3.4 Software Driver Installation

This is one area that is already essentially an end-user 'self-installation' process but because the modem and its associated software is supplied by the ISP it is still often perceived as their responsibility. This is further compounded if the customer purchases the modem and has to input the settings specific to the UK. This can cause significant levels of customer support calls which are handled in the first instance by the ISP. (This is not normally an issue with PSTN modems as the end-user will contact the equipment supplier to resolve technical problems.) Another big advantage for moving to self-installation is that when end users are free to buy their own modems and fit them themselves, the network provider will no longer have to provide first-line support on CPE installation. In future, the modem will be available from BT outlets or via the BTopenworld Web site. This should avoid most of these issues as the product will have been preconfigured to work with the service.

4.3.5 Self-Installation Today

Currently self-install is available with the ISP providing the modem and drivers. The modem is delivered with typically two micro-filters and the customer can purchase more if required. The future expectation is that, with growth in the ADSL modem market, it will become a commodity item just as V.90 modems are today. In future, ADSL modems will be purchased from BT outlets, the BTopenworld Web site [4], or other retailers, and will be complete with software drivers and micro-filters as part of the package.

4.4 Distribution of Broadband Within the Home

To really exploit the true value of an 'always-on' broadband connection it soon becomes apparent that some form of distribution within the home is needed. While the number of homes with PCs is rising all the time, so is the number with two or more machines. This is usually as a result of replacing the original PC and allowing other family members to use the 'old' machine or wanting to connect to other PCs such as laptops. In considering a home network, the consumer is faced with several, somewhat conflicting technologies.

4.4.1 Wireless

For speed of installation, wireless LAN technology seems to be the obvious choice. Indeed when looking at the specifications of the latest IEEE802.11b standards, the data rate of 11 Mbit/s looks very attractive compared to legacy 10 Mbit/s 10baseT wired systems. It is for this reason that wireless seems to be the preferred option for valued-added resellers wanting to supply bundled home-networking solutions.

However, there are some 'downsides' to this technology.

- Cost

  The first and most obvious is the cost compared to wired solutions. A typical system to network four computers via an access point (similar to a network hub) would cost at current prices about £500 although this could become much cheaper if the technology really takes off in the home.

- Performance

  Secondly, the performance may not be quite as good as expected. Real-world data rates are less than 5 Mbit/s and may be even lower than 2 Mbit/s depending on the PC's proximity and types of wall/floor materials, etc. Interference is also a problem, particularly from devices such as a microwave oven, which can radiate comparable levels of microwave energy around the same spectral frequencies that the wireless LAN uses.

- Security

  Finally, the lack of security may be a major oversight by the user who is unaware of the issues surrounding wireless LANs. Either way, some form of wireless encryption should be used and this again can impact on data rates.

The truth is that wireless LANs have yet to be installed in large numbers particularly in the home environment and issues of interference and security may yet become more significant as the installed base rises. However, products are currently available and BTopenworld already offers a wireless LAN product as an easy and fast way to encourage end users to network their PCs and potentially make better use of their broadband connection.

4.4.2 Mains-Borne Signalling

"Every home has a ready-made network with access points as convenient as the nearest power outlet ..." or so proponents of mains-borne signalling would have us believe.

The basic problem with using mains power wiring for networking is that the system was never intended to be used for high-speed data — that is not to say that the system cannot be made to work. Indeed current systems have been shown to work with data rates as high as 10 Mbit/s, but such complexity comes at a price and again there are issues of interference and security to deal with. Even the innocuous fridge can be a major source of high impulsive noise on the mains as the compressor switches in and out. Obvious sources of more serious interference are found with any device which 'chops' the mains voltage to derive another voltage, e.g. fluorescent lighting, light dimmers, etc. Such problems can be overcome but will probably result in a data throughput significantly lower than the theoretical maximum. To date there do not seem to be many off-the-shelf mains networking products.

4.4.3 Home Telephone Wiring

It seems obvious that, if the home telephone wiring is being used to broadcast the ADSL signal about the home, then the same wiring could be used to carry network traffic. This approach was first developed in the USA several years ago and resulted in the Home Phoneline Networking Alliance (H-PNA) [5, 6]. Indeed, the original draft specifications were intentionally changed as xDSL technologies evolved, so that H-PNA systems could co-exist without interference as the first proposals used similar carrier frequency ranges. Recently a new version has become available,

H-PNA2, which increases the typical network data rate from 1 Mbit/s to 10 Mbit/s, i.e. comparable to 10baseT systems.

The attraction of these systems is that they exploit bandwidth in the home telephone wiring that is above that used by the xDSL carriers, and so, by using this network, offer the end user a ready-made home network. This is, of course, true, provided the telephone extension sockets are conveniently located near the required points. In practice, this is more often the case in the United States where a large number of telephone outlets in the home is more common than in the UK. Thus unless the sockets are conveniently located it probably makes more sense to install purpose-made category-5 (Cat5) network cable rather than category-3 (Cat3) telephone wire.

This leads to probably the last option available and that is using conventional Cat5 wiring.

4.4.4 Traditional Cat5 Wired Networking

In terms of performance and cost this represents the best value for money option for the consumer. For as little as £100 it is possible to install a complete network including hub, network cards (NICs) and cable that is capable of networking, say four PCs at speeds of typically 100 Mbit/s.

Already products are appearing that will push speeds up to 1 Gbit/s in the near future. The downside in this case is the time and effort to install the Cat5 cabling and for this reason this networking technology, while the preserve of most offices and businesses, has failed to really take off in the home.

While the price for this technology may be attractive, most consumers are not prepared to invest the several hours required installing and terminating cables and tolerate the potential cosmetic damage that may result from installing cables around the home. This assumes that they have the necessary skills (and tools?) to even undertake the work. While home networking kits do exist that contain Ethernet cables and network interface cards (NICs) to network several PCs, these kits are mainly for the small office home office (SOHO) end of the market and not for the home DIY installation.

It is conceivable that a better 'home DIY' kit could be produced with user-friendly instructions, RJ45 wall sockets with screw terminals, white (rather than grey) cable, etc, supplied in the same way as home telephone extension kits with throw-away plastic IDC tools. However, current thinking is that such an approach is still only of interest to the 'early adopter' and not the average broadband user.

4.5 The Need for a Home Gateway

Once the technology has been selected for the home network and the system installed, it soon becomes apparent that some form of home gateway is required (see

Chapter 5 for further details). Even without broadband this becomes an issue, since it is often desirable for all the users on the network to be able to share resources such as printer(s), files and, more significantly, Internet access. The latest versions of Microsoft Windows operating system make home networking more evident to the home user. Special utilities are now provided with Windows ME, 2000, XP, etc, to automate the configuration of PCs on the network and, in particular, configure shared Internet access. While this is even of benefit to narrowband users, it is particularly beneficial to broadband users with always-on connections. Each user on the network has potentially a permanent Internet connection providing the gateway PC is running.

This, however, presents another problem in itself. What if the gateway PC is not being used? It is wasteful just to run a PC to provide Internet access for others. With the 'USB connected PC' described above, this is the only way to provide multiple access to the Internet, as the modem must be hosted off one PC and that PC is then the Internet gateway. A better solution would be a dedicated ADSL modem/router that could act as the home gateway to any PC. This is effectively what is provided by the business Ethernet product as supplied by BT Wholesale and shown in Fig 4.5. This product provides a direct Ethernet presentation to the end user and has an in-built 4-port hub to allow direct connection of up to four PCs.

Fig 4.5 Business Ethernet modem/router product supplied by BT Wholesale.

The reason that the ADSL modem/router is not more widely used in the home environment is simply down to the price of this item compared to the consumer-oriented USB product.

In spite of this, however, a low-cost ADSL modem/router acting as the home gateway is essential if the ADSL service is ever to migrate from just high-speed Internet to encompass other value-added services such as VoD. As has been mentioned above, while the USB-connected solution affords low-cost entry level CPE, it is not really the preferred way to route different services into the home.

4.6 Additional Services that a Home Gateway Could Provide

From a network or service provider's perspective, it is desirable to be able to supply further value-added services on top of high-speed Internet-connection sharing that makes the decision to buy or rent a home gateway even more appealing. Indeed, the secret to holding on to the customer in the face of increasing competition and so reduce churn is being able to provide distinct product offerings that differentiate your service from others. Many of these potential services are relatively straightforward to provide once the customer has a broadband connection and makes justification of the monthly rental that much easier!

4.6.1 Simple Home-Network/End-Device Management

Probably a major obstacle to home networking taking off in the short term is the perceived complexity of configuring the PCs to use the home network. Any user that has been forced to make changes to their PC 'Network Neighbourhood' or 'File and Printer Sharing' will be dissuaded from trying to configure the settings themselves. Indeed, while some users may have a vague understanding of what an IP address is, they perhaps have no idea how to set the subnet mask or DNS/gateway settings. Again, this is a service the home gateway could offer by acting as a DHCP server, assigning these values automatically and leaving the user to plug in the cables or switch on the wireless link. Even the issue of network address translation (NAT) could be fully automated, thus allowing the end user to exploit the maximum leverage from a single fixed IP number, which would allow the gateway to carry out the IP translations automatically. Without this functionality, the user is potentially forced to set individual IP numbers to ensure correct operation of their equipment. NAT is employed to 'reuse' IP numbers locally as there are not sufficient numbers available with the current 32-bit numbering scheme to allocate to every device its own unique number.

Until IPv6 is sufficiently widespread to make it possible to pre-assign IP addresses in a similar way to media access control (MAC) addresses, there will always be a need for gateways that provide IP mapping services to ensure the most efficient use of the existing IP number pool. For this reason, these types of protocol are probably best handled in a home gateway rather than in connected PCs or other Internet devices. There is also opportunity for the gateway supplier to provide special 'customised' versions of the home gateway kernel specially adapted to particular home networking scenarios/devices.

4.6.2 VoIP/VoDSL

Voice over IP or DSL is probably more applicable to business customers but could be a service which, if appropriately priced, could make a home gateway connected to broadband very cost effective compared to traditional telephony services. It is very likely that the more feature-rich home gateways will provide some form of derived POTS that could be used as a second line/third line, etc, for low-cost telephony.

To really exploit voice telephony using DSL, it makes more sense to use symmetric DSL (SDSL) where the data rates upstream and downstream are comparable and ideally 1 to 2 Mbit/s. With such a symmetric line speed it is quite easy to provide several (4, 8, 16, etc) voice channels as required, providing the ISP/network provider supports this type of service since the voice traffic is extracted at the DSLAM and routed over standard voice circuits. The advantage of this approach is that it offers voice services of the standard currently provided by normal POTS.

Alternatively, lower cost home gateways could provide VoIP as a bundled application (with or without video using a USB-connected camera) providing a lower quality but cheaper voice service for longer distance or international calls where the cost savings could justify the lower performance/time delays. In effect, what the user gets is the ability to make many more simultaneous voice or multimedia calls without the need for extra lines.

4.6.3 Game Consoles

We have yet to see on-line games really take off on consoles. While it is quite common for many current PC games to support Internet gaming, this is less common on the console. The reason for this is that even the latest consoles do not always include some form of on-line connectivity as standard. While the Sega Dreamcast was one of the first consoles to include a modem as standard, the promise of on-line gaming failed to make this product a real success and it has subsequently ceased production.

Arguably, if the Sony PS2 was designed today, it is likely that some form of broadband connection would have been added as standard. This has now been recognised as a requirement and Sony are promising a broadband add-on which provides an Ethernet connection. This is also to counter the threat of the Microsoft X-box which provides a network connection as standard.

Perhaps broadband will be the final catalyst which makes on-line gaming really take off, with the promise of an always-on connection and sufficient bandwidth for even the most intensive of games.

The home gateway could again be key to this connectivity providing the required Ethernet (or wireless) interface and routing functionality without the necessity for a dedicated PC. This is particularly true for the Sony PS2 which could be supported via its native USB connection thus eliminating any need for the broadband adapter at all. This in itself could be quite an incentive to purchase a home gateway if consoles such as the PS2 were supported directly for Internet gaming without the need for any additional hardware.

4.6.4 Home Security

When looking at security, various aspects should be considered.

- Home network security

  Security is becoming increasingly important since always-on connections could afford hackers much more opportunity. There is a market for applications that protect the home network. While home gateways should include some form of firewall software, the ability to simply and effectively configure this functionality as the customer wants will be key to a successful product.

- Surveillance

  The potential security benefits can go further. By providing an interface such as USB and bundling a USB camera and suitable software, the home gateway can double as a low-cost home surveillance system. Only with an always-on broadband connection does it become cost effective in the home environment to run an in-built Web server application to which you could log in from anywhere around the world to check your home (or indeed the baby sitter).

  Other lower data-rate sensors such as passive infra-red movement detectors and door/window switches could similarly be networked (perhaps via wireless or the mains) back to the home gateway to complement the information provided via the camera. Indeed, the whole market for valued-added telemetry services in the home becomes a viable product proposition again now that an IP-aware data service is continuously available in the home. Instead of having to design a bespoke telemetry system just for a particular application, e.g. utility meter reading, it is now possible to use the same home gateway to interface to a range of applications. Also gone are the issues of how to interface to the telephone line and how to avoid interference from other CPE (i.e. telephones, faxes, answering machines, etc) which are also monitoring the line for incoming calls.

Now, via the home gateway's Ethernet, USB, wireless, mains, POTS or traditional serial interfaces, comes the ability to connect a host of products that can interwork and offer real benefits to the end user.

Perhaps the ability to remotely close the curtains, programme the video, switch on the oven, turn on the central heating via a simple HTML Web application running on the home gateway will finally become a reality!

4.7 Summary

We are now beginning to see more home-oriented gateway devices which are designed to provide the services required by the domestic user, offering routing functionality to support multiple PCs, firewalls to protect against unauthorised access, a file server, and potential support to non-PC devices such as TVs. It is important to think 'outside the box' when deciding the functionality that could be supported by the home gateway as traditional 'Wintel' solutions using PC-like hardware and software may not be the most cost-effective solution or the most appropriate for a system without a keyboard and screen. Perhaps now more than ever, the timing is right for other suppliers, such as consumer equipment suppliers, to seize the initiative and, using open standards and software (such as LINUX), produce highly featured products without high royalty costs and win the battle for the home gateway product of choice.

Despite many attempts to supply products that combine the functionality of the TV with the PC, such as the 'Internet Access Device' from Access Devices [7], these two consumer devices are still kept separate by most users. Broadband connectivity via the home gateway could at last provide the services to make this convergence happen. The bandwidth available using the broadband service coupled with recent advances in MPEG encoding and the on-going cost reduction in decoders provides the possibility that the home gateway could do much more than Internet access. The unit could also support specialist TV services which otherwise are not commercially viable. Such an approach could even assist in the Government's plans to advance the digital TV take-up by providing an alternate delivery mechanism to satellite in areas where digital coverage is problematical.

To achieve this and also provide other higher bit-rate services will almost certainly require VDSL data rates of 8 or even 12 Mbit/s. This is, of course, possible, but the expected reach may necessitate fibre-to-the-cabinet to ensure reliable performance and therefore the cost of supplying the service could be prohibitive when compared against the expected revenues.

Again, the technology used for home distribution may become significant since a wireless LAN supporting data rates of 5 Mbit/s may not be sufficient for good-quality digital television distribution without some form of quality-of-service (QoS) protocols to avoid 'glitching' when other traffic is put on the network by other users. New versions of the wireless standard (IEEE802.11a) take the underlying data rate to 54 Mbit/s and therefore should be sufficient to avoid collisions. Alternatively, the humble wired LAN, offering 100 Mbit/s today and 1 Gbit/s soon on Cat5E cable, may be more than enough for today's and tomorrow's services brought to the home via broadband and distributed via a home gateway for the foreseeable future.

References

1 ITU-T —http://www.itu.int/publications/

2 On-line index for BT SINs and SPINs, etc — http://www.sinet.bt.com/

3 USB specification — http://www.usb.org/

4 BTopenworld — http://www.btopenworld.com

5 Home Phone Networking Alliance — http://www.homepna.org/

6 HPNA products — http://www.homepna.com/

7 Access Devices — http://www.accessdevices.co.uk/

5

RESIDENTIAL GATEWAYS

P M Bull, P R Benyon and R Limb

5.1 Introduction

A significant proportion of homes now contain computers connected to the Internet, and an 'always-on' connection is becoming a reality for many. Increasingly, people have more than one computer and are starting to network them. There is a steady move towards convergence of entertainment and computing, with set-top boxes offering Internet access and PCs configured as entertainment centres. Looking ahead, it is envisaged that there will be substantial numbers of networked appliances and embedded devices around the home. This offers enormous business opportunities for service providers, but the infrastructure must be as reliable as the existing utility services and just as easy to use. The residential gateway (RG) is seen as a key component in the home network, but appreciation of the form and function of the RG varies quite widely.

The first section considers the evolution path for the RG as it emerges from a simple hub through to the services platform of the future. Then the alternative proposals for RG location are discussed, from the traditional telco view that it should be part of the core network, through to the consumer view that it is a domestic appliance that they own. The key management issue of autoconfiguration is looked at from the perspective of a future where devices are becoming pervasive and range down to the very small. Then the management of applications on the RG are examined, with particular reference to the Open Services Gateway Initiative (OSGi). Finally some thoughts about the future are presented.

5.2 Evolution of the Gateway

For many people, the first stepping stone into home networking is driven by the need to make two (or more) computers share resources such as files, printer(s) and a single Internet access link, commonly using modem-sharing facilities on one of the PCs (Fig 5.1). This can work reasonably well for a few machines, although conflict can arise between the use of the 'gateway' machine as a PC, e.g. to run games, and its reliable operation as a network access point. It is also likely that PCs are not left operating continuously, and therefore the gateway may not be running when needed. As this configuration usually does not address security concerns either, it becomes necessary for the user to install and maintain personal firewalls on each connected machine.

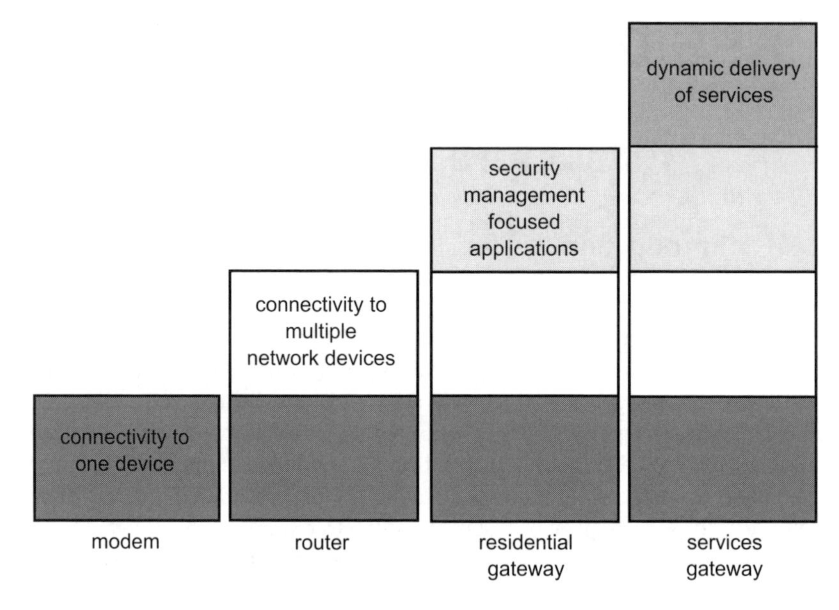

Fig 5.1 Gateway evolution.

Network device vendors have responded to this by enhancing existing products to deliver devices offering some or all of the following:

- modem (ISDN, DSL, cable, etc);
- multi-port hub (usually 10/100BaseT Ethernet);
- basic firewall, plus address/port translation;
- dynamic IP address allocation (DHCP) services.

These have the benefit of being small, self-contained units that are more likely to be left powered and provide some measure of network security, but, as they require some expertise to set up, they tend to be better suited to the more technical user. They are unlikely to have any capability for remote management or diagnostics and are often more accurately described as 'home hubs' rather than residential gateways. They may be connected to ISDN services offering 'near-always-on' access, or to xDSL or cable services that provide 'always-on' access.

Although applications such as home automation and security are frequently cited as key applications for home networks, they are unlikely to be prime drivers for installing a home network infrastructure. Broadband data and especially entertainment services, are seen by service providers as offering more compelling business opportunities, leading to the emergence of a new range of devices aimed primarily at service providers and gateway operators. These offer additional features over the home hub, such as:

- automatic service detection, e.g. for DSL;
- support for multiple physical network technologies, e.g. Cat5 10/100BaseT Ethernet, Phoneline and wireless;
- derived voice telephony facilities — some offering quite sophisticated PBX functions;
- network security/firewall functions;
- remote management by gateway operator (GO).

Some of these, in the form of a set-top box (STB) with digital TV decoding, are aimed at the broadcast entertainment audience. Others focus more on broadband data and voice, but can pass entertainment data through. These boxes are envisaged as being provided as part of a service package and managed by the service provider. Because these devices are to be designed to fit the stringent cost targets of the service providers, they tend to have limited expansibility and capacity for additional applications.

As the RG is starting to emerge, attention is already turning to the services gateway (SG). The SG is a platform on to which applications and services can be securely delivered and managed. It is desirable for this platform to be standardised in some way so that applications are not specific to any particular product. One solution to this is being developed by the OSGi, which is described in more detail in section 5.7.

5.3 Inside, Outside or Centralised

It might be assumed that the gateway must be located within the customer's premises, but some propose that the gateway should be external to the premises and even be located in the network itself. A key business driver is that of gaining 'ownership' and control of the territory — the customer premises.

A traditional telco approach would be to assume that there will be a single provider of services (service aggregator) who delivers services via a single access route and who controls the gateway (gateway operator). The service aggregator and gateway operator may actually be separate organisations, but appear to the customer as an integrated service provider. A benefit to the customer is that they are protected from the complexities of installing and managing the operation of their network and from the risk of service interaction problems. Security and application updates can be managed remotely and in the background. The gateway box, which will, of necessity, be built to a target cost, may then be provided in the price of the subscription or on a lease basis. Some of the RG devices are clearly being designed with this in mind and the vendors are developing gateway management systems (GMSs). Indeed, this model and the various operator roles are explicitly addressed in the OSGi specification [1]. Telcos are natural contenders in this market, since they have experience of operating reliable mass-market telecommunications infrastructures.

Some propose that, in order to control quality of service, the customer should not be allowed to interfere with the gateway, in the same way that historically they were not allowed to interfere with the primary telephone point. Instead, the gateway should be located and secured on an outer wall of the customer's premises. Some further propose that applications and functions such as security should reside on servers in the service provider's central office so that the customer does not have to worry about upgrades, capacity limits, etc. The hardware required in the customer's premises can then be simpler and of lower cost. The service provider benefits from economies of scale, and contention ratios, through centralisation, potentially lower the cost of entry for the customer.

It is likely that the gateway will reside at the customer's premises and that gateway functions will be distributed, rather than performed by a single device. There is a new generation of 'disruptive users' [2] who have become used to the Internet culture. They are accustomed to subscribing to services from a number of providers, and to downloading and using new applications, and they may use multiple ISPs. They have control and demand rapid response to changes. They are also less tolerant of a supplier imposing changes on their environment. There will also be multiple gateways and access paths provided by many different operators. For example, in our own homes, we already have digital satellite TV via a set-top box, a gateway to the Internet and telephony via our telco, terrestrial TV feed (digital TV services do not scale very well in a family home yet), mobile telephones, etc. There is no reason why our cars should not contain a gateway that can connect to the home network when in proximity to the home and the OSGi vision extends to include this.

In principle, the home network has some similarities with our existing utility systems. The gas supplier owns and maintains the gas meter with which, though installed in our home, we may not interfere. The gas boiler, which produces hot water and heats the house, is owned and controlled by us.

5.4 Management Issues

From the perspective of a service provider, the RG should have a blend of:

- autoconfiguration — to minimise 'truck-rolls' and support calls;
- simple user interface for local management;

- remote management capability to enable diagnostics, upgrades and service management.

On installation, it would be desirable for the customer to simply plug in the gateway, have it detect service presence and connect to the gateway operator's management centre to register itself. This is not necessarily a straightforward process at present. For example, a DSL service may be presented at the network termination in one of a range of configurations, requiring the gateway interface to be appropriately set.

Once registered, the gateway management system could:

- initiate and check results of any self-tests to verify correct operation;
- perform any required software updates;
- enable customer-specific service options supported by the gateway, including things like security configuration.

Figure 5.2 is a schematic representation of possible components of an RG system.

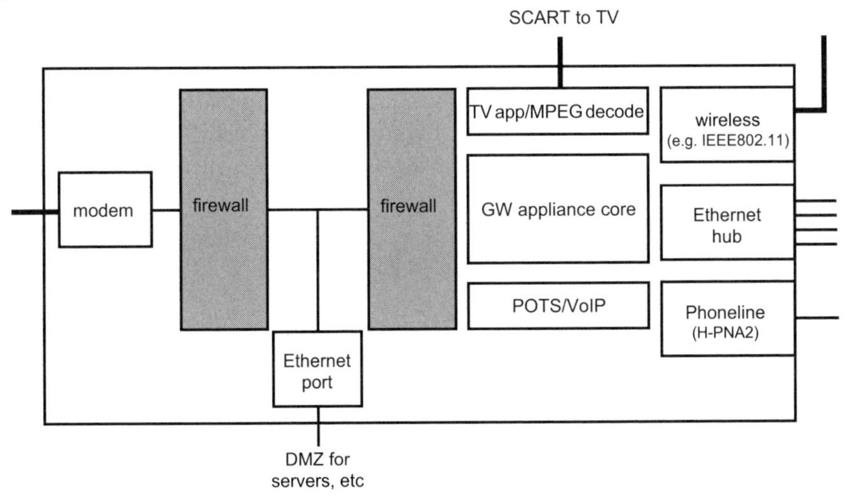

Fig 5.2 Schematic of a gateway.

The access network connects to a modem (DSL, cable, etc). Immediately behind this is a firewall providing network security, possibly extending to the application level. Rather than feeding directly to the core functions of the gateway, Fig 5.2 shows two firewalls separated by a demilitarised zone (DMZ) in a fashion similar to that commonly adopted for enterprise security. To the DMZ may be connected externally facing servers, application proxies and maybe even wireless LAN access. This may seem over-complex for a home environment, but there are some prototype products with this facility.

The core functionality of the gateway is next and will vary according to the target market for the specific product. The schematic in Fig 5.2 shows an STB system that contains TV decoding and connects directly to a TV. It also contains a network bridge that allows various types of physical LAN to be handled, feeding into a 10/100BaseT Ethernet hub, Phoneline (e.g. H-PNA2 [3]) interface and wireless LAN hub, for example.

A secure management channel is required for the GO to maintain the firewall(s) and the core application services of the RG.

Looking at the home side of the RG, the problems of configuration and management become somewhat more interesting. The customer is likely to own and have control of a wide range of devices. The RG may offer services such as DHCP to automatically assign addresses for IP devices, but the topology of the network and the wide range of devices that could be connected to it are arbitrarily complex. There is a need for an infrastructure that supports autoconfiguration.

5.5 Devices and Autoconfiguration

One of the drivers for having a residential gateway is the vision that in the future there will be an increasing number of 'smart', networked devices around the home. This is beginning to happen today with personal digital assistants (PDAs), networked printers and cameras with built-in Web servers. Some white goods manufacturers are also starting to seriously consider network-enabling their products.

It is one thing to enable these devices to be connected, but quite another to get the devices to communicate with each other and/or central service(s). Some of the issues are:

- knowing what protocol the device is using;
- knowing where the device is in a network (both in a logical sense and in a physical sense);
- knowing when the device is connected and available and when not;
- knowing something about the capabilities of the device.

Not surprisingly, there are a number of proposals and initiatives under way to address one or more of these issues. The following sections provide a brief overview of some of them.

5.5.1 UPnP (Universal Plug and Play)

Over the last few years, people have become familiar with the idea of 'plug and play'. This has enabled them to plug devices into their computers and have them automatically detected by the operating system and the appropriate device drivers are then loaded. Microsoft and others are now building on the idea (although it is much more than a simple extension) with a framework called Universal Plug and Play (UPnP) [4]. The idea is that even quite simple devices, provided that they are capable of communicating using TCP/IP and HTTP (the basis of Web server communication), can be automatically discovered on a network, be configured and even be controlled remotely, if that is appropriate.

The concept extends to component services within a composite system. For example, a digital television contains components including a display, MPEG decoder, audio amplifier, etc. A Webpad might contain a display and headphone output. Each of these components can be published as UPnP services, so that a user can run an application on the Webpad that uses the MPEG decoder in the television to deliver video to the Webpad display and audio to the Webpad headphone output.

It is beyond the scope of this chapter to fully describe UPnP, but, briefly, UPnP defines a set of protocols that devices use to describe their services and capabilities. These are 'advertised', or broadcast, over the network for other UPnP devices or control points to discover and then access. Alternatively, control points can search the network for UPnP-enabled devices. UPnP is based on existing protocols, using TCP and UDP and, above that, an extension to HTTP. Information is passed using XML, which is now well established, via a protocol known as SOAP (simple object access protocol) [5, 6].

One aspect of UPnP which does cause the authors (and others) concern is security.

The current specification does not really address this at all. There are no measures a user or developer can take within the UPnP specification to prevent access to devices or authorise their use. The original view was that UPnP would only be used in the relatively benign, closed network in the home and, as such, there was little risk. However, with the increase in 'always-on' connections from the home to the Internet via cable or xDSL, it is now realised by the UPnP Forum that security must be addressed. One can expect that as the specification matures, some security measures will be added and hopefully implemented. Currently there are very few UPnP devices available and, while more can be expected in the near future, it may be that such devices will not proliferate until a more secure environment can be demonstrated.

5.5.2 Jini

Jini [7] is another discovery framework. Based on the Java platform, Jini, from Sun Microsystems, will not only discover devices, but services as well. This is important because as devices become smarter, they not only provide some basic functionality, but may well provide one or more services to other devices or applications on the network. While UPnP is limited to transfer of (XML) messages to invoke remote actions, Jini allows the movement of Java objects around the network. This is very powerful, but can become quite complex and has the possible limitation in that a Java platform is required. However, it is not imperative that objects are transferred;

there are interfaces to SOAP and CORBA [8], so that bridges can be built to other protocols.

5.5.3 Salutation

Salutation [9] is an open architecture that again aims to provide a framework for service discovery and session management. The aims are very similar to those of Jini, but attempt to overcome one of the perceived limitations of Jini — the dependence on the Java platform. Salutation claims to be OS, platform and language independent. The architecture is open and completely licence and royalty free. There is a 'Lite' implementation that is Open Source and is targeted at small footprint devices. A number of manufacturers are already supporting Salutation in some of their products, including Canon, Ricoh and Fuji Xerox. The Salutation architecture defines Salutation Managers (SLMs) which co-ordinate with each other and mediate between devices. The SLMs present the Salutation API and drive the protocol on the network. Services are decomposed into functional units, which are an abstraction of the functions provided by an application or device. One of the goals of Salutation is not to be tied to a particular protocol. While a Salutation Manager Protocol is defined, other open protocols may be used. In the short or medium term, it does not seem likely that any one of these protocols (or some new protocol) will become an outright winner. Devices and systems will be available for some time that will use one or more protocols. Our view, shared by others in the industry, is that a gateway can mediate between these protocols and the devices using them. The next section describes this in more detail with some simple illustrative scenarios.

5.6 Pervasive Computing

A related driver for a residential gateway, and a vision for the future of computing and devices, is an area of research variously known as 'Pervasive Computing' [10] 'Ubiquitous Computing' [11] and even the 'Invisible Computer' [12]. The common theme through all these areas of research is the idea that computers (anything from the latest PC to a small computational device embedded in an appliance) will be in many places and available for interaction, but more importantly will be networked and will co-operate with each other. Thus your PDA will no longer just be a 'small piece of neat technology' that you carry around to hold your diary and personal data. For example, it will interact with your local environment to become an audiovisual remote control when you enter the lounge of your home. Not only that, but it will display what programmes are on now and what you may want to see (or record) based on your personal pre-set and learned requirements. To achieve this scenario, a number of devices and technologies may well be active — the PDA, location-sensing devices, TV listings server, personalisation server, a data network, IrDA

(Infra-red Data Association) devices, an audiovisual recording device. However, the user does not wish (nor need) to know this. What the user wants is for it to 'just work'. Another way of looking at this is that computers, sensors and devices will all become part of the infrastructure.

It is possible to provide solutions to the scenarios just described that encompass all the sensors, devices, software, etc — a bespoke system. But what of the users that already own half of the technology required? They will not be interested in buying more of what they already have, but would rather extend the capabilities. Perhaps they have a network and some sensors that could contribute to the solution but use a different protocol. In this case a central system could be employed to make use of the existing technology, translating protocols as appropriate, adding software and configurations where required, managing the auto-detection and configuration of devices, etc. Such roles could be attributed to a residential gateway, or, more particularly, a services gateway such as is provided by the OSGi platform.

5.7 OSGi — The Open Services Gateway Initiative

5.7.1 What is OSGi?

The OSGi is an independent, non-profit corporation working to define and promote open specifications for the delivery of services over wide-area networks to local networks and devices. The extensive list of member companies includes service providers, network operators, utilities, software developers, box vendors, device manufacturers and research institutions.

The technical work of the OSGi is carried out in six expert groups looking at the core platform, architecture, devices, remote management, security, and vehicle applications. Much of this work is targeted at creating APIs for new services that could be deployed upon a service gateway. Several working OSGi software frameworks have been produced and released as products.

5.7.2 Specifications

The OSGi service gateway specification defines a Java-based software architecture designed to support the development and deployment of services on to a services gateway. By defining an open, common architecture, the intention is to provide a framework that enables service providers, developers, software vendors, gateway operators and equipment vendors to deploy and manage the services on a gateway such as a set-top box, cable modem, DSL modem, PC or other dedicated box.

OSGi specifications consist of a set of Java APIs that, when combined with a working service framework, provide the necessary environment for downloading and managing services. The first OSGi specification (service gateway 1.0) was released publicly in May 2000 and included the framework and three basic service specifications for logging, Web serving and device access. Subsequently, the OSGi service platform 2.0 [1] has improved and extended the existing APIs by defining services for user administration, configuration management, and bundle preferences management.

5.7.3 Background

At present, service providers all have their own wires into the home or business. As the variety of communications technologies used by the different utility, telephone, cable and TV companies diversifies, the infrastructure becomes more complex. As a result, many opportunities to integrate the hardware and software to provide services such as home monitoring, telephony and audio/video are being lost.

The OSGi approach is to run a service gateway on the network of each home or business to provide an integration layer — a configurable software framework which allows managed software modules, possibly from different vendors, to communicate with each other using agreed service APIs. Normally, the framework software would run on hardware which also incorporates a modem, firewall and other Internet-related applications — although, if this is done, it is clearly very important to ensure that the OSGi platform itself is protected from external attack.

The platform neutrality and 'network-aware' nature of Java makes it a natural choice for such an initiative. The availability of interfaces to other devices, which conform to existing standards (e.g. Jini, HAVi, UPnP), to perform other tasks also favours Java.

5.7.4 Business Model

The business model associated with deployment of services based upon OSGi is still at its formative stages. Several roles have been envisaged:

- framework provider who supplies the basic framework software;
- gateway manufacturer who builds the hardware;
- gateway operator who installs and manages the gateway;
- service provider who designs and provides the software for a service;
- service aggregator who integrates services from different providers;
- service operator who installs and manages the software for a service.

Of course, it is likely that for a given installation, one company will play multiple roles. The overall aim is to allow services to be dynamically loaded and upgraded eliminating the need for 'truck rolls' by enabling networks and devices to share a common gateway.

In effect, the OSGi project is attempting to lower the deployment and life-cycle costs of delivering services to homes and businesses through the use of common standards embodied in the service gateway. In principle, this should also mean that services can be run on any gateway that complies with a given version of these standards.

5.7.5 Framework Fundamentals

As outlined above, an OSGi framework provides an environment for electronic downloading of services (Fig 5.3). This environment includes a Java run-time system and support for life-cycle management (starting, stopping, etc), persistent data storage, version management, and a service registry.

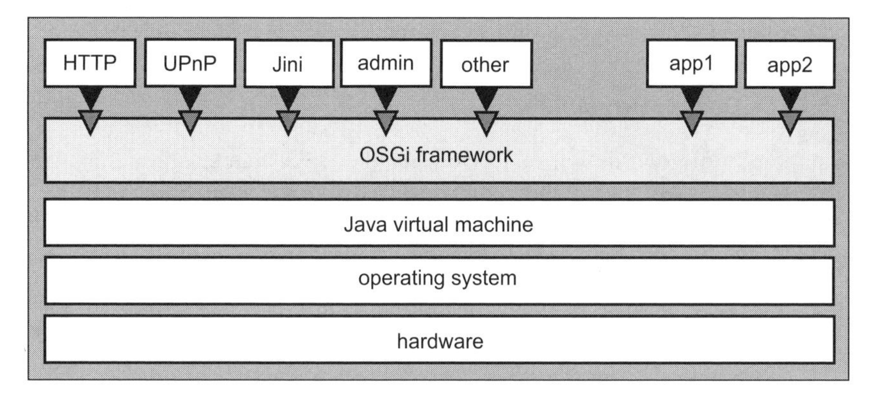

Fig 5.3 OSGi framework environment.

Once this basic framework has been set up on a customer's network, services may be added to it through the managed download of software components called 'bundles'. In practical terms, a bundle consists of a Java archive (jar) file that can be downloaded from the Internet. Java archive files are identical to 'zip' files except that they contain a special file called a manifest that tells the framework about the services that the bundle implements.

As well as Java class files, a bundle can contain other resource files, such as Web pages, data and images, that can be used by themselves or other bundles. Each bundle has its own Java class loader which is restricted to loading its own classes or packages of classes explicitly 'exported' to the framework by other bundles as defined in their manifest files. This provides the developer with a degree of isolation not normally achieved by manipulating the Java CLASSPATH. The class export mechanism also supports versioning in that a bundle can export or import particular versions of a Java class package.

Once a bundle has been successfully installed, it can take on several different states (installed, resolved, uninstalled, starting, stopping and active) under the control of the framework and its user interface. The framework tracks the state of each bundle, ensures that interdependencies are managed and if necessary restores bundle state, for example, when the framework is restarted. The framework also manages the passing of 'events' between bundles as their state changes. Events are generated when services are registered/unregistered or when the framework is launched or has a problem. If the installation of a new bundle fails, the framework can revert to an old version of the bundle if it exists.

A bundle may register ('publish') one or more services with the framework which effectively makes objects available for other bundles to call and use. Once a service has been registered, another bundle can query the framework and look it up using a set of lightweight directory access protocol (LDAP) criteria [13]. If it finds the service it wants, it can get an object reference and call methods. If a service becomes unavailable for some reason, the framework can inform a dependent bundle which can then take appropriate action.

5.7.6 Design Principles

The principles of good design of services for the OSGi framework are similar to those associated with any other component-based model such as CORBA or DCOM. While the details of how to design and implement OSGi services are beyond the scope of this chapter (but see Chen and Gong [14]), some basic principles can be identified:

- separate interface from implementation by placing them in different Java packages;
- try to design interfaces that grasp the intrinsic functionality of a service;
- factor out any common functionality into an independent service;
- exclude configuration aspects of a service from the interface;
- when the service interface has to change, try to maintain backwards compatibility;
- do not reinvent existing service APIs.

5.7.7 Device Access

The OSGi device access component allows bundles to interact with devices as objects in the Java virtual machine, attempting to integrate all devices into a single coherent framework. It assumes that low-level device drivers have already been installed and can be used to detect the connection of new devices, which the device manager can register as services, as they appear.

For instance, connecting a printer to the parallel port of a PC should result in a device with a parallel service interface being detected and registered with the framework. Since it may not always be desirable to talk to a printer via such a generic interface, a process of interface refinement is undertaken by the device manager to find better abstractions for the service. It does this by asking existing services registered with the framework if they can refine the interface, given its properties and characteristics. This will usually result in additional services being registered, providing bundles with the opportunity to talk to a particular device in a number of different ways.

5.8 Looking Ahead

It is tempting to look forward with a vision of the 'house of the future' containing a full-blown services gateway. As a utility appliance, it would be modular, allowing it to be expanded and upgraded over time as required. One can imagine it becoming the central 'controller' for the home, providing media serving, security, tele-medical services and a wide range of other functions that can be downloaded and maintained by a range of service providers.

However, there are a number of reasons why this seems unlikely to happen. Such technology is relatively expensive. A major home utility appliance, such as a gas central heating boiler, costs around £500 and is expected to have a useful life of some 15-20 years. A service gateway as outlined above could be expected to cost £1000+ and might have a maximum useful life of 5 years given the current rate of technology churn. Commercial drivers indicate that for some years ahead the RG will be a cost-constrained device of focused functionality that is supplied as part of a service agreement. We have already indicated that it is likely that there may be several access routes, each with its own gateway and service applications. From a practical standpoint, it is undesirable to concentrate the whole household functionality at a single point of failure; it makes more sense to distribute the functionality across a number of nodes.

Norman [12] argues that for products to move beyond the early adopters, technology must disappear to be replaced by simple, easy-to-use appliances. The likely scenario for at least the next few years is that there will be a range of simple (to operate), remotely manageable, appliance gateways providing access termination, some specific service function(s), and a network hub. The customer will have one (or more) of these. Behind these are likely to be network-attached appliances, such as a media server, games appliance, etc, so that the customer can build an environment with the combination of facilities desired and with manageable cost increments. These appliances may also have remotely managed updates and services through which the gateway appliance must pass. There will be issues about how these products co-exist and co-operate as vendors and service providers strive to gain territory and maintain their market; but, ultimately, customers will demand more open interworking so that their networks provide more benefit than the sum of the individual components.

If the longer-term future is to bring the dawn of pervasive, 'invisible' information appliances, where does this leave the services gateway and what are the issues of interoperability? While many people are willing to buy integrated solutions from one supplier, many, even in the mainstream, will want to select from a number of suppliers. One of the possible roles of the services gateway could be to resolve and manage the myriad of devices and their associated protocols. However, perhaps more likely in the long term, as devices become smarter and more autonomous, is that the services gateway functionality becomes distributed and the physical appliance disappears. Will the truly intelligent, autonomous device become a reality? How will people interact with it? What are the interface and social implications of such devices appearing in the home? These and related questions are currently exercising the research minds of industry and academia.

References

1 The Open Services Gateway Initiative: '*OSGi — the managed services specification*', — http://www.osgi.org/

2 Gillett, S. E. et al: '*The disruptive user — Internet appliances and the management of complexity*', BT Technol J, **19**(4), pp 40-45 (October 2001).

3 Home PNA — http://www.homepna.org/

4 UPnP Forum — http://www.upnp.org/

5 SOAP Forum — http://www.w3.org/TR/SOAP/

6 Selkirk, A.: '*Using XML Security Mechanisms*', in Regnault, J. and Temple, R. (Eds): '*Internet and Wireless Security*', Institution of Electrical Engineers, London, pp 45-62 (2002).

7 Jini Specification — http://www.sun.com/jini/

8 CORBA Specification — http://www.omg.org/gettingstarted/overview.htm

9 Salutation — http://www.salutation.org/

10 IBM Pervasive Computing — http://www-3.ibm.com/pvc/

11 Weiser, M.: '*Some Computer Science Problems in Ubiquitous Computing*', Communications of the ACM (July 1993).

12 Norman, D. A.: '*The Invisible Computer*', Cambridge, MA, MIT Press (1998).

13 Howes, T.: '*A string representation of LDAP search filters*', (1996) — http://www.ietf.org/rfc/rfc1960.txt

14 Chen, K. and Gong, L.: '*Programming Open Service Gateways with Java Embedded Server Technology*', Addison Wesley (August 2001).

6

HOME AREA NETWORK TECHNOLOGIES

C E Adams

6.1 Home Area Networking Overview

We have come a long way from the days when all we needed to do was connect two PCs together using an RS232 cable in order to share data, or to play against each other within a game of Doom! Today we want to share music and video files, share a common resource like a printer, or share our high-speed connection to the Internet.

Some of this, of course, has always been possible, but in the past has been too difficult, expensive or too restrictive in bandwidth terms. People are now realising the potential of sharing their digital content on a much wider scale using the power of the Internet.

Networking has moved on considerably and we are no longer limited to one or two technologies to fulfil the above needs. A whole array of both wired and wireless solutions now exists, the data rates of which range from a few hundred kilobytes of throughput, to a hundred megabytes and beyond. It is predicted that, in the future, bandwidth limitations will not be the main problem. The key issue will be the ease with which you can connect up your digital home.

6.2 The Connection Technologies

The range of home network interconnection technologies can be broken down into three distinct categories:

- new wires;
- no new wires;
- wireless.

Each of these categories brings with it a number of key advantages and disadvantages making it difficult to see if one individual solution will win the day. In reality the future of the home network will probably be a range of these technologies in co-existence.

6.2.1 New Wires

The thought of tearing apart your house to install a completely new network must horrify even the most DIY-proficient people. Ethernet and other similar wired technologies require point-to-point cabling which, if installed during the initial building stages of a house, presents less of a problem, but, if you need to retrofit in an older property, things can become complicated. Speed and reliability are key issues for justification of such an installation with other technologies currently not matching wired performance. Physical wires are currently the most reliable way of distributing high bandwidth applications around the home environment.

6.2.1.1 Ethernet

Ethernet is a reliable networking technology, which has been in use for many years and is probably the best known and most widely used networking technology today. It originated from the Xerox Corporation around 1974 and was developed by Dr Robert Metcalfe; it was originally designed to link Xerox Alto workstations and other related equipment. Xerox defined physical standards for connecting this equipment along with the communication protocols for use on the physical medium, the whole system having an overall bandwidth of about 2.94 Mbit/s. It was soon realised that this method of inter-workstation connection could be transferred to support any computer system and was adopted by a multi-vendor consortium, to become known as the DEC-Intel-Xerox (DIX) standard in 1980. This in turn was eventually adopted by the Institute of Electrical and Electronic Engineers (IEEE) to finally become the IEEE802.3 standard in 1985, which today most people know as Ethernet.

Being a close proximity system, the connection between nodes should be no more than 100 metres. It originally used coaxial cable to interconnect devices, but this has been replaced by unshielded twisted pair (UTP) cabling normally made up internally of four twisted pairs of wires. Data rates of 10 Mbit/s and 100 Mbit/s can be transported across inexpensive Cat5 UTP cable, but, in the future, the connection medium may well be fibre optic or Cat5e UTP cabling carrying Gigabit Ethernet with a 1 Gbit/s data rate. Costs of standard 100baseT Ethernet network interface cards (NICs) have fallen dramatically over the past few years and it looks like Gigabit NICs are following this trend.

The Ethernet physical layer is responsible for getting data on to the physical medium or cable. Sitting above this is the link layer which is split into two areas, the first being the logical link control (LLC) sub-layer which performs error checking, the second sub-layer being the media access control (MAC) which is responsible for getting data on and off the network and access to it by each connected node. This has to be carried out in a fair manner making sure that no single node can lock out others — this process is called carrier sense multiple access/collision detection (CSMA/CD).

Each device on the network waits until there is no signal present on the network (carrier sense), and when this is true it can start to transmit data. If another device is transmitting, it must wait until the network is quiet before attempting to transmit. All devices are equal on the network and thus no single device can take priority at any time (multiple access). If two devices do attempt to transmit at exactly the same moment in time, this will result in a collision. When this occurs, the two devices will stop transmitting and will attempt to retransmit after a random time period (collision detection).

The transmitted data is sent serially from the network interface in the form of frames made up of various fields (see Fig 6.1). These include 48-bit source and destination address fields, a data field and an error-checking field (CRC). These unique 48-bit addresses (equating to a HEX code) are pre-assigned and hard coded into the network interfaces by the manufacturers and are referred to as the MAC address. They act as an individual hardware identity for each node on the network. All network interfaces on the system look at the frames passing and only read an entire frame if the destination address matches its own; it will then pass this data up to the network layer for further processing. Two other types of address are used, multicast and broadcast, the former being used to deliver data to a group of nodes previously given a specific multicast address, unlike the broadcast address which will be interpreted by all network interfaces.

preamble 7 bytes	start frame delimiter 1 byte	destination address 2 or 6 bytes	source address 2 or 6 bytes	length LEN 4 bytes	data ¦ PAD 46 to 1500 bytes	CRC 4 bytes

Fig 6.1 IEEE802.3 frame.

The data field carries high-level protocol packets, such as transmission control protocol/Internet protocol (TCP/IP), between computers and is used for networking operations across the Internet. At the IP (Internet protocol v4) level a 32-bit address scheme is used to identify the node in use which is aware of its own 48-bit MAC address, but not the MAC addresses of other nodes on the wider internetwork. These are identified using a protocol called the address resolution protocol (ARP). The Ethernet topology is a linear bus made up of segments or LANs, which can grow by connecting repeaters. These repeaters, or hubs as they are more commonly referred to, normally have multiple ports and the data from the input is repeated to all outputs. The maximum number of hubs (normally 5) and segments must meet round-trip timing requirements for the network to function correctly. The network should also have two distinct ends and not be connected in a loop, which will cause operational problems. Finally, one more important device to mention is a switching hub that allows further individual segments to be connected and operate as separate LAN entities, thus reducing the number of nodes in each broadcast domain. This is achieved by packet switching electronics within the hub and is used to control the flow of data between designated ports on the device, which helps to manage data traffic on the network and reduce the risk of contention.

Extensions have been added for quality of service (QoS), via IEEE802.1p, in order to map traffic types on to IEEE802.1p priority codes, and VLAN support via IEEE802.1q. Switching devices can then use this information to determine the scheduling class to which a packet belongs, e.g. for mixed services such as video, voice and data.

This is only a brief review of how Ethernet works and there are many more complex issues not covered here. For further reference please see Metcalfe and Boggs [1] who give a more in-depth explanation of Ethernet and related subjects.

- Advantages of Ethernet:

  — established, proven and well-supported technology;

  — reliable medium;

  — high speed (e.g. 100 Mbit/s);

  — 100 m between nodes;

  — supports high-bandwidth applications;

  — high security;

  — low cost;

  — easy bridging to other networking technologies (e.g. HomePNA and Wireless Ethernet).

- Disadvantages of Ethernet:

  — requires new cables to be installed for home network;

  — may require some configuration by user.

Ethernet has been around for nearly twenty years now and is slowly making the transition from a corporate environment to the domestic arena. Formerly expensive networking equipment (e.g. hubs, switches and cabling) and the PCs themselves have dropped to affordable consumer prices making wired home networks feasible. Unfortunately it still requires a reasonable understanding of networking to get computers to talk to each other using this solution and there are many pitfalls for a novice user who just wants to be able to 'plug and go'. Attempts have been made to ease this problem with the introduction of simple networking kits containing network adapters, cabling and simple software installation wizards, but industry analysts, IDC, predict that by 2004 Ethernet-based home networks will only account for 15% of the total installed market.

6.2.1.2 IEEE1394 (Firewire®/iLink®)

The Firewire bus was originally developed by Apple as a high-speed serial bus to be used to connect printers, scanners, external hard drives and other associated hardware to PCs. The IEEE adopted Firewire (also known as iLink on Sony products) and in 1994 set up a trade association to help promote this new technology and to standardise a specification, eventually to be known as IEEE1394. In 1995 a specification was released, but early interoperability problems fuelled work on the next revision. By 1998 the IEEE1394a standard was born, giving data rates up to 400 Mbit/s and following this, a year later, the IEEE1394b specification became available, promising data rates of up to 3.2 Gbit/s and full backwards compatibility with earlier standards.

The HAVi (Home Audio Video Interoperability) consortium [2] has been involved in defining IEEE1394. This is a group of leading consumer electronics manufacturers, including Sony, Panasonic and Philips, whose aim is to provide a seamless and easy connection medium for new digital consumer products. IEEE1394 was chosen for its bandwidth capability, inherent QoS mechanisms, isochronous/asynchronous data flows and its support for digital copy protection. All are seen as crucial requirements for the delivery of future entertainment services over home networks. IEEE1394 has also been endorsed by the European Digital Video Broadcasting (DVB) Standards Forum as the standard interface for digital television in the future.

Being a peer-to-peer-based system (see Fig 6.2), it does not require a PC to act as a host on the network, unlike the current implementation of the universal serial bus (USB). This means that individual devices may be hot-plugged together in pairs or multiples, normally in a daisy-chain fashion, allowing instant connectivity for data transfer. The bus uses 16-bit addressing, giving provision for up to 64k nodes in total. A maximum of 16 hops is achievable in one chain using a standard 4.5 m cable length, giving a maximum network length of 72 m. Up to 64 devices may be connected to each bus segment as IEEE1394 uses a 6-bit physical ID address. Bus bridges are used to extend the network from a single segment and the use of a 10-bit bus ID allows for the connection of a maximum of 1024 bus segments. Devices can either be powered independently or draw their power from the bus, which can provide between 8-40 V DC at up to 1.5 amps. Cables are similar to Ethernet, containing two twisted pairs for data and two wires for power connectivity.

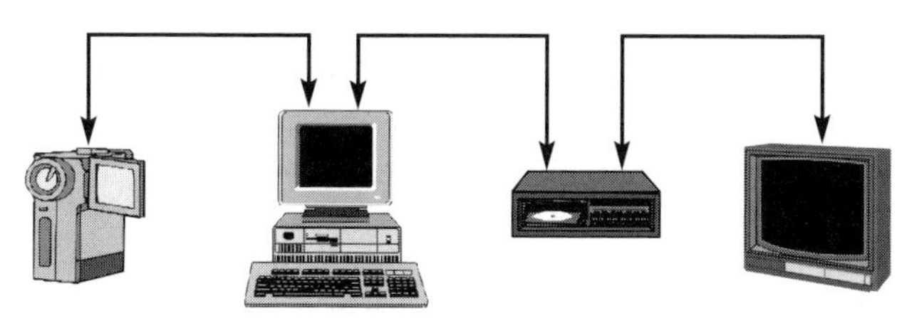

Fig 6.2 IEEE1394 peer-to-peer topology.

Like Ethernet, IEEE1394 has a layered structure (see Fig 6.3). There are three main layers and a serial bus management controller interfacing at all layers. The physical layer is responsible for the actual physical connections to the network (e.g. connectors) and the electrical signals carried. It performs both data encode and decode, along with arbitration of the actual devices connected to the bus. Sitting above this is the link layer that takes translated serial bus data from the physical layer and formats it into two types of IEEE1394 packets — isochronous and asynchronous. All packet transmission, reception and cycle control for isochronous channels are handled at this level, as well as error checking. The transaction layer handles several tasks including asynchronous transactions with the host controller and IEEE1394 network.

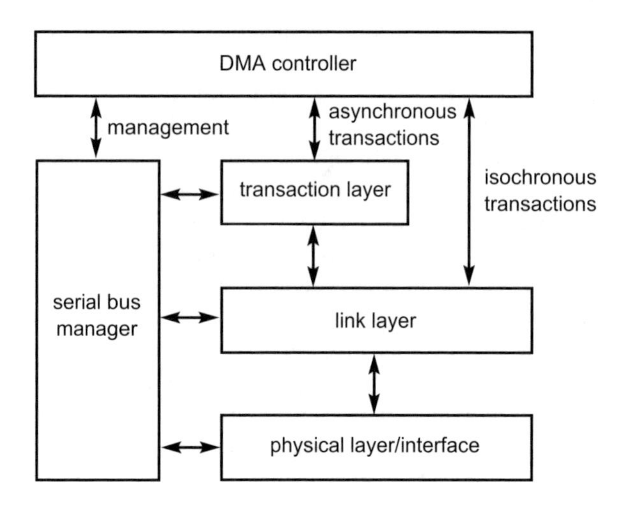

Fig 6.3 IEEE1394 protocol stack.

Finally we come to the bus manager which allows the connection of a wide variety of hardware devices. It dispenses with the need for a PC or other bus controller to be present and acts as the bus master in this instance. If any devices are connected which support isochronous communications, the bus controller will act as an isochronous resource manager.

Isochronous transfer of data is more suited to the delivery of video and time-critical applications, where timing is more crucial than the actual delivery of the packets. Here no error correction and no re-transmission of packets takes place. This is in contrast to asynchronous transfer that aims to guarantee the delivery of the data rather than the actual timing, making this more suitable to applications such as data transfer from disk drives.

USB is a similar networking system to IEEE1394 and is seen by many as a competitor, but in reality they are complementary, as each system has different bandwidth, cost and application targets. USB is more suited for PC-to-peripheral connection, i.e. scanners, printers, keyboards, mice, etc, whereas IEEE1394 is typically aimed at high-quality digital-video-based equipment, requiring high-bandwidth connectivity, such as the digital camcorder. It is probable that both systems will coexist in equipment in the near future. This is evident today if you examine what is being integrated into modern multimedia PCs. Even with the advent of the faster USB 2.0 specification, the differentiation between USB and IEEE1394 is unlikely to change.

- Advantages of IEEE1394:
  - — supported by the IT/consumer product industry;
  - — endorsed by standards bodies (e.g. DVB);
  - — HDTV interface standard for the USA;
  - — does not require PC to act as host to network (i.e. peer to peer);
  - — relatively low cabling and implementation costs;
  - — high speed, up to 400 Mbit/s with evolution to 3.2 Gbit/s;
  - — devices 'hot swappable', plug and play;
  - — support for up to 64 devices per segment;
  - — single port connection for all devices;
  - — power provision through cables with standardised connectors;
  - — non-proprietary open and royalty-free standard;
  - — supported 'out of the box' in Microsoft's XP operating system;
  - — suited to home entertainment AV equipment applications.
- Disadvantages of IEEE1394:
  - — short reach between nodes (4.5 m) means local cluster use more probable;
  - — IEEE1394a is unsuitable for full home networking.

The development of new, high-data-rate connection systems like IEEE1394 is driven by the need to connect devices such as digital camcorders, VCRs, digital satellite and cable set-top boxes, DVD players, games consoles and home theatre/ entertainment systems. Market predictions from IDC indicate that by 2003 there will be nearly 50 million IEEE1394-compliant information/entertainment devices world-wide, making IEEE1394 an important variable in the future of the home-networking equation. It is more likely that IEEE1394 will be used in a local cluster mode for AV equipment, rather than a full home network, because of the limitations of the short cable lengths. This may change in the future with the take-up of the IEEE1394b specification and its ability to transmit over greater distances using twisted pair or fibre optic cabling. More information on IEEE1394 can be obtained from the Web site [3].

6.2.1.3 USB (Universal Serial Bus)

The USB was originally developed in 1995 by a consortium of PC manufacturers together with Microsoft and other software companies, as a modern replacement for the slow and cumbersome RS232 serial bus and IEEE1284 parallel port. The slow arrival and high cost of IEEE1394 originally helped to spur on the development of the USB. The consortium's goal was to provide a high-speed bus for the connectivity of modern peripherals like scanners, printers, keyboards and mice. In 1995 the USB Implementors' Forum (USB IF) was set up to ensure a rapid introduction of USB equipment to the consumer market by providing support and guidance to manufacturers. The first real acceptance of USB came with the large-scale success of the Apple iMac in 1998, which included USB connectivity for its keyboard, mouse and other peripheral devices.

USB 1.1 provides two speeds of operation — 1.5 Mbit/s and 12 Mbit/s. It uses standardised cables up to 5 metres in length and a maximum of six hops from the host is achievable, giving a network length of 30 m. Up to 127 devices can run on the bus simultaneously and power is provided at 5 V and up to 500 mA in total, if required by the connected devices. If there is a need to extend the network, additional hubs can be connected in a tiered fashion by daisy-chaining up to a total of five hubs (see Fig 6.4); these come in both powered and unpowered formats. The cable comes in a four-wire format, two of the wires being used for power and a single twisted pair for data. Unlike IEEE1394, USB currently requires a host controller (normally a PC) to function but new silicon is now available from Philips to allow any peripheral device to act as a master, providing peer-to-peer connectivity. PC control can be used to advantage as it can place some devices in sleep mode if running power-saving settings. USB 1.1 also adheres to the Microsoft

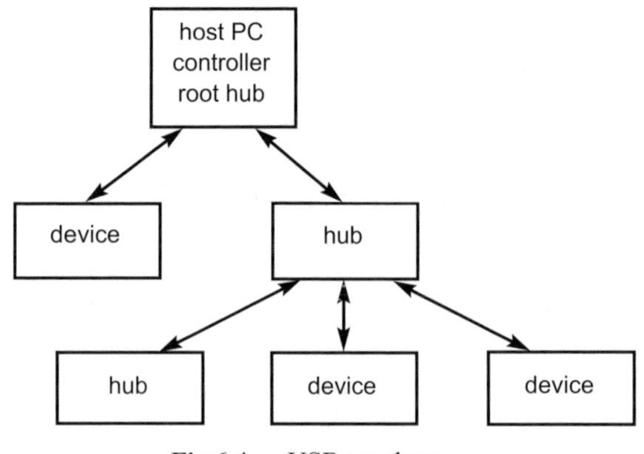

Fig 6.4 USB topology.

plug and play (PnP) specification, allowing devices to be configured on connection and hot swapped without the need to reboot the host PC.

USB supports three modes of data transfer between the host and connected peripherals. Firstly, the isochronous mode is used to connect hardware like audio and video devices, where the full-uninterrupted delivery data rate is required for particular applications (e.g. streaming). Secondly, bulk transfer mode is used where blocks of data need to be transferred and verified. These could be data transfers between the host and hardware such as disk drives or printers. Finally, interrupt transfer mode is used for items like keyboards and mice which run at the slower 1.5 Mbit/s data rate. On connection to the bus or at power-up, all devices are enumerated (assigned a 7-bit address) by the host controller, which also interrogates the device to see which of the above transfer modes is required. Total bandwidth consumed by isochronous and interrupt devices can be up to 90% of the full 12 Mbit/s (although in reality a payload of 8 Mbit/s can be sustained). Above this point the host will deny any further access to the bus by these types of device. The final 10% is used for bulk transfer and control information. The host sets up and controls pipes between applications and individual peripherals over which data is streamed and control information is sent. The data on these pipes is divided up into frames of 1500 bytes and starts every 1 ms. With multiple pipes on the bus, time-domain multiplexing is used under host control to keep things in order.

USB 2.0 is much the same as USB 1.1 and even uses the same cables and connectors, but has an increased bandwidth of 480 Mbit/s. It is backwards compatible with USB 1.1-compliant devices and is aimed at the next generation of high-bandwidth peripherals. A whole new range of applications is being or has been developed to exploit the USB port now common on PCs. USB memory sticks (personal drives), Web application authentication keys, software dongles and encryption keys are but a few of the new wave of USB devices.

Future developments for USB include the 'On-The-Go' (OTG) [4] supplement, which addresses the limitations of USB in a portable device environment. Many new portable devices like MP3 players and telephones may require peer-to-peer connection without the need for a full host controller. Also limitations of the current connector size and reduced power requirement issues are being addressed, driven by today's thirst for mobility. With Bluetooth looking like the future standard for wireless connectivity of portable devices, it is hard to see the need for a wired solution at all. Time will tell, but for today the greater bandwidth of USB offers speed advantage over Bluetooth's wireless capability and low data rate in certain target applications.

- Advantages of USB:
  - — supported by the IT/consumer product industry;
  - — standardised system used on most modern PCs and peripherals;
  - — low cabling and implementation costs;

  — devices are hot swappable/plug and play;
  — up to 127 devices connectable to bus;
  — single port connection for all devices;
  — power provision through cables with standardised connectors;
  — good OS support.

- Disadvantages of USB:
  — short reach between nodes (5 m) (hubs required to extend);
  — efficiency is dependent on implementation;
  — USB 1.1 limited to theoretical 12 Mbit/s shared bandwidth;
  — currently computer must act as host in a USB network;
  — currently an unsuitable technology for full home networking.

There are over 1 billion devices in use today that are enabled for USB connectivity. The current broadband service delivered in the UK by BT uses a USB modem for the domestic customer connection. It is a proven technology and although it utilises wires, it is easy to install. Due to the short reach of USB it is unlikely that it will ever be used for full home networking and like IEEE1394 is more suited to local cluster use, primarily peripheral connection. Further information on the USB is available on the Web site [5].

6.2.2 No New Wires

Reuse of existing networks within the home provides many benefits over installing a completely new network and the pain that goes with it. Take a look at what networks already exist within the home. Power cabling and telephony spring to mind straight away. It is now possible to take advantage of these legacy networks with the introduction over the last few years of some exciting new technologies. Although in their infancy they will become increasingly more important as these technologies mature.

6.2.2.1 HomePNA (Existing Phonelines)

HomePNA (Home Phoneline Networking Alliance) is an alliance of over 150 companies, led by 11 major industry players including 3com, HP and Motorola. The alliance is a non-profit-making organisation that aims to provide a standardised phoneline networking solution and associated products. It was launched in 1998 and within the same year produced the HomePNA 1.0 specification, which provided 1 Mbit/s networking over existing telephone wiring without affecting the consumer's telephone service. By December 1998 the first HomePNA 1.0 products started to appear. In July 1999 HomePNA announced a proposed technology for a 10 Mbit/s version of its networking solution which would be fully backwards compatible with version 1.0. By November 1999 the version 1.0 specification had been certified for use in Europe, greatly widening the market for network and product sales. One month later version 2.0 of the HomePNA networking solution was complete providing 10 Mbit/s bandwidth and QoS functionality. By the end of Q1 2000 more than 2 million units of both v1.0 and v2.0 of the HomePNA product had been shipped world-wide.

The original HomePNA 1.0 product was based on a technology developed by Tut systems with backing from Microsoft, giving 1 Mbit/s bandwidth and co-existing with other telephone line services like voice, ISDN and ADSL. Using frequency division multiplexing (FDM), it splits the bandwidth on the telephone line into separate frequencies, each allocated to different functions, i.e. POTS, ADSL and data networking (see Fig 6.5). Data networking sits in the 5.5 to 9.5 MHz frequency range and uses a carrier frequency of 7.5 MHz. Like Ethernet, HomePNA 1.0 uses IEEE802.3 framing, addressing and the CSMA/CD mechanism for multiple device access to a common network. Unlike Ethernet, the telephone line, as a physical medium, presents many more challenges as a high-speed data network. The average telephone wiring within a house can be made up from completely random topologies, introducing high levels of signal attenuation. Signalling, ringing and other telephone equipment connected to the line can also introduce problems. The wiring is subject to differing levels of noise from within the home environment, not only from consumer appliances, but interference from other RF sources. To cope with some of these problems and to maximise data throughput, as much data as possible needs to be encoded in each signal pulse on the line. Thus the Tut system uses larger packets within the Ethernet frames and a patented pulse-modulation technology to encode multiple bits into each pulse to achieve reasonable data rates in such a hostile environment. Equalising filters have also been used within nodes to help improve line quality, along with a cyclic redundancy check (CRC) thereby improving the overall reliability of the technology.

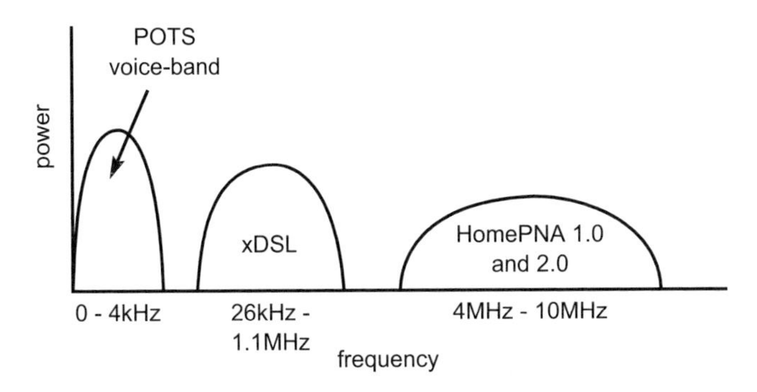

Fig 6.5 Frequency division multiplexing (FDM) of phoneline bandwidth.

With the 1 Mbit/s bandwidth limitation of HomePNA 1.0, a newer and faster version was required to satisfy the needs of the more bandwidth-hungry applications (e.g. streamed video) and to share the now faster access technologies available to the home. The HomePNA 2.0 specification was based on Epigram's patent pending technology, using QAM and FDQAM (frequency-diversity quadrature amplitude modulation). This is a rate-adaptive technology that varies the packet encoding from 2 to 8 bits and also employs two operating baud rates:

- 2-Mbaud, which is its normal operating mode, provides peak data rates between 4 and 16 Mbit/s;
- 4-Mbaud is used when line conditions are favourable giving possible peak transfers of up to 32 Mbit/s.

Training for the correct rate adaptation mode is handled by an 8-bit code within the frame control field of the HomePNA frame, which also contains preamble, frame type, CRC and Ethernet packet information (see Fig 6.6). A passband of between 4 and 10 MHz is used, with a carrier frequency of 7 MHz; this ensures backward compatibility with v1.0 of the HomePNA standard. FDQAM, which is used in the 2-Mbaud mode, sends two redundant copies of the baseband signal in separate frequency areas, thus helping to ensure at least one will be received if the original becomes corrupted, increasing the reliability of the system. To accommodate the need to carry streaming and other time critical/bandwidth-dependent applications, quality-of-service provision was included, giving eight priority levels to traffic, and a new collision resolution algorithm was employed for fair access to the network. In these respects HomePNA 2.0 is superior to Ethernet which does not provide fair access or priorities. To combat impulse noise a fast retransmission system called limited automatic repeat request (LAPQ) is implemented at the link layer in software.

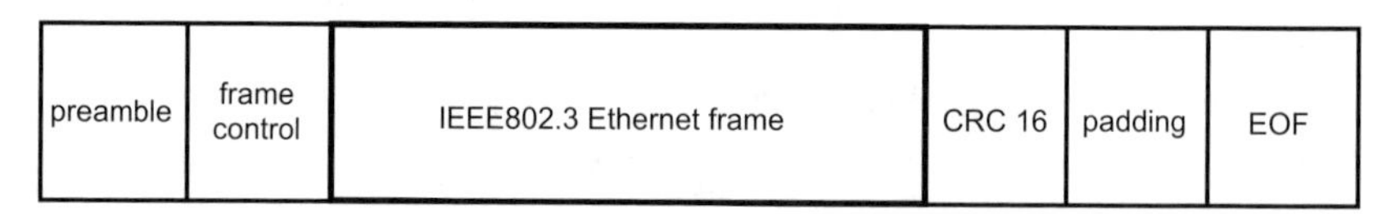

Fig 6.6 HomePNA 2.0 frame format.

HomePNA 2.0 has a line rate of 16 Mbit/s but the actual achievable user data payload is 10 Mbit/s, accounting for all overheads. If the peak line rate of 32 Mbit/s is achievable under favourable line conditions, then the network should be capable of producing a 21-Mbit/s user data payload. A third generation is being developed, that will continue to support QoS, but raise the bandwidth to 100 Mbit/s, enabling it to support high-bandwidth streamed entertainment services. The HomePNA has also standardised a method for voice over HPNA (VoHPNA) which allows multiple telephone lines to be carried digitally over a single cable. This can be used with home gateways to provide PBX features and for additional voice lines derived via

DSL. There are a few non-PC-based devices available today that have HomePNA embedded and it is hoped that v3.0 will be incorporated into more silicon solutions to enable a more diverse range of high-speed HomePNA consumer appliances.

- Advantages of HomePNA 2.0:

  — supported by many vendors;

  — has been standardised by ITU-T as Recommendation G.989.1;

  — 10 Mbit/s user data payload (16 Mbit/s line rate);

  — up to 50 devices connectable to network simultaneously;

  — no hubs or routers required;

  — 300 metres between nodes (maximum network span between furthest end-points);

  — co-existence with POTS/ADSL technologies;

  — backward compatible with HomePNA 1.0;

  — requires no new wiring for network (dependent on existing telephone wiring);

  — easy to add to wiring using extension kits if required (Cat3 cables);

  — quality of service implemented;

  — good EMC profile;

  — low implementation costs;

  — secure environment;

  — consumer friendly.

- Disadvantages of HomePNA 2.0:

  — uncontrolled network characteristics may affect performance;

  — some problems encountered with legacy telephone equipment;

  — co-existence of both v1.0 and v2.0 on network may cause performance hits.

Limitations on the number of telephone sockets within homes may have an impact on the proliferation of HomePNA as a viable home networking technology, compared to the cost of implementing a wireless solution. That said, HomePNA still offers some significant benefits over other solutions that users may find appealing, like superior security and good data rates. If a more diverse range of high-speed HomePNA-enabled devices become a reality, this may help to stimulate the market, but ease of installation will always be a stumbling block for most wired solutions. In the consumer world, users just want to be able to plug and play without the need for extra wiring in their homes; a case in point is the success of DECT-based cordless telephony in Europe. More information on HomePNA is available on the Web site [6].

6.2.2.2 HomePlug (Existing Powerlines)

The HomePlug consortium is a non-profit organisation. It was originally founded in March 2000 by thirteen leading IT companies who have a mutual interest in technologies that enable the networking of data over existing electrical power cabling within a dwelling. 3com, Cisco, Compaq, Conexant, Enikia and Intellon form part of the consortium of over 80 companies keen on powerline technology, emanating from a variety of industry fields like semi-conductor manufacturers, hardware/software suppliers, and service and technology specialists. Their goal is to create an open specification for this networking technology and to promote new products to accelerate its adoption. Several of the companies above had been working on their own proprietary systems, but by June 2001 the HomePlug v1.0 specifications were published and Intellon's PowerPacket technology was chosen as its baseline. In May 2001 trials covering 500 homes in the USA and Canada were carried out to field test this new technology. Some 10 000 outlet-to-outlet tests took place, resulting in 80% of all outlet pairs achieving at least 4.8 Mbit/s throughput and 98% achieving 1.4 Mbit/s.

Other technologies exist within the powerline-networking arena, but this chapter will concentrate on Intellon's PowerPacket system, as adopted by the HomePlug consortium. The HomePlug solution sits within the 4.5-21 MHz band and uses orthogonal frequency division multiplexing (OFDM) digital modulation to split the signal up over the 84 available narrowband sub-carriers. The data bit-streams are modulated on to the sub-carriers using differential quadrature phase shift keying (DQPSK) and inverse fast Fourier transform (IFFT) techniques to produce individual channel waveforms for simultaneous transmission over the electrical network. This whole process is reversed at the receiving end. Forward fast Fourier transform (FFFT) and demodulation is used to recover the original data. This solution helps to deliver large quantities of data over a very short period of time, a necessity in noisy environments; it also yields data rates of about 14 Mbit/s at the MAC level if all carriers are usable. To further aid any interference/impulse noise problems the Viterbi and Reed-Solomon forward error correction (FEC) system is used at the transmitter, helping to recover data when any sub-carriers encounter noise-induced attenuation during a transmission.

In operation, two nodes will initially agree which sub-carriers should or should not be used to transmit the signals — this is derived from the transmission characteristics of each sub-carrier (some may be affected by interference, rendering them almost unusable), with those that are attenuated too highly being discarded. This dynamic adaptation helps to ensure reliable communication from node to node, by travelling around the noise, rather than trying to push through it. A second, stronger system of modulation can be applied, that uses differential binary phase shift keying (DBPSK), but throughput is reduced from 152 bits per symbol to 76 bits. HomePlug can switch between the two systems as required. In the most extreme line conditions ROBO (robust orthogonal frequency division multiplexing) is implemented which defaults to DBPSK for all sub-carriers.

Again, like HomePNA, HomePlug uses Ethernet framing and addressing schemes and the CSMA/CA protocol for multiple device access to a common network; it also implements a process called virtual carrier sense to help reduce collisions. When a node sees a valid frame on the network, it will then read the payload length at the start of the frame and work out how long the network will be in use. The node can then set a time to postpone the transmission until the network is clear. QoS is also implemented via a priority scheme using priority bits for frames requesting access to the network. Four levels of priority exist within this scheme with the highest being allocated to time-critical applications like streaming audio and video. To prevent the network from becoming congested, HomePlug breaks up any packets that have long transmission times; this is beneficial to packets with a high priority and allows for queue jumping. If the network is quiet, however, a long packet can be transmitted in its entirety.

A HomePlug frame contains start and end of frame delimiters, both containing a preamble, followed by a frame control field of 25 bits. Sandwiched between the delimiters, is a 17-byte frame header and payload made up of between 20 to 160 symbols. Finally the HomePlug frame includes a 2-byte frame check sequence (see Fig 6.7). The preamble is a known bit pattern that can be detected by all nodes, even in the worst of line conditions and the frame control field contains packet status and length information as used for virtual carrier sensing, described above.

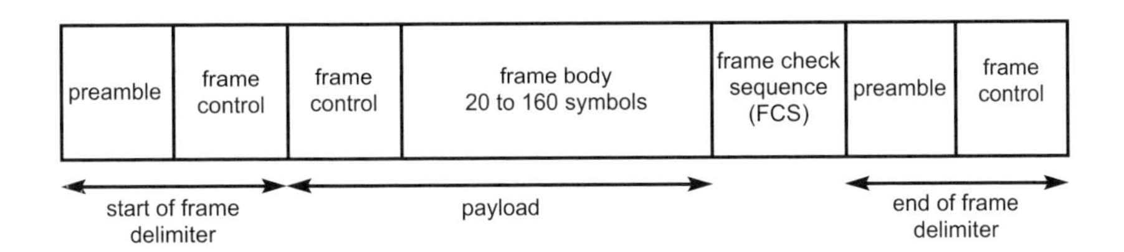

Fig 6.7 HomePlug frame format.

As a powerline network transmits data over existing electrical wiring, the HomePlug network needs to be secure and not interfere with adjacent properties. A security system needs to be implemented to stop data being decoded on networks within other buildings. The former issue is handled by lowering the power output of the transmitter, allowing just enough power for nodes to communicate within the internal network, but preventing it from being snooped from an external source. HomePlug also implements a 56-bit data encryption standard (DES) key and a password over the internal home network, adding an additional layer of security and ensuring any data arriving at a neighbour's house is unintelligible.

- Advantages of HomePlug:

  — currently supported by many mass-market IT vendors for the USA;

  — 14 Mbit/s MAC rate (actual payload 1.5-5 Mbit/s);

  — power outlets most common connection within a dwelling (every room normally serviced);

  — full coverage of a home, compared to limitations of wireless solutions (thick walls);

  — works independently to line voltage and frequency of current;

  — will co-exist with other powerline technologies (i.e. X10, CEBus and Lonworks);

  — security issues handled by encryption using 56-bit key and password;

  — consumer friendly and easy to set up;

  — supports QoS;

  — evolution to 100 Mbit/s bandwidth in roadmap.

- Disadvantages of HomePlug:

  — not a cheap solution currently;

  — not suitable for portable devices unless used with a gateway;

  — competing technologies to HomePlug standard being proposed;

  — leakage to external networks may be an issue;

  — consumers sceptical about using mains for data networking;

  — full data rates may not be achievable within European regulations on EMC emissions.

Powerline networking seems to be the most appealing of all current home networking technologies. Although expensive at present, the costs will fall with demand. To be able to take products out of the box and simply plug them into existing mains outlets, giving almost instant network connectivity seems like the dream solution for mass-market adoption — the reality is a little way off at the moment. With other standards being proposed to compete with HomePlug and consumers still sceptical about using the electrical network to connect computers, it may be a while before we see powerline accepted as a viable technology in the mass market. Another factor affecting the adoption of powerline networking world-wide is regulation on electromagnetic emissions, which is stricter in Europe than in the USA. These regulations are subject to much controversy and lobbying by the powerline communications industry, but it is still not clear whether the promised data rates will be achieved within European regulations. One option is to reduce emissions by lowering the transmitted data rates and increase the risk of networking being unavailable at some power outlets. Until some of these issues are resolved, high-bandwidth powerline networking is unlikely to take off in Europe in the near future. It is still very early days for powerline networking, but companies like Netgear, Linksys and SMC are leading the way by launching powerline products based around the Intellon INT5130 chipset for the USA market. Phonex Broadband also entered into this fledgling product area, late in 2001, with their QX-201 NeverWire 14 product. This device allows the user to connect up to 16 Ethernet enabled devices through its Ethernet port over existing power wiring.

If regulatory and technical issues can be overcome and the bandwidth significantly increased, a mixture of powerline and wireless technologies could help to rapidly increase the adoption of home networking to the mass market by making it a truly plug and play environment. Powerline will be a technology to watch in the future. For further information on HomePlug, please refer to the Web site [7].

6.2.3 Wireless

The ability to connect devices to a home network without having to worry about cabling has tremendous appeal. Looking at the market for IEEE802.11b wireless equipment, which is growing at a considerable rate as prices continue to fall, proves this. Wireless technology is moving from the corporate to the domestic environment, saving user installation time and disruption to the fabric of the home. The limitations on bandwidth are finally being removed with new, faster variants appearing, opening up this technology to many more multimedia applications. Wireless looks like it will be one of the strongest contenders in the home networking market of the future and predictions from the Cahners In-Stat Group show that it will account for 48 per cent of all in-home networking installations by 2006.

6.2.3.1 IEEE802.11 and Derivatives

Originating from the original IEEE802.3 standard for wired Ethernet, the next natural progression was for a wireless system based around this well-proven technology. The IEEE initially started to look at a new standard for wireless in 1990 and by June 1997 the initial standard for IEEE802.11 was finalised, with the technology based in the 2.4-GHz ISM (industrial, scientific and medical) band. At the physical layer, two formats of spread spectrum modulation techniques were employed, direct sequence spread spectrum (DSSS) and frequency hopping spread spectrum (FHSS). These modulation schemes are not compatible. IEEE802.11 provided data rates of 1 or 2 Mbit/s for all implementations. Alongside these radio-based systems, an IR (infra-red) based solution was proposed sitting in the 300-428 GHz band and using pulse position modulation. The RF systems are by far more popular than the IR solution, as line of sight is required for correct IR data transmission.

The initial take-up of the RF-based LANs was slow, fuelled by high costs and the fact that users had become used to the higher throughput of wired Ethernet. By the end of 1999 the IEEE had published two new supplements, IEEE802.11b and IEEE802.11a which dealt with the bandwidth constraints. The former operated in the same band as the original specification, but now adding 5.5 and 11 Mbit/s (air rate) to the original two data rates and the use of FHSS was dropped in favour of DSSS. In contrast, IEEE802.11a offers up to 54 Mbit/s and sits in the less cluttered 5.2 GHz band, but for current use in the UK users will require a Test and Development licence and must limit the output power and RF bands used.

DSSS works by spreading out the transmitted data over a wider band. At the physical layer a high rate 11-bit Barker sequence (chipping code) is modulated on to each bit to spread the data before transmission — this helps to reduce the effects of interference and enable recovery of any lost data. At the receiving end, the RF signal is de-spread to recover the original information. For IEEE802.11b, the overall bandwidth used by DSSS is in the region of 20 MHz and therefore three non-overlapping channels can co-exist in the 2.4-GHz ISM band.

FHSS in contrast hops between frequencies using 22 patterns known to both ends of the system. Like Bluetooth the frequency band is split into 79 1-MHz channels but the hop rate is far slower at 2.5 hops/sec for IEEE802.11. This hopping technique makes FHSS far less susceptible to interference, as the data is only on each frequency for a fraction of a second (normally stated as the dwell time). DSSS uses DBPSK and DQPSK modulation techniques for the 1 and 2 Mbit/s data rates respectively. For the higher two rates it switches to complementary code keying using 4-bits per carrier for the 5.5 Mbit/s data rate and 8-bits for 11 Mbit/s. However, two and four level Gaussian frequency shift keying (GFSK) modulation is implemented for FHSS.

Unlike wired Ethernet, IEEE802.11b uses a slightly different approach to network access and collisions. CSMA/CA is applied, as it is difficult to detect collisions within an RF-based network. Carrier sense is used at the MAC level to detect if any of the RF channels are usable. The physical layer passes information to the MAC layer concerning RF strength for each channel and thus it can be said to be in the clear or in use. All nodes listen for traffic and will transmit if a channel is free, but, like Ethernet, if a channel is busy, a node will back off for a random period to stop the network from becoming tied up (see Kleinrock and Tobagi [8] for further reference to CSMA in a wireless environment).

Two modes of operation exist for IEEE802.11b — infrastructure and *ad hoc*. The former consists normally of at least one access point (AP) connected to a fixed LAN or distribution system of some description. The infrastructure-mode basic service set (BSS) provides a path for communication for all wireless nodes via an access point rather than directly between each other. The extended service set (ESS) may consist of one or more access points on the same fixed LAN or distribution system forwarding traffic between themselves. This allows wireless nodes to roam seamlessly from one access point to another. Access points can also be used to extend the range between nodes. In *ad hoc* mode (e.g. peer-to-peer) all wireless nodes can communicate directly with each other if in range, without the need to traverse an access point.

Unlike Ethernet LANs, wireless LANs cannot always detect all stations on the network and a problem known as the 'hidden node' arises. This occurs when one node cannot detect transmissions from another node within the network, making it impossible to accurately tell if the physical medium is busy. To help alleviate this problem in the infrastructure mode, virtual carrier sense is used, allowing nodes to reserve time on the physical medium using a 4-way handshake procedure (see Fig 6.8). A node must initially send a 'request to send' (RTS) frame if it wants to transmit data on the network (node A). The RTS contains address information and message length; this information is used by the network allocation vector (NAV) in all nodes to warn other nodes not to send data while the current node is transmitting. The sent RTS is received by an access point which will respond with a 'clear to send' (CTS) frame, which also contains address information and message length. If the hidden node (node B) did not hear the original RTS from the transmitting node (node A), it will have heard the CTS response from the access point and will update its NAV — thus a collision can be avoided.

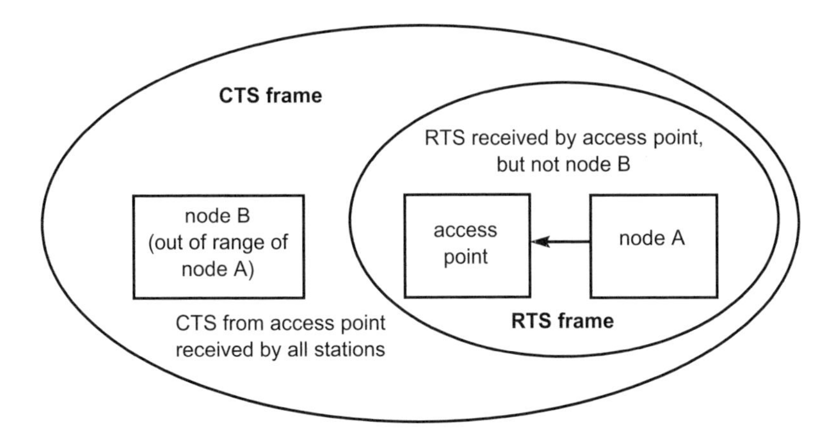

Fig 6.8 Virtual carrier sense using RTS/CTS.

As the addressing scheme from IEEE802 was maintained for IEEE802.11, the 48-bit address used within Ethernet is compatible from the wired LAN right through to the wireless segment. The frames within IEEE802.11 (see Fig 6.9) are very similar to standard Ethernet packets containing a header, payload and a 32-bit CRC. It also contains a frame control field that encompasses the protocol version and frame type (management, data or control). The duration/ID field (used by NAV) follows this, along with address fields and a sequence control field used for packet fragmentation process. Packet fragmentation is used to break up large packets for transmission, reducing the chances of corruption and the need to retransmit; if this is necessary only a small fragment will need to be retransmitted thus making the process much faster.

frame control 2 bytes	ID 2 bytes	address 1 6 bytes	address 2 6 bytes	address 3 6 bytes	sequence control 2 bytes	address 4 6 bytes	data 0 to 2312 bytes	CRC 4 bytes

Fig 6.9 IEEE802.11 frame format.

Wireless LANs have had a considerable amount of bad press over the last few years, with the knowledge that the embedded security system WEP (wired equivalent privacy) has been easily compromised. Newer versions of the specification will include improvements to security such as stronger encryption, dynamic key exchange and key-hopping technologies.

Currently the IEEE are working on a new standard, IEEE802.11g which resides in the 2.4-GHz ISM band and is expected to deliver in excess of 20 Mbit/s throughput. Using DSSS to achieve rates up to 20 Mbit/s and OFDM technology to raise this to 54 Mbit/s will result in this implementation ending up in competition with IEEE802.11a. The advantages of IEEE802.11g are that it will be backward compatible with IEEE802.11b, while residing in the 2.4-GHz spectrum will make this technology far more accessible to many more countries. This may all be overcome if vendors decide to support both formats within the same adapter or full ratification of IEEE802.11a for use in Europe.

- Advantages of wireless LANs (IEEE802.11):
  — widely adopted by many manufacturers;
  — 11 Mbit/s air rate (user data payload 4.5 Mbit/s);
  — progression to faster implementations (IEEE802.11a and IEEE802.11g);
  — IEEE802.11b components may be upgradable in firmware to IEEE80211g;
  — reliable technology with fall-back to slower speeds when necessary;
  — backwards compatible with older DSSS implementations;
  — *ad hoc* networks supported;
  — long range (25-500 m);
  — Wi-Fi standard for IEEE802.11b product compliance;
  — ideal for home networking where installer requires minimal wiring.

- Disadvantages of wireless LANs (IEEE802.11):
  - — not a user-friendly technology (being overcome in Windows XP);
  - — QoS issues with IEEE802.11b (being addressed in IEEE802.11e task group);
  - — interference can drastically affect throughput;
  - — physical environment may affect performance (e.g. thick walls);
  - — shared ISM band may cause performance hits;
  - — original WEP security easily compromised.

Wireless will most definitely become one of the most popular home networking technologies for the future, with falling prices and evolution to higher bandwidths. Although initially expensive, IEEE802.11a products have started to hit outlets and, as the price falls, the increased bandwidth of up to 54 Mbit/s will be very appealing to those users who want to deliver multimedia applications around their dwellings. It is an appealing technology to the home networker who can quickly install a network with minimal disruption to their premises, but it will require a competent understanding of computers and networks to configure it correctly. This may be overcome in the future with the support for the IEEE802.11 standard within Microsoft's Windows XP operating system. As an adapter is plugged in, the operating system will automatically look for any available wireless networks, without the need for manual configuration. A factor that may help to drive up adoption of wireless LANs and reduce costs further is the deployment of public wireless LANs in the not too distant future. This is a hotly debated topic on both sides of the Atlantic with some commercial successes and failures in the USA. Further information on wireless LANs is available on the Web site [9].

6.2.3.2 HomeRF

HomeRF, like other technologies mentioned within this chapter, owes much of its existence to other already established wired and wireless technologies. Its support for data using CSMA/CA and voice using time division multiple access (TDMA) are derived from IEEE802.11 and DECT standards. In March 1998 the HomeRF Working Group (HRFWG) was formed and, like other forums of this nature, its aim was to promote the development and adoption of HomeRF v1.0. In November 2001 the European arm of HRFWG was launched as rule changes from ETSI enabled certification of HomeRF v2.0 in Europe. Today the working group has in excess of 90 active members involved in promotion and future developments of the shared wireless access protocol (SWAP), including the now ratified HomeRF v2.0. Completed in March 2002, version 2.0 greatly improves many areas of the original specification, such as security, QoS and speed, and products are becoming available from vendors like Proxim.

HomeRF is a technology that has been developed specifically with the home user in mind, and, designed to promote home networking, it also includes voice transmission capabilities. Version 1.0 had support for four simultaneous voice lines at near-wire-line quality; this has been doubled in version 2.0 providing up to eight simultaneous calls. The inclusion of voice has facilitated features such as call forwarding to multiple cordless handsets, voice mail or other telephony equipment (e.g. fax). This adoption of embedded voice has led to co-operation between the HRFWG and DECT forum to try to promote HomeRF as the next generation of DECT (Global DECT standard). It sits in the more world-wide friendly 2.4 GHz ISM band, as opposed to the original 1.88-1.90 GHz band of DECT technology, currently limiting its adoption to European countries.

Originally only supporting a data rate of 1.6 Mbit/s, HomeRF struggled to compete against IEEE802.11b's superior throughput and falling costs. The new specification has increased the data rate to a more respectable 10 Mbit/s with fallback rates of 5/1.6 and 0.8 Mbit/s and the HRFWG are supporting a roadmap to 25 Mbit/s. Based in the 2.4-GHz ISM band and using FHSS, over the same 79 channels as IEEE802.11, it provides superior interference avoidance and security by frequency hopping at rates of up to 50 hops/sec using 22 different hop patterns. The security is enhanced by the inclusion of 128-bit Blowfish encryption and a 24-bit network ID required by devices before they can participate within the network. With v2.0's support for QoS being an integral part of the protocol and increased bandwidth, it is far more suited to multimedia applications than the earlier version. Isochronous data is given priority over the asynchronous traffic, useful both in supporting up to eight isochronous simultaneous audio/video streamed media sessions and in helping to make sure near-wire-line quality is maintained for the voice channels.

Like IEEE802.11, HomeRF can work in a point-to-point mode, thus dispensing with the need for an expensive access point. Proxim has developed the Symphony base-station (gateway) which gives connectivity to a DSL modem, firewalling, roaming, DHCP and NAT all in one unit, along with HomeRF v2.0 compatibility. Other manufacturers are now producing bridges for connection between HomeRF and other networking technologies such as wired Ethernet or HomePNA. Some manufacturers have even taken the step of embedding HomeRF technology within their new wave of devices. Siemens produces the Gigaset range of products and Simple Devices has launched products such as the Simple Clock and SimpleFi, the former being an Internet-connected clock/scheduler, the latter a device for playing back streamed audio over a wireless connection.

- Advantages of HomeRF:
  - — supported by some of the leading players in the IT industry;
  - — support for 127 nodes;
  - — no access point required;

— eight separate/simultaneous voice lines;

— good security and noise immunity;

— reasonably fast 10 Mbit/s (v2.0) with roadmap to faster rates;

— user friendly technology;

— inclusion of QoS for multimedia/voice applications;

— support for multicast streaming;

— range of 50 m;

— low power (Rx 120 mA, Tx 250 mA, standby 3 mA);

— backwards compatible with original HomeRF specification;

— costs falling;

— a usable solution for home networking.

- Disadvantages of HomeRF:

— slow take-up due to deployment delays and original slow data rate;

— competing against established IEEE802.11b technology;

— environment may limit range and performance (thick walls).

HomeRF has had a very bumpy start, but with the ratification of version 2.0 and the push for it to be adopted as the Global DECT standard it still remains a viable home networking solution. It will have to go head to head with the stronger IEEE802.11 standard, but contains many elements today (e.g. QoS/voice support) that its rival has yet to support, or is in the process of implementing. The key to the mass-market home network adoption has to be ease of installation with minimal configuration; HomeRF has had a head start in this area over IEEE802.11, but Windows XP is helping the latter standard to catch up. In a market dominated by cheap IEEE802.11 components and a push for the deployment of public wireless LANs using IEEE802.11 technology, it is hard to see a future for HomeRF. Time will tell and the underdog has gone on before to succeed. Compare this to Bluetooth's chequered history over the last few years — only now is it being accepted as a workable standard for portable device wireless interconnection. Further information can be obtained from the HomeRF Web site [10].

6.2.3.3 Bluetooth

Bluetooth began life in 1994 as a project originating from Ericsson mobile communications, set up to investigate ways of replacing cables interconnecting their mobile telephones to other accessories, by using a wireless solution. IrDA was already being used by other players in this market to perform the same function, but it was felt that it did not meet all the required needs. Radio was seen as the next best technology, overcoming some of the shortfalls of IrDA, such as line-of-sight communication issues. By 1998 other major players such as Intel, Toshiba, Nokia and IBM, who all shared in their vision of an open standard for a short-range wireless system for interconnecting devices, had joined Ericsson. They collectively formed the Bluetooth special interest group (SIG) in February of the same year, agreeing to work towards a specification for Bluetooth technologies and by July 1999 Version 1.0 of the Bluetooth specification was released. Within a few months several other leading vendors, including 3com, Lucent, Motorola and significantly Microsoft, joined the original members. Several ratifications of the specification followed, correcting some problems and issues thrown up by the original, and by March 2001 Version 1.1 was made public. The Bluetooth SIG now has in excess of 1900 members from all fields of industry, helping to ensure that this technology will be accepted as the *de facto* standard for connection of short-range mobile personal area networks (PANs) in the future.

Bluetooth resides in the unlicensed 2.4-GHz ISM band. This is the same part of the spectrum already used by devices like baby monitors, cordless audio and video accessories and even the microwave oven in the kitchen. It also shares its bandwidth with other networking technologies, such as IEEE802.11b wireless LAN and HomeRF. Bluetooth only transmits at very low power output levels (1 mW), giving it a short range of operation, normally about 10 metres. Potential problems and initial concern with interference between IEEE802.11b and Bluetooth were discovered and studies have shown that performance degradation begins to occur when devices are collocated within 3 m of each other. Various solutions are being examined to overcome simultaneous close proximity use of these two technologies. Bluetooth uses a packet-switching protocol based on the FHSS technique. The band is split into 79 separate frequencies (this is reduced for some countries due to out-of-band regulations) spaced at 1-MHz intervals, allowing the signal to hop at up to 1600 hops/sec between them. This helps to give a high level of immunity to noise and interference within the used spectrum and allows the co-existence of many devices within the same location to function simultaneously by hopping momentarily from one frequency to the next. GSFK is a spectrally efficient modulation scheme used by Bluetooth where the binary data is represented by positive and negative shifts in frequency, a binary '1' being associated with a positive shift.

Data communication takes place over an asynchronous channel that can support up to 721 kbit/s in asymmetric mode and still support 57 kbit/s in the return direction. When in symmetric mode, a data rate of 433 kbit/s is achievable bidirectionally. Alternatively up to three simultaneous synchronous voice paths are available each supporting 64 kbit/s in both directions. A final configuration may also be set up to support both asynchronous data and voice simultaneously, priority being given to synchronous voice packets.

One of the criteria for Bluetooth functionality was to allow for *ad hoc* networking. When two or more devices come into close proximity, communication takes place between them and a personal area network, sometimes referred to as a piconet, is established between all participating devices (Fig 6.10). This all happens without any user intervention and is an automatic process. Once devices become members of this *ad hoc* network they may wish to share information or perform a control function on another device.

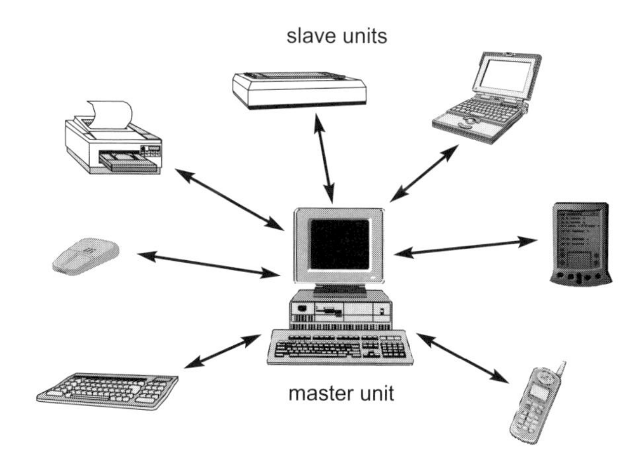

Fig 6.10 Bluetooth piconet.

Every device will have a unique embedded 48-bit address (derived from IEEE802 standards) that corresponds to a range of addresses pre-set for a particular device category, normally agreed among manufacturers. Initially when a device is turned on (assigning itself as the master), it will send out either an enquiry or page message (dependent on whether the destination address of another device is known or not) to see if any similar devices (slaves) are listening and within range. If the master device does not receive a response it will then send a beacon sequence to wake up any slave units that are in standby mode. Bluetooth also implements FEC to help to overcome noise/interference-induced data loss and to reduce the need to retransmit information.

User protection and information privacy is implemented at the physical layer and employs a challenge-response algorithm and stream cipher encryption using secret key lengths of 0, 40 and 64 bits. As Bluetooth is a very low power technology, it has three power-saving modes, including 'hold' mode, used when no data needs to be sent but quick wake-up may be required, followed by 'sniff' mode. Slave devices that need to listen (or sniff) into the piconet only use this mode — both 'hold' and 'sniff' modes retain their MAC addresses. Finally 'park' mode, being the most energy conscious, keeps the device locked to the piconet, but does not pass any traffic.

- Advantages of Bluetooth:

  — wide support for standard across several industry sectors (large SIG membership);

  — low cost for implementation (targets of $5 per device);

  — very low power consumption (0.3 mA standby, 30 mA max during data transfer);

  — small form factor, ideal for portable devices;

  — enables *ad hoc* networks;

  — supports both data and voice connections;

  — world-wide market currently growing at an ever-increasing rate, increasing adoption;

  — primarily a cable replacement technology, but has many other applications;

  — enabled in many new products.

- Disadvantages of Bluetooth:

  — low speed, 721 kbit/s (but suited to target applications);

  — short range, 10 m (higher power version giving 100 m);

  — potential to clash with other technologies within 2.4-GHz ISM band;

  — not suited for full home networking solution (better suited to local PAN).

In its current format Bluetooth is still very much aimed at personal area networking of portable devices and accessories. It is not well suited for whole home networking and indeed this was never its goal. With the proliferation of portable devices into our lives, we will need a simple zero configuration method to interconnect these terminals to our fixed in-home networks and this is Bluetooth territory. Current limitations on bandwidth will be overcome with the Bluetooth 2.0 specification, allowing many new possibilities such as multimedia services enabled to portable terminals. Bluetooth will also make inroads into new areas not traditionally associated with networking, but which may benefit from being enabled by this technology. Areas like the automotive industry, vending/ATM machines and even domestic white goods are prime candidates for the new wave of Bluetooth applications with the new JSR-82 specification allowing development with Java. Recent developments will also allow the building of cellular-like networks, with support for seamless handover, using both voice and data. Bluetooth now has full support from Microsoft and will certainly figure as part of BT's future public access hotspot strategy. There is no question that Bluetooth is here to stay and will play a big part in people's lives in the future — they just may not know it! Further information on Bluetooth can be found on the Web site [11].

6.2.3.4 DECT (Digital Enhanced Cordless Telecommunications)

Originally designed as a digital replacement for analogue cordless phones, DECT overcame many of the inadequacies of its analogue counterparts. Higher quality audio, better security and roaming between base-stations were just some of the features that helped drive this new technology. It started in 1987 with the development in the UK and Sweden of the CT2 and CT3 cordless standards. ETSI amalgamated and enhanced the two technologies in 1988 with a replacement standard called DECT. In 1992 the completed DECT standard was made available and the first products began to appear. Initially the DECT market was seen to grow dramatically in Germany, but as costs fell this technology found its way into many other European countries.

During 1994 the general access profile (GAP) was completed. By implementing the GAP protocol manufacturers can ensure interoperability, for basic telephony operations, between handset and base-stations of other manufacturers. Over the past 10 years we have seen a market explosion in the DECT arena with a vast number of products becoming available. They range from simple cordless telephones to complete wireless PABXs and with the proliferation of mobile telephony, dual function DECT/GSM units are also available. DECT has also been used for wireless local loop (WLL) applications in several countries.

Primarily a micro-cellular radio system DECT provides connections between fixed parts (base-stations) and portable parts. The portable parts could be cordless handsets or data terminals. It sits in the 1.880-1.900 GHz band and uses a GFSK modulation scheme. Other frequencies have been adopted where necessary to suit country-specific regulatory requirements.

MC/TDMA/TDD (multi-carrier, time division multiple access, time division duplex) is the technique used to select channels for radio access. In the frequency domain, ten RF carriers exist within the DECT band, but, within the time domain, 10 ms timeframes are used and split into 24 separate time-slots (see Fig 6.11). This results in a possible total of 120 duplex channels available for use. A full duplex voice transmission of 32 kbit/s will require a pair of 2 time-slots (used for Rx and Tx) resulting in a total of 12 simultaneous voice calls being accommodated over the entire band. For data transmission, channels can be bonded together to give a theoretical maximum of 552 kbit/s throughput allowing for all overheads, but, in practice, this may be considerably lower.

Base-stations continuously transmit a beacon signal on one channel containing base-station information, paging messages and access rights. This allows portable parts to lock on if they are allowed access and to check for free radio capacity to initiate communication links. A list of free channels is maintained by regular scanning of all idle channels by the portable part for RF activity and takes into account any that are affected by interference, ensuring that any radio links will use the best path. If during a transmission the used channel degrades, handover to another channel will take place. Radio links are always initiated from the portable

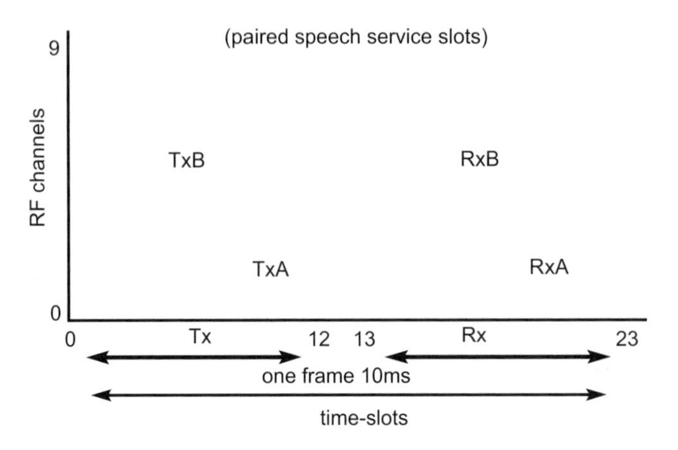

Fig 6.11 DECT frequency/time spectrum.

part even when receiving incoming calls; this is achieved by using paging messages from the base-station to portable parts with the correct access privileges.

Using any wireless link requires some form of security system; DECT employs a challenge/response key authentication system for mobile to base-station connectivity along with strong encryption of data across the radio link.

- Advantages of DECT:
  - — widely adopted within Europe for wireless voice communications;
  - — co-existence with ISM band;
  - — range 50 m indoors, up to 300 m outdoors;
  - — high immunity to interference;
  - — good security;
  - — near-wire-line speech quality;
  - — low cost.
- Disadvantages of DECT:
  - — low speed (1152 kbit/s);
  - — not currently suitable for full high-speed home network.

Although the GAP profile was primarily aimed at wireless voice applications a number of data profiles now exist, with DECT being seen as a medium for narrowband data transmission. It is not currently suited to high-speed home networking due to data rate limitations (1152 kbit/s), but consumer products (Fig 6.12) are beginning to appear for low-speed networking of PCs using its wireless capabilities. An example is the BT On-Air 1800 product which primarily is a wireless modem with digital answering machine functionality. It contains a v90 modem within the base-station and comes equipped with a DECT cordless handset for voice calls. A USB-enabled DECT wireless dongle allows the user to connect their portable or desktop PC wirelessly to the Internet via a dial-up PSTN connection. Another similar product is the BT Airway wireless telephone and Internet sharing system. Basically it is a wireless PABX using either a 2- or 4-line controller that may be connected to standard PSTN lines or, with the use of an adapter, ISDN. Cordless DECT handsets can be used with the controller or wired telephones via DECT wireless adapters. Additionally a special DECT dongle can be used to connect your PC's serial port wirelessly to the controller's data network, enabling file sharing and printing with other PCs or connection to the Internet.

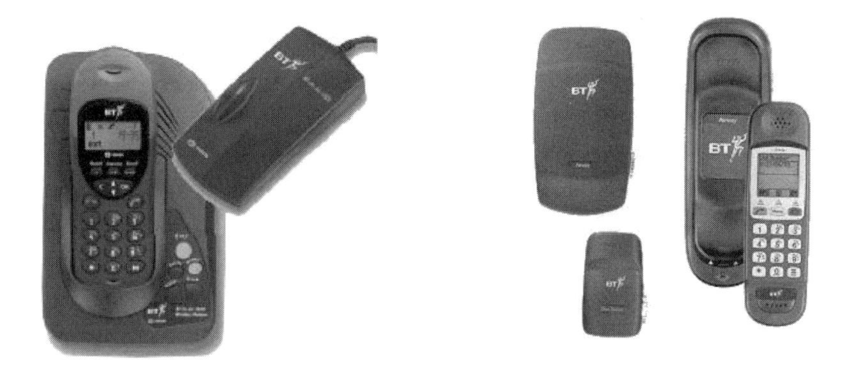

Fig 6.12 BT On-Air 1800 and Airway products.

The DECT Forum predicts that by 2003 over 200 million DECT products will have been shipped world-wide, proving DECT is now well established; however, it remains to be seen where and how DECT will be used in the future. Although this technology is currently not well suited for high-bandwidth applications, there are still many areas where DECT wireless data networking solutions could be implemented. Further information on DECT is available on the Web site [12].

6.3 The Role of the Residential Gateway

The residential gateway (RG) is an extremely important part of the home-networking story. In its simplest form its function is to provide, on its public facing side, connectivity to the Internet. It may use some form of integrated access technology (e.g. xDSL, ISDN, PSTN), or connect externally via Ethernet to a separate access modem. On the home side some form of home networking technology (e.g. Ethernet, Wireless, HomePNA, Powerline) is used to interconnect a range of devices to the single Internet connection and to each other.

With the need to connect many devices to the Internet, each with an IP address, and the belief of some that the four billion IPv4 IP addresses will rapidly be used up over the next few years, various solutions are being employed to combat this issue. Dynamically assigned addresses from the ISP have, up to now, helped to reduce concern about IP address shortages, but this is in a world of temporary dial-up connections. As more users see the benefits of ADSL and its always-on availability the address shortage problem will surely once again rear its head. The residential gateway uses functionality to help rationalise IP address use and to provide a simple layer of protection for devices residing on the home network. NAT assigns devices connected on the home side with private IP addresses (normally from an address pool in the 192.168.x.x range) and translates these to a single visible public side IP address, thus hiding the connected device's real identities. This public address can be statically set, but is often dynamically assigned, using DHCP by either the ISP or the connected access modem. To ease configuration of the devices on the home side of the network, the residential gateway may well also be running a DHCP server to automatically allocate IP addresses. Unfortunately NAT does present problems to some applications and in certain instances stops them working completely. Video-conferencing, streamed video/audio and even FTP can fall foul of NAT's address translation functionality. Another function rapidly becoming seriously important is the addition of security features such as firewalls to help prevent intrusion by hackers from the public side.

Opinions vary on where the residential gateway should sit within the home network. It can be integrated into a consumer device saving on overall cost, e.g. a set-top box, gaming console or even a home entertainment centre, containing access, gateway and home networking functionality alongside its normal applications. Alternatively, as with the majority of today's commercially available residential gateways, it can be a separate device that is connected to the access modem, networking all the other devices together.

There are many companies producing stand-alone residential gateway devices ranging in price and functionality. 2Wire markets its high-specification HomePortal range [13]. The HomePortal 100 has all the minimal functionality you would expect such as DHCP, NAT and firewall (e.g. port filtering and forwarding). It also includes support for other networking technologies besides Ethernet on the home network. HPNA v2.0, USB and optional wireless IEEE802.11b networking are all supported and the last provides connection to portable devices, like laptop computers and PDAs. As with many gateway devices the unit is configured via a password protected Web browser interface. Extra functionality includes virtual private networking (VPN) with passthrough for both point-to-point tunnelling protocol (PPTP) and IPsec. Features that have been omitted from this device but can be found on other units include buffered printer ports and demilitarised zone (DMZ).

This is a port on the outside of the firewall providing a connection for applications that cannot transverse NAT or firewalls (e.g. a personal Web server needing visibility to the outside world). Multiple port hubs and switches are becoming more commonly integrated as standard, saving the expense of separate external units.

To gain true acceptance in the mass market the residential gateway will have to reside within a consumer device providing much more to the user than just home networking and Internet connectivity. Likely scenarios include set-top boxes giving a broadband connection providing perhaps integrated interactive TV and voice services in addition to high-speed Internet. Moxi Digital Inc, with the development of their licensable design for the Moxi media centre, envisages integration with home entertainment multimedia servers [14]. This is a device which not only provides cable or satellite connectivity and an Internet gateway, but also DVD, PVR, networking and home server capabilities for the storage of MP3, photo and video content on its internal hard drive.

There is also a vision of a high-powered PC-based central media server accessed by low-cost PC clients and Internet appliances. The high costs of Internet appliances today (e.g. Webpads, Internet audio players) could be reduced as they connect by wired or wireless means to a central server using its computing resources and applications. Microsoft's Windows XP operating system already supports the required client/server structure and remote desktop technology to achieve this and is being exploited by Microsoft's Mira project.

During the last few years, products like Microsoft's Internet Connection Sharing (ICS) have attempted to make home networking more acceptable in the mass market by integrating functionality into the operating system. Using ICS basically turns your PC into the gateway resulting in a downside, in that you will have to leave the main PC powered up and running, whenever you want to connect any other devices in your network to the Internet. ICS still requires an element of computer/network understanding to set up, which is normally beyond the scope of the average residential user. The stand-alone RG is much more suitable in this instance as it can be left powered up and devices turned on and off as they are required.

It can be clearly seen that to connect multiple devices, using a home network, requires some knowledge of computing and networking practices and has only really been attempted over the past few years by experienced computer users and gamers, normally found working within the computer industry. In the domestic market for home networking, consumers will not have the luxury of a network administrator or computer expert but will rely on equipment to be simple to plug together and requiring zero or minimal configuration. Without this it is difficult to see how home networking will gain wider adoption. Residential gateways are devices and solutions that will play an important role in achieving this next step for home networking (see also Chapter 5).

6.4 Devices, Appliances, Services and Applications

Although several companies cut back their home networking development during 2000, there are products available that make use of the always-on Internet connection. The following sections review approaches and devices in the IT and entertainment industries. Voice over IP (VoIP) will develop as an alternative to PSTN-derived voice services and eventually the Internet and telephone network will become one and the same. VoIP is a key technology for voice services over broadband — for the technical details, see Swale [15]. There is also a good Web site for general information on home networking [16].

6.4.1 Entertainment/Information

Home entertainment centres will enable content sharing over the home network and bring the entertainment and computer industries closer. SonicBlue [17] has launched the Rio Central audio centre in the USA enabling users to store and catalogue large CD collections in MP3 format held locally on its internal hard drive. If required the content can then be burned on to the unit's integral CD-RW drive. The HomePNA 2.0 output facilitates networking of the content from the home PC and around the home to other HPNA enabled devices. Several companies such as Turtle Beach and Philips are now producing similar devices.

Another concept is to use your home PC as the storage server for all content. Motorola has produced the SimpleFi product [18], which links the PC to a remote set-top unit via HomeRF technology. The remote end is simply plugged into a hi-fi and decodes MP3 content sent from the PC over the wireless link.

Although no longer available, the Kerbango radio was launched as the world's first Internet-enabled radio appliance. Basically a hardware device, it connected to the Internet to receive streamed audio broadcasts, using links to the Kerbango Web site to achieve tuning/channel selection. A number of companies are still producing similar devices like iRAD and Smartmedia with its DDL Player [19].

Stand-alone PVR technology is beginning to surface in the consumer market, with a number of units available. Panasonic, Sony and Thompson all provide hardware, but the services used by these units are operated by companies like TiVo [20].

Next-generation DVD players will have Internet access. The iDVDBox, being launched in the USA during 2002, will be available in Europe at the end of 2002, and will merge MP3, interactive DVD, interactive TV and Internet technologies. Internet access is via a 56 kbit/s modem or USB for broadband Internet [21].

The gaming console market has always been associated with stand-alone units, but in recent years multiplayer gaming has taken off dramatically. Some units have allowed interconnection of multiple consoles to enable multiplayer action, but this requires close proximity of the players. The PC gaming market has captured a whole new culture of gamers by using the power of the Internet to bring together remotely located players and console technology is now catching up. The next logical step has finally been taken by bringing broadband connectivity to these platforms. Both Sony's PS2 and Microsoft, with its Xbox, are launching Internet-based multiplayer gaming services, using Ethernet technology for connection to broadband services.

A number of Webpads or tablets have appeared in the market-place but, at around £1000, three times as expensive as PDAs, prices need to fall for the mass market to adopt them. RSC, FIC, Viewsonic, Philips and a host of other suppliers are marketing solutions in the Webpad arena — although aimed at vertical market applications the Webpad could eventually find its way into the living room. Microsoft shares this vision with the development of a tablet under the former 'Mira' project name [22]. The concept is of a mobile monitor, which can be detached from the main PC and used remotely over an IEEE802.11b wireless connection. In the remote mode this monitor could be used as a Webtablet using a stylus to input data on its touchscreen and giving access to applications anywhere around the home. Although aimed initially at portable monitors an array of other devices could benefit from this approach, e.g. remote audio and video players. Appliance costs could be reduced making such devices more affordable to consumers. Alternatively, the advance of PDAs may threaten the Webpad concept's future as mass-market portable devices offering similar features at a much lower price enter the market.

Simple Devices promotes a product called the Simple Clock, which as well as acting as a standard clock can also act as an Internet clock with personal schedules synchronisable with Microsoft Outlook. The unit can display information about local weather, news, etc; it can also play MP3 and access Internet radio stations along with standard AM/FM broadcasts. Another nice feature is the ability to build communities by allowing users to send wake-up greetings and messages.

Photo albums (sending pictures from a storage server to a device for display) and picture storage (storage of pictures on a server in the network, which can be accessed by friends and family) are growing areas. The Ceiva Logic company offers an end-to-end service, hosting the server where the photograph is stored and transmitting to the address of the relevant photoframe for display [23].

6.4.2 Home Control, Security and Safety

A number of vendors offer home control and automation devices. For example, Honeywell offers a series of Internet-enabled devices to control heating [24]. Although based on PSTN, a broadband-enabled version of the Honeywell telephone access module will follow.

The current system uses a telephone access module to provide secure remote control of the heating, ventilation and air-conditioning. The Honeywell Home & Away System is a home network that covers security, control and Internet access.

Even the white goods industry has not escaped the Internet age. Ariston and Merloni Elettrodomestici both offer a new generation of household appliances that are network and Internet aware. Through the adoption of the Web-ready appliances protocol (WRAP), some of these new products embed the technology required to promote new services including remote assistance, consumption management and Internet connectivity. They transmit their data across the mains outlets using powerline-networking technology and can connect to the outside world through currently available residential gateways using the Telelink interface device.

A number of companies offer a series of network-enabled cameras that can be used for monitoring and security from remote locations. Their functionality ranges from purely networked security cameras, to those with built-in digital video recorder (DVR) technology right through to complete multi-camera solutions [25].

6.5 Summary

Are we ready for home networking? There are a number of factors affecting the adoption and development of home networking and home technologies from both the industry and user perspectives.There are three key industry elements that affect end-user behaviour and thence adoption.

- What can the network itself actually deliver?
- What devices will be available to place on the end of this network?
- What services will these devices and the network enable?

The home user will only adopt the technology when they see answers in four crucial areas.

- Benefit — is there a reason and definite use, are there exciting applications and services?
- Low cost — considering the benefits, is it affordable?
- Easy to install — does it involve disruption to décor, does it come with minimal set-up/configuration?
- Size, look and feel of the technology — does it blend with the furniture?

The current slowdown in the telecommunications, IT and Internet markets has been a major factor that has held back the adoption of home networking and Internet appliances.

But there are a number of drivers that will stimulate the home networking market:

- the take-up of higher bandwidth access technologies (e.g. ADSL);
- the need to share this single point of access to the Internet;

- digital content available from the Internet (i.e. music, videos);
- replacement of old analogue consumer products by digital equivalents;
- new low-cost Internet appliances and technologies;
- advances in wireless technologies (e.g. bandwidth, security);
- zero or minimal configuration of networks and devices (e.g. plug and play).

Today's domestic consumer equipment is moving towards a networking ethos, but the digital home with a network to support entertainment, information and communication may take a little longer to become a reality. When it does arrive it is likely to consist of a number of the technologies described in this chapter co-existing together.

But what is the outlook for home networking technologies?

In the short term, the stand-alone residential gateway will continue to be popular as an enabler for multi-user broadband connectivity and home networking technologies. The devices are getting cheaper and they alleviate the need for a user to leave their PC switched on 24 hours a day, seven days a week. Couple this with the emergence of faster, affordable wireless-based networks and we will see a steady growth within the domestic and SME environments for home networking.

Many users may wish to take the PC-centric route, having already purchased a high-specification machine, capable of acting as the networking hub and household PC. Again, coupled to a wireless-based networking solution, the user will be able to set up and run a home network and extract the full power of their investment.

In the longer term, as the Internet continues to evolve, we are more likely to see the home entertainment server become the networking hub in the home.

In order for home networking to achieve mass-market adoption, the industry needs to steer towards the point where the Internet becomes invisible to the end user. Devices will need to connect simply without configuration and applications and services will need to be compelling and easy to use. For example, connecting an Internet radio appliance will need to be as easy as plugging in a telephone while tuning will need to be as simple as an FM radio is today.

It is currently very difficult for a user to decide on the best solution for a home network — which network to choose, which standard to adopt. This will change if the industry eventually picks a standard to back, but this may be some time off as we are still in the early days of this technology race. An early leader looks like wireless networking, although Powerline, which has many problems to overcome, shows great potential. Plug and play may be a phrase that has been over used in the computer industry, but is one that reflects what the home user really wants — minimal hassle with maximum benefit.

References

1 Meltcalfe, R. and Boggs, D.: '*Ethernet: distributed packet switching for local computer networks*', Communications of the ACM, **19**(7), pp 395-404 (July 1976).

2 HAVi — http://www.havi.org/

3 IEEE1394 — http://www.1394ta.org/

4 USB OTG — http://www.usb.org/developers/onthego/

5 Universal serial bus — http://www.usb.org/

6 HomePNA — http://www.homepna.org/

7 HomePlug — http:// www.homeplug.org/

8 Kleinrock, L. and Tobagi F.: '*Packet switching in radio channels: Part 1 — carrier sense multiple-access and their throughput-delay characteristics*', IEEE Trans on Communications, **23**(12), pp 1400-1416 (December 1975).

9 Wireless LANs — http://www.wlana.com/

10 HomeRF — http://www.homerf.org/

11 Bluetooth — http://www.bluetooth.org/

12 DECT — http://www.dectweb.com/

13 2Wire — http://www.2wire.com/

14 Moxi Digital Inc — http://www.moxi.com/

15 Swale, R. P. (Ed): '*Voice over IP: systems and solutions*', The Institution of Electrical Engineers, London (2001).

16 Home networking — http://www.homepcnetwork.com/

17 SonicBlue — http://www.sonicblue.com/audio/rio/rio_audiocenter.asp

18 Motorola — http://www.gi.com/noflash/simplefi.html

19 Smartmedia — http://www.smartmediaonline.com/index.asp

20 TiVo — http://www.tivo.com/home_flash.asp

21 iDVDBox — http://www.idvdbox.com/

22 Microsoft — http://www.microsoft.com/windowsxp/smartdisplay/

23 Ceiva Logic — http:/www.ceiva.com/

24 Honeywell — http://content.honeywell.com/yourhome/zoning/tam.htm

25 Axis Communications — http://www.axis.com/

7

STANDARDS FOR BROADBAND CUSTOMER PREMISES EQUIPMENT

R I Galbraith

7.1 Introduction

The creation of standards for customer premises equipment (CPE) is considered as a key driver to enabling the availability of appropriate low-cost CPE platforms to connect to BT's future broadband networks. These standards will allow providers to offer a wide range of new services and applications to be delivered to a connected broadband home.

CPE standards will help to drive the convergence of the broadcast, telecommunications and computer technologies, enabling service providers to offer bundled service offerings (including video, data and voice) to the consumer. It is essential that the emerging standards for broadband CPE are compatible with standards for broadband xDSL multiservice networks.

These standards will benefit all sectors of the broadband industry, including consumers, equipment suppliers, network operators, broadband service providers and regulators. Of course, it is also important that these standards do not stifle innovation, and that suppliers can still innovate through product and service differentiation.

This chapter describes current BT standards activities in the areas of broadband CPE, in particular standards for broadband set-top boxes (STBs) and in-home networking. A number of activities in relevant standards forums are described, in particular CPE activities in the Digital Video Broadcasting (DVB) Forum [1] and the Full Service-VDSL (FS-VDSL) Forum [2].

7.2 Digital Video Broadcasting (DVB) Forum

The DVB Forum is an industry-led consortium of over 300 broadcasters, manufacturers, network operators, software developers, regulatory bodies and others in over 35 countries committed to designing global standards for the delivery of digital television and data services.

DVB systems are developed through consensus in the working groups of the Technical Module, which approves the technical output from the various working groups. The DVB Project works to strict commercial requirements established by member organisations. Once standards have been published, through the European Telecommunications Standards Institute (ETSI), they are available at a nominal cost to anyone, world-wide. All manufacturers making systems that are compliant with these open standards are therefore able to guarantee that their DVB equipment will work with other manufacturers' DVB equipment. Another advantage is that, since the standards are designed with a maximum amount of commonality, and based on the universal MPEG-2 coding system, they may be effortlessly carried from one medium to another — a feature that is frequently needed in today's complex signal distribution environment. DVB signals can move easily and inexpensively from satellite to cable, or from cable to terrestrial.

The DVB's first phase of development was focused on a number of open standards for digital video broadcasting, to enable broadcast and interactive services over all transmission networks including satellite, cable, terrestrial and microwave systems.

More recently the DVB recognised the need to define a standardised interactive 'middleware' architecture for digital STBs, known as the multimedia home platform (MHP) [3]. This is described in section 7.2.1.

Phase 2 of DVB has now commenced and is focused on convergence of DVB services with the broadband Internet. This includes building specifications for the transport of DVB services over the Internet infrastructure, and would apply to distribution over telco networks such as ADSL and VDSL. This work is ongoing in a group called DVB Internet Protocol Infrastructures (DVB-IPI). Other relevant activities are in copy protection (DVB-CP) and wireless in-home networks (DVB-WIN). These activities are discussed in the following sections.

7.2.1 DVB Multimedia Home Platform

7.2.1.1 MHP Background and Scope

At the start of digital broadcasting, the concept of an 'open' application programming interface (API) developed slowly. The first set-top boxes were designed to receive digital television broadcasts of pictures and sound, together with teletext information. There was very little spare memory or processing power to run an electronic programme guide (EPG) or any interactive applications.

The first DVB platforms were all vertical market operations, where the multiplex operator had control of the whole chain, the complete end-to-end system, including STBs and which API is to be used. The initial phases have resulted in the deployment of a large number of proprietary middleware platforms deployed across the world. This has resulted in a number of authoring problems for the operators, as interactive applications have to be authored in multiple proprietary formats. The DAVIC [4] organisation was the first public body to recognise the need for a standardised API to convey text and graphics and to run applications beyond the EPG function. DAVIC proposed that MHEG5 should be adopted as a presentation engine, with an optional Java Virtual Machine for more advanced applications.

Following the work of DAVIC, the DVB Project became aware of the need for a fully standardised API in order to offer a universal API under the same fair, reasonable and non-discriminatory terms as other DVB specifications. The aim was to develop one standard middleware for digital interactive TV and Internet access, thus overcoming market fragmentation and allowing for a more horizontal market, where one STB could receive applications, services and interactive content authored from any service provider. This led to the formation of the DVB Multimedia Home Platform (MHP) project in 1997.

7.2.1.2 MHP Requirements

Key requirements for the MHP project included:

- interoperability;
- scalability;
- upgradability;
- separation of data and applications;
- support for conditional access systems;
- security, conformance and interoperability testing.

The scope of the project [5] is shown in Fig 7.1.

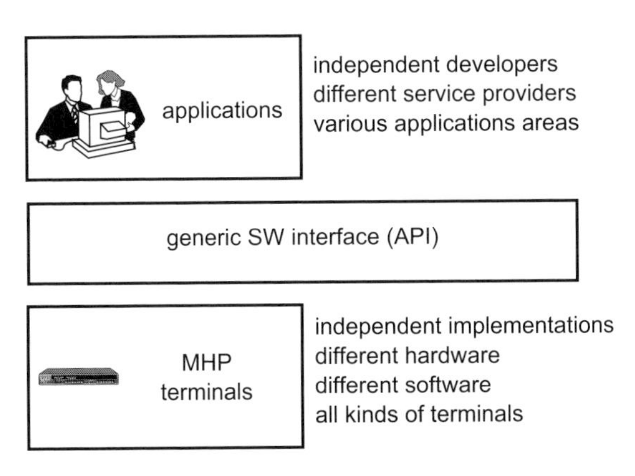

Fig 7.1 MHP project scope.

In the horizontal market scenario, any digital content provider can address any type of terminal ranging from low-end to high-end STBs, integrated TV sets and multimedia PCs, through applications delivered to a defined set of APIs. The APIs provide a platform-independent interface between applications from different service providers and the manufacturer-specific hardware and software implementation.

Examples of MHP applications include pay and free-to-air TV, enhanced teletext, games, eCommerce (e.g. home shopping, banking), interactive advertising, and Internet access.

7.2.1.3 MHP Architecture

MHP Solution Layers

The DVB MHP receives content from a DVB broadcast channel and can use an interaction channel. The content is normally displayed on a TV screen and operated by a remote control. Figure 7.2 illustrates the basic MHP layers [6].

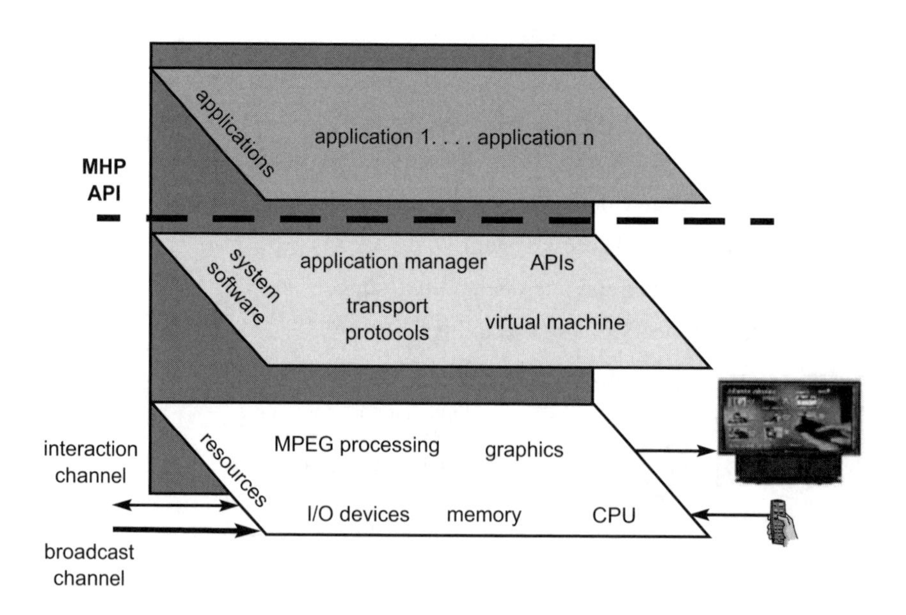

Fig 7.2 Basic MHP layers.

Resources include MPEG processing, memory, I/O devices, CPU and graphics processing. The system software includes an application manager and support for a number of transport protocols for the broadcast and interaction channels. The system software also includes support for a virtual machine, which abstracts the manufacturer-specific hardware and software, e.g. OS and drivers, from the standardised APIs. The applications can only access the platform via these standard APIs.

The core elements of the MHP are:

- the general architecture;
- transport protocols (DSM-CC object carousel for the broadcast channel and IP for the interactive channel);
- content formats (e.g. support for PNG, JPEG, MPEG-2 video/audio, subtitles, fonts);
- the Java Virtual Machine;
- DVB-J APIs;
- optional DVB-HTML APIs;
- the application life cycle and signalling;
- security and authentication;
- detailed profile definitions;
- minimum platform capabilities;
- the conformance regime;
- authoring guidelines.

MHP Profiles

Because of the wide range of MHP applications, the DVB MHP group defined three profiles based on the application area. In this way it is ensured that applications developed for a particular profile will run on MHP boxes which are compliant with that particular profile.

Figure 7.3 illustrates the MHP profiles. The base profile is the Enhanced Broadcasting (EB1) which supports access to broadcast (one-way) services and local interactivity. The Interactive Broadcasting (IB1) profile includes EB1 plus support for interactivity via a return channel. The initial version of the MHP specification (MHP1.0) is aimed at definition of the first two profiles and is based on a DVB-J platform. The third profile, called Internet Access (IA1), adds support for content based on DVB-HTML and is specified in the second version of the MHP specification (MHP1.1). It should be noted that MHP1.1 also describes optional additions based on DVB-HTML for profiles EB1 and IB1.

DVB-J Platform (and MHP1.0)

MHP1.0 specifies an extensive application execution environment for digital interactive terminals, independent of the underlying vendor-specific hardware. The execution environment is based on the use of a Java virtual machine and the

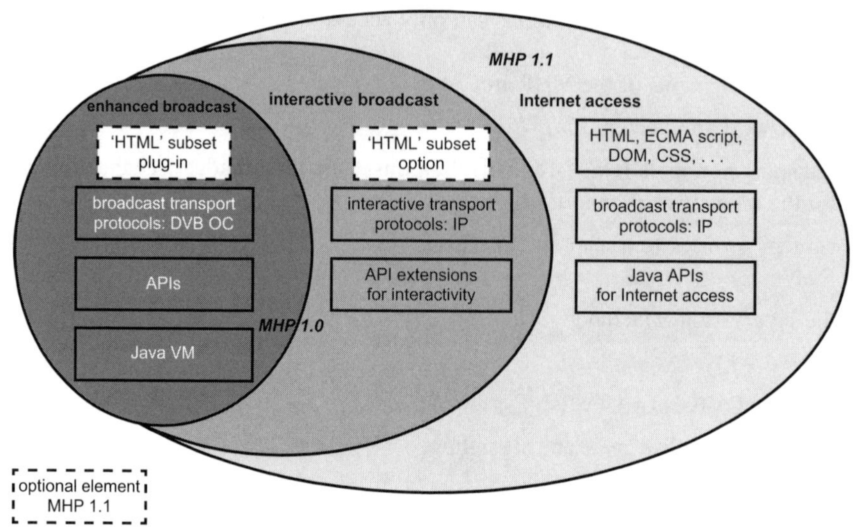

Fig 7.3 MHP profiles.

definition of generic APIs that provide access to the terminal's resources and facilities. A Java application using those generic APIs is called a DVB-J application. The DVB MoU ensures, for implementors of DVB-J, fair, reasonable and non-discriminatory terms for the use of any IPR owned by another DVB member.

Figure 7.4 shows the structure of the DVB-J platform. The application manager controls the operation and configuration of the MHP. The flexible software architecture enables a manufacturer to differentiate itself with specific user interfaces and features, which is considered important for the horizontal market.

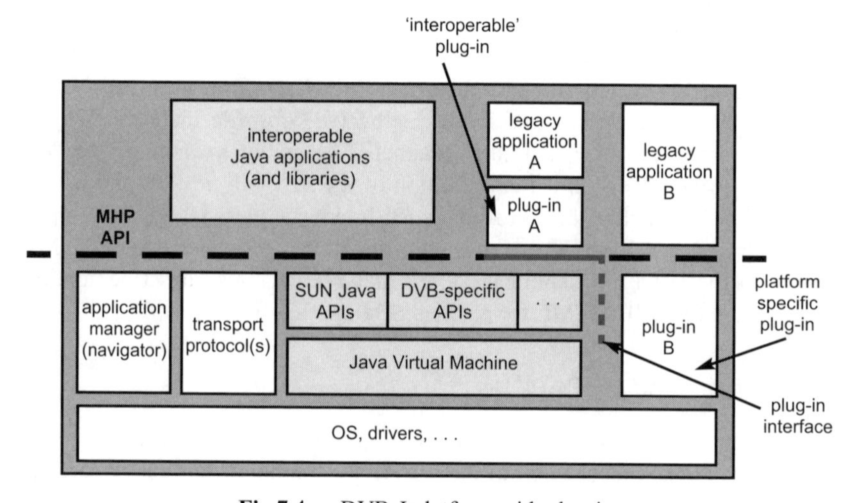

Fig 7.4 DVB-J platform with plug-ins.

The DVB-J APIs are divided into the following categories.

- Java APIs defined by Sun:
  - — fundamental Java APIs (lang, util, beans ...);
  - — presentation APIs (awt, JMF);
  - — service selection APIs (JavaTV).
- APIs defined by HAVi (Home Audio Video Interoperability Forum):
  - — presentation/GUI APIs.
- APIs defined by DAVIC:
  - — CA APIs;
  - — tuning API.
- APIs defined by DVB:
  - — extensions to Java APIs;
  - — data access APIs;
  - — service information and selection APIs;
  - — I/O device APIs;
  - — security APIs.

As there are a number of deployed DVB systems with proprietary APIs, to ease migration the DVB-J platform provides support for plug-ins for legacy APIs. These plug-ins can be implemented in DVB-J or can be implemented directly on the system software which would require a platform-specific plug-in, e.g. MHEG or OpenTV.

DVB-HTML Platform (and MHP1.1)

MHP1.1 specifies the Internet access profile, in which applications can control the basic operations of open Internet resident clients (e.g. Web browser, e-mail and news client). In addition to the DVB-J application format, an optional application type has been defined called DVB-HTML [7].

Much of the existing HTML content today has been primarily designed for the PC browser — the layout and interaction mechanisms are very different for a TV environment than for a PC. This underlined the need for a completely new declarative language, based on the W3C [8] specifications and complying with the DVB requirements.

Current browsers do not comply with the W3C [8] specifications and the same content may be rendered differently in different browsers. The DVB-HTML language is based on XML and a subset of the W3C specifications has been selected for the TV environment.

It should be noted that, due to its XML nature, DVB-HTML may be displayed on a PC WWW browser (although the result may be unpredicatable if the browser is not conformant to the DVB-HTML specification), but, for display on an MHP device, the Web content will be rejected if it has not been transcoded into DVB-HTML prior to transmission.

The DVB-HTML language is based on XML1.0, a metalanguage used for the definition of mark-up languages. The rules of an XML language are described in a Document Type Definition (DTD). Therefore DVB-HTML requires the validation of all documents that are signalled as being of the DVB-HTML application type. Any application which violates a constraint in the DVB-HTML DTD will be rejected by an MHP platform (note that proprietary extensions to the DVB-HTML language may be permitted, but only if they are in a private XML namespace). DVB-HTML builds on existing W3C specifications, such as modules from XHTML and cascading style sheets, DOM and ECMAScript [8].

For a content provider to target different platforms (TV sets, PCs, mobile devices, etc) and to be able to differentiate the TV environment, the 'dvb-tv' media type was created. By decoupling style information from the document structure, content providers can address the displays of different platforms by encapsulating, in the associated stylesheet, the style settings with those media rules (e.g. screen for desktop viewing, dvb-tv for TV viewing). For example, a desktop device could present itself as a TV device and emulate TV experience and vice versa.

7.2.1.4 MHP Status

In February 2000, DVB approved the first release of the Multimedia Home Platform (MHP) specification (MHP1.0). In July 2000, the specification was formally accepted by ETSI as Technical Specification TS101812 [9]. MHP1.1 was published by ETSI in November 2001.

An appropriate licensing and conformance-testing regime is currently being defined in DVB, where the MHP logo will be granted to a manufacturer only after successfully passing a set of defined test suites, with use of a self-certification regime.

Many STB manufacturers and leading interactive TV middleware providers are building DVB-MHP systems. MHP applications have been demonstrated on a number of platforms at a number of international consumer electronics shows in 2001/2002 [10].

There have been recent announcements that a number of Scandanavian countries are committed to MHP, and the MHP is also due to be launched in Germany. Also CableLabs, which sets standards for the North American Cable Industry, has adopted DVB-MHP.

MHP paves the way for the deployment of the open standard API, which will facilitate convergence of the broadcast, telecommunications and computer technologies.

This will make a huge impact on the set-top-box and digital-TV market. We will now see content from different providers available through a single device that uses the MHP common API, thus making it completely independent of the hardware platform on which it is running. It will enable a truly horizontal market in the content, applications and services that the consumer can choose.

The business implications are enormous, as new and exciting content and applications will stimulate the expansion of the broadcasting industry into the multimedia age, and link the broadcasting and Internet worlds together, enabling portability of interactive TV (iTV) content.

7.2.1.5 Future MHP

The next steps in MHP, will include integration of a number of ongoing DVB activities, such as:

- broadband Internet;
- standards for storage and personal video recorder (PVR) — an example of a PVR box today is TiVo, which is currently on sale in consumer outlets;
- MHP as a home gateway;
- integration of copy protection technologies;
- support for future content formats for media streaming, e.g. MPEG4.

Some of these activities are described in the next sections.

7.2.2 DVB Internet Protocol Infrastructures

7.2.2.1 Introduction to IPI

As well as traditional DVB delivery mechanisms, future DVB content could be transported by a broadband IP network operator and delivered over a network such as xDSL.

A group established in mid-2000 has been set up to define standards for the consumer terminal equipment to allow a receiver connected into an in-home network to receive unicast and multicast content delivered over IP. This would include applications such as video-on-demand (VoD) and also delivery of broadcast channels over IP and Internet carousels.

The group is attended by many of the European network operators, TV service providers, STB manufacturers, TV head-end manufacturers, network vendors and middleware vendors.

7.2.2.2 IPI Requirements

The DVB-IPI specifications will be aimed at delivery of a range of services over IP, such as:

- entertainment — broadcast TV, PayTV, VoD, music, picture download, games;
- general information — advertising, sports news, entertainment news, emergency information, general news, travel information, stock exchange information;
- educational — distance learning, computer-based training;
- messaging — e-mail, multimedia messaging;
- communication — videotelephony, VoIP;
- service information — electronic programme guide, service discovery and selection.

The DVB-IPI group will deliver a number of documents, which are referenced within an Architectural Framework — the baseline document for delivery of DVB services over IP-based networks [11]. This document introduces the DVB-IPI reference model and basic service class descriptions. The DVB-IPI architecture is applicable to all system and service implementations (using integrated receiver decoders, TV sets and multimedia PCs), as well as to clusters of such devices connected to home networks. The architecture describes an abstract layer model and a home network architecture.

7.2.2.3 IPI Architecture

Layer Model

The high-level reference model for DVB services on IP is shown in Fig 7.5. The various entities in the layer model are described briefly below.

- Content provider

  This is the entity that owns or is licensed to sell content (note that a direct logical connection may exist between content provider and home client, e.g. for rights management and protection).

- Service provider

  This is the entity providing a service to the client. Different types of service provider will exist for DVB services on IP, e.g. simple ISPs and content service providers (CSPs). The CSP acquires/licenses content from content providers and packages this into a service. A CSP could offer an IP-based TV multicast service to the end user, or a unicast service such as VoD.

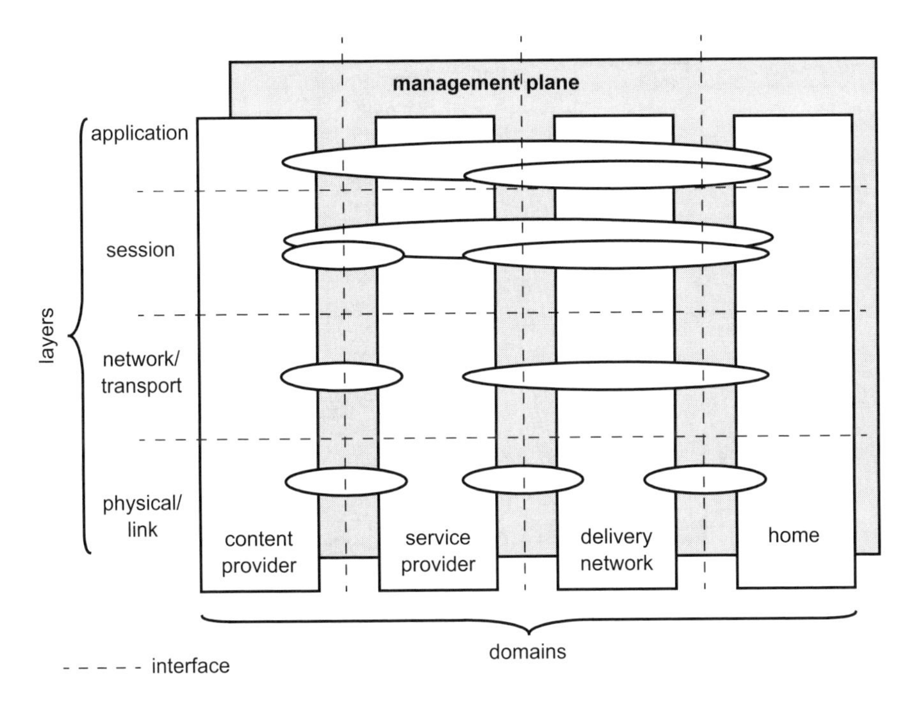

Fig 7.5 DVB-IPI layer model.

- Delivery network

  This is the entity connecting clients and service providers. The delivery system is usually composed of access networks and a core network, using a number of network technologies, e.g. xDSL. The delivery network is transparent to the IP traffic.

- Home

  This is the domain where the audiovisual services are consumed. Either a single terminal or a network of terminals may be used. It should be noted that the IP network layer enables an elementary end-to-end logical link by providing routing and packet forwarding. In DVB, the application layer is specified as the MHP.

Home Reference Model

The DVB-IPI home reference model is shown in Fig 7.6. The elements depicted in Fig 7.6 are described below.

- Delivery network gateway (DNG)

  This device connects one or multiple delivery networks with one or multiple home network segments. It could act as a bridge or router interconnecting

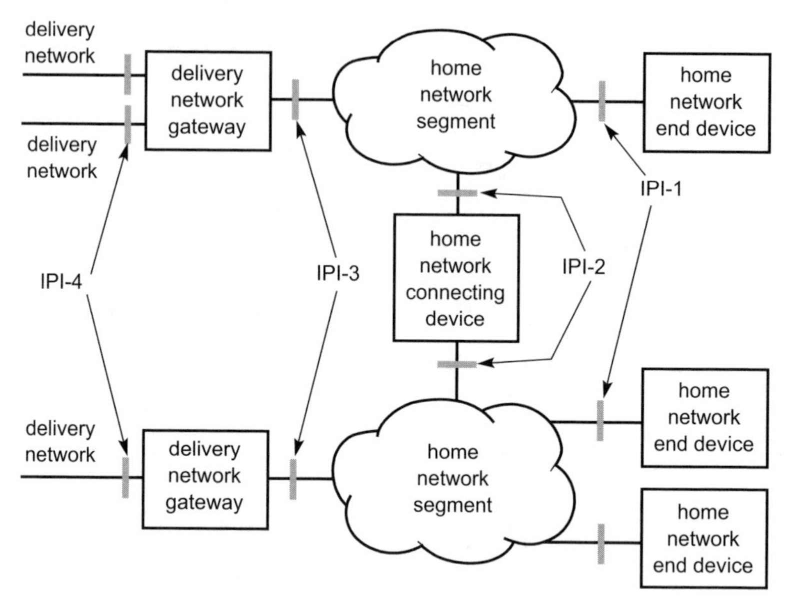

Fig 7.6 DVB-IPI home reference model.

different link-layer technologies or could act as an intelligent gateway providing functionality above OSI layer 4.

- Home network segment (HNS)

  This segment consists of a single link-layer technology and provides a layer 2 connection between the home network end devices and connecting components. Home network segments in DVB include wired technologies based on 100baseT Ethernet and IEEE1394, and wireless technologies, e.g. HiperLAN, IEEE802.11.

- Home network connecting device (HNCD)

  This is a device which connects two or more home network segments with each other, and functions as a bridge, router or gateway.

- Home network end device (HNED)

  This device is connected to the home network and typically terminates the IP-based flow, e.g. an STB.

The prime target for standardisation by DVB is the interface (IPI-1) to the home network end devices, to enable high-volume, low-cost equipment. IPI-1 describes the necessary protocols required for the delivery of DVB services over IP-based networks. The interface description will be independent of the physical-layer and link-layer technologies used in the home network.

The IPI group plans to deliver a number of specifications, in addition to the architecture.

- Transport of DVB services over IP-based networks

  This describes how DVB services are transported over IP, and defines both how the service is initiated and what the requirements are that will be needed on the network for correct and timely delivery of the service. It also specifies how the multicast and unicast services can be tuned and the mechanisms required for IP QoS.

- Service discovery and service selection

  This describes how an HNED discovers the available service providers, which services/channels they provide, and the mechanism to select those services.

- Network provisioning and IP addressing

  This describes how an HNED obtains an IP address and other basic IP services, and how the device is managed.

- IEEE1394

  This will cover the use of IEEE1394 in the home network segment.

- Ethernet

  This document will deal with how Ethernet is used within the home network segment (based on 100BaseT Ethernet).

- Security

  This will cover all aspects of network security in the home.

Many of the DVB-IPI group documents are already at an advanced stage, and it is likely that they will be completed before the end of 2002.

7.2.3 DVB Copy Protection

The DVB has issued commercial requirements for a new DVB content protection and copy management (DVB CPCM) system to provide a common framework for the protection of content beyond a boundary point of a DVB conditional access (CA) system (e.g. the card in existing satellite receivers). This would include in-home digital networks and personal video recorder technologies. A consumer domain architecture example [12] is shown in Fig 7.7.

Content is delivered by content providers, through service providers, to the consumer. This is carried out using a variety of delivery means which include protected and unprotected delivery methods (i.e. conditional access and/or digital rights management). When content enters the consumer domain it may move from a residential gateway (e.g. connected to an xDSL access network) or STB, using DVB CA, into a DVB CPCM environment. The aim of the DVB system is to provide end-to-end protection of content from the content provider through to the point of consumption of the content by the end user.

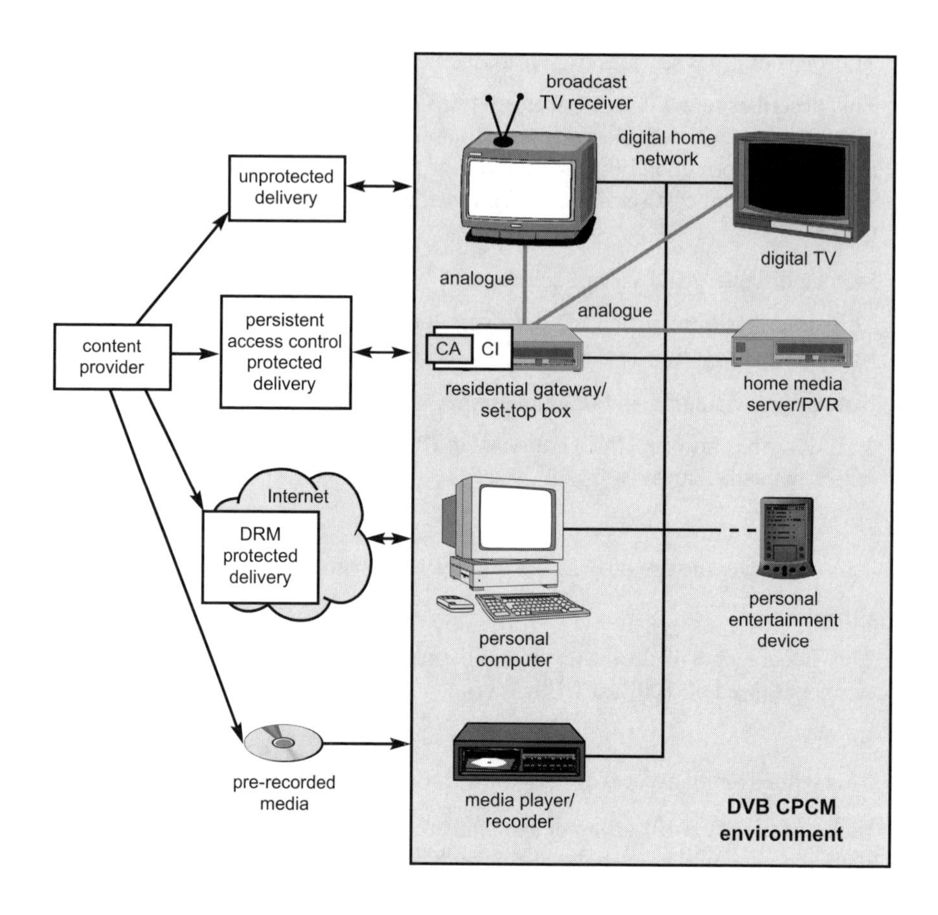

Fig 7.7 DVB-CP consumer domain example architecture.

A number of copy control states could be possible, e.g. copy once, copy never. This protection of content must also be supported by analogue and digital interfaces between devices in the consumer domain. The DVB CPCM system will define and support a standardised digital interface, at a logical level, establishing 'trust' between devices that allow protected content, usage state information, and the control of content usage, to be securely exchanged between two or more compliant devices.

The DVB Technical Module has issued a call for proposals [12] for technologies to meet these requirements. Over 20 companies have responded, including a large number of STB/silicon manufacturers and CA providers. The DVB aims to agree a specification for a common system. A key requirement here is the acceptance of the system by high-value content owners, and low-cost implementation in devices. As there are many issues to be resolved about such a system (e.g. licensing and management), it is likely to be mid-2003 before the system is agreed.

7.2.4 DVB In-Home Wireless

The DVB wireless in-home networking (DVB-WIN) technical group has agreed commercial requirements for wireless in-home networks primarily targeted at domestic applications but they will also allow for small office home office (SOHO) applications. This should address wireless distribution of multiple channels of video and associated services from a home gateway to portable television sets and other devices in the home.

The technical group are currently evaluating a number of in-home wireless technologies, such as IEEE802.11 and ETSI HiperLAN2, to meet the commercial requirements. Issues include QoS, frequency bands and channels, convergence layers, security and power management.

7.3 Customer Premises Equipment for FS-VDSL

The full service-VDSL (very high speed digital subscriber line) is a large group of over 70 companies representing network operators (over 14 representing North America, Europe and Asia) and vendors (including CPE) who share a common vision for an end-to-end multiservice video delivery platform [2]. The aim is to deliver a number of specifications to enable a common low-cost end-to-end video-centric multiservice network to enable offers of bundled services, including video, Internet and voice, to be delivered over a telco broadband xDSL network.

There is significant BT Group representation and active participation in FS-VDSL which has enhanced the success of the forum. In addition, the forum is under BTexact Technologies technical directorship.

7.3.1 FS-VDSL CPE Requirements

The aim of the FS-VDSL CPE specification [13] is to facilitate the standardisation of VDSL system requirements for mass deployment, in order, by maximising commonality of equipment, to achieve low-cost CPE.

The requirements address the distribution of video, data and voice services in an in-home environment having a high bit rate xDSL broadband access connectivity. The CPE architecture aims to enable the provision of a bundle of services in a reliable way, with minimal user intervention, respecting the key requirements of security and conditional access to the contents and at a cost compatible with mass-market deployment.

7.3.2 FS-VDSL CPE Architecture

The FS-VDSL CPE architecture allows for physical implementations which may perform all processing and decoding in a single customer premises device ('centralised architecture') or distribute functional processing and decoding into two or more customer premises devices. Within a premises, typically the following functions may exist:

- a service splitter, which separates the VDSL signals from other low-frequency services (such as POTS or ISDN);
- a VDSL modem termination function — sometimes termed a VDSL transceiver unit remote terminal (VTU-R);
- protocol processing in a 'home gateway' function to distribute the video, data and voice;
- in-home distribution interfaces and networks — this could include a combination of wired and wireless technologies;
- MPEG decoding functions for the viewing of broadcast TV and VoD;
- home appliances such as PCs and any other customer premises device that connects to IP data services;
- analogue voice interfaces to connect to VoATM or VoIP services.

The distributed and centralised architecture configurations are shown in Figs 7.8 and 7.9.

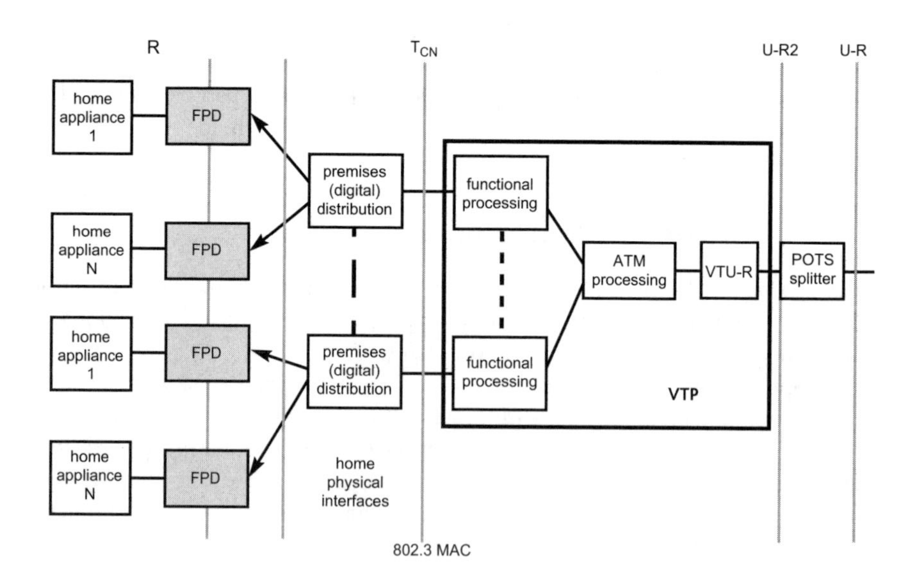

Fig 7.8 Distributed CPE — VTP with distributed approach.

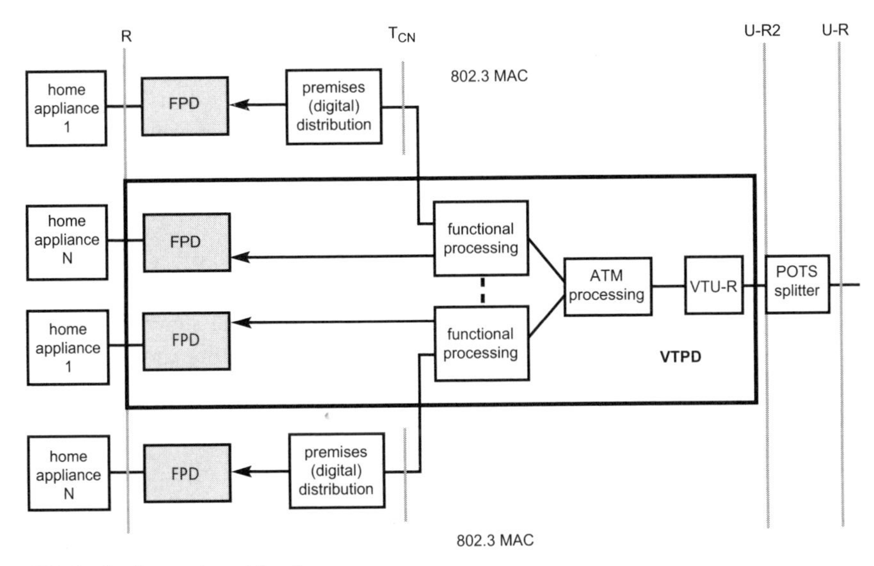

Fig 7.9 Centralised CPE — VTPD with centralised approach (with optional distributed CPE).

VDSL termination processing (VTP) refers to the function of VDSL modem termination, and protocol processing. A device that implements the VTP function includes interfaces to the in-home network. So the VTP is similar to an xDSL broadband home gateway which performs the function of modem termination and home-network distribution.

VTP and decoding (VTPD) refers to both the functions of VTP and of video decoding. An example VTPD would be the integration of an xDSL broadband home gateway inside a DVB STB. The FS-VDSL CPE architecture also defines a number of reference points as shown in Table 7.1.

Table 7.1 Reference architecture reference points.

Reference point	Description
U-R	Interface at the POTS splitter
U-R2	Interface at the input to the VDSL modem (before the VDSL PHY) (includes cabling)
TCN	Interface at the input/output of the digital customer premises distribution network
R	Interface at the output of FDP to the home applicance — the actual location of this interface is difficult to define and varies

The VTP (U-R2 interface) carries the VDSL line code. The interfaces to the home network (TCN) are based on an IEEE802.3 compatible MAC layer but the PHY layer is not specified in FS-VDSL, but left to the operator decision. Example interfaces are 100baseT Ethernet, HomePNA, IEEE802.11 wireless and Powerline.

Typical example interfaces at the home appliance (R) could include analogue video interfaces, such as SCART and interfaces such as USB, Ethernet and IEEE1394. Functional processing associated with the home appliance could include digital decoding (e.g. video/audio decoding, application processing, voice processing).

The model allows for a number of business scenarios from the network operators, dependent on the service offerings and who manages the CPE, for example the VTP or VTPD could be supplied by:

- the network operator;
- a retailer (or service provider).

Some examples of business scenarios are shown in Fig 7.10. A network provider may include the equipment providing the VTP or VTPD functions as part of their service offering (see Figs 7.10(a) and 7.10(c)). Alternatively, network providers may offer a service at the U-R or U-R2 differentiation point, and therefore the VTP or VTPD functions are supplied by the service providers (see Fig 7.10 (b)). Interface U-R2 includes the cabling from the splitter U-R interface up to the VTP connector. Control by the provider for QoS and management extends to the TCN interface. The home service delivery control extends through the FPD. The demarcation points may vary based on network operator and regulatory requirements.

The FS-VDSL CPE specifications define a number of mandatory and optional functions and flows based on the above architectures. These include:

- packet flows for the service types (video, voice, data) at the U-R2 interface:
  - — NAT (network address translation) flow;
  - — routing flow;
  - — bridging flow;
  - — broadcast TV flow;
  - — channel zapping flow via IP multicast;
  - — VoATM (optional);
- VTP/D functional processing;
- functional processing and decoding (FPD) functional specifications.

It should be noted that the FS-VDSL CPE specifications are primarily aimed at defining CPE specifications to ensure compatibility with the FS-VDSL network

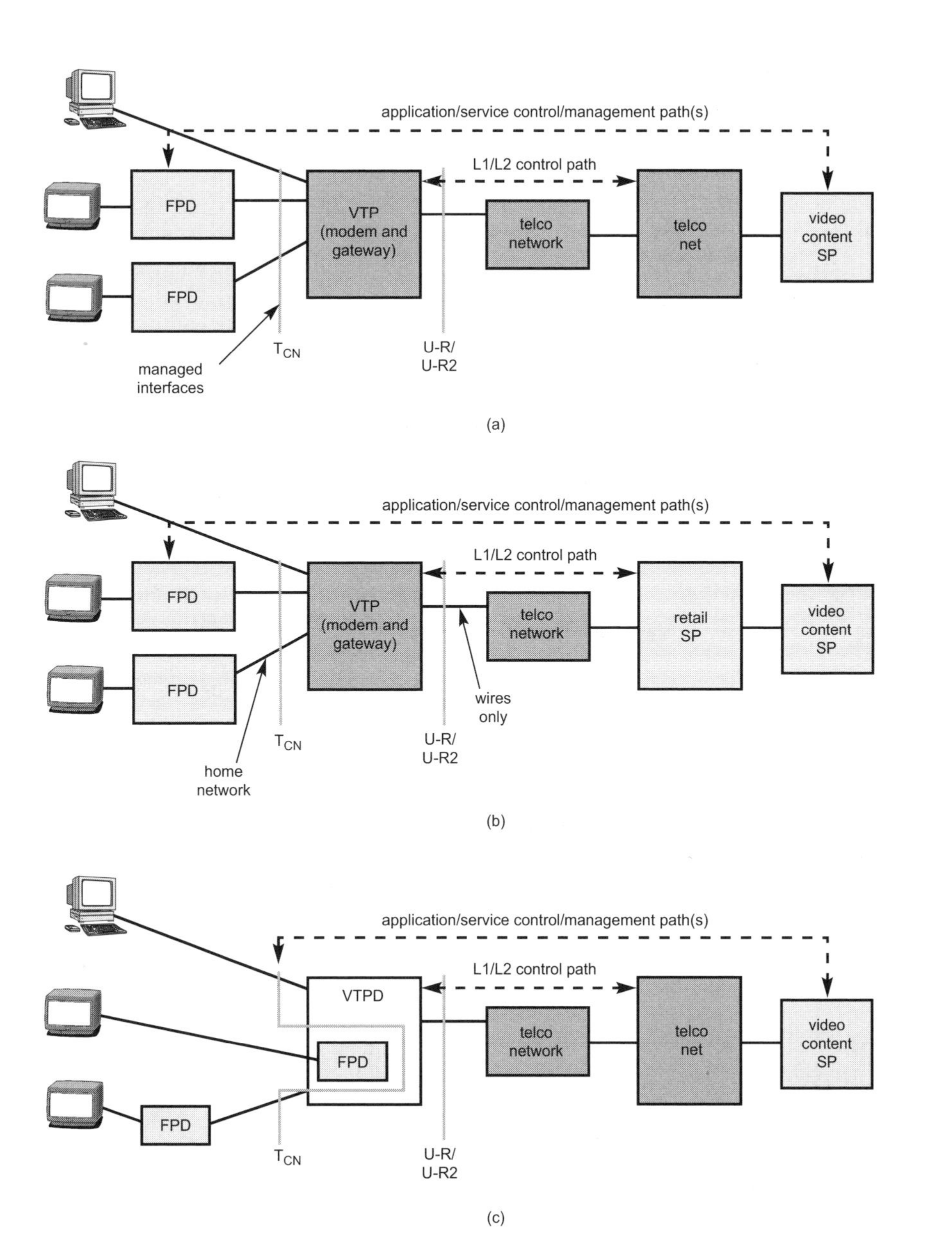

Fig 7.10 Reference model showing example business solutions.

system architecture. Therefore FS-VDSL has not specified a CPE middleware architecture or any copy protection mechanism, as these are likely to be specified by a service provider. FS-VDSL is considering digital rights management (DRM) from an end-to-end perspective to ensure that the network architecture is secure.

There is ongoing verification of the FS-VDSL specifications via a number of international demonstrators hosted by operators (including BT). A number of access vendors, middleware and CPE vendors have been selected by network operators to validate the FS-VDSL system specifications, and to enable a demonstration platform for FS-VDSL. This has evolved into the broadband-home/broadband-office demonstrator, administered by BTexact at Adastral Park.

7.3.3 FS-VDSL CPE Status

The schedule for approval and publication of the FS-VDSL specifications has been finalised. The FS-VDSL management committee approved the documents in June 2002, to allow for publication and feedback from the wider industry. It is planned to downstream the FS-VDSL specifications into the ITU.

7.4 Home Networking Forums

There are a large number of home networking industry forums that are progressing standards for home network physical layer technologies. These include wired and wireless technologies and it is likely that a number of these technologies will be seen being deployed in the home on a range of consumer products. A broadband gateway device may support a number of these technologies (e.g. copper wire, coaxial, wireless, powerline) for distribution of video, voice and data within the broadband home. Some of these forums are listed below.

- Wired
  - — Home Phoneline Networking Alliance (H-PNA) (http://www.homepna.org);
  - — HomePlug (home powerline networking — http://www.homeplug.org);
  - — IEEE802.3 Ethernet (http://www.ieee.org);
  - — USB (universal serial bus — http://www.usb.org/);
  - — IEEE1394 (http://www.1394ta.org/).
- Wireless
  - — HomeRF (http://www.homerf.org/);
  - — IEEE802.11 (http://www.ieee.org);
  - — HiperLAN (http://www.hiperlan2.com);

— DECT (http://www.dectweb.com);

— Bluetooth (http://www.bluetooth.com).

There are also a number of forums that are developing standards from a management and application-sharing perspective for home gateways and consumer equipment:

- HAVi (Home Audio Video Interoperability — http://www.havi.org);
- UPnP (Universal Plug and Play Forum — http://www.upnp.org/);
- OSGi (Open Services Gateway Initiative — http://www.osgi.org);
- Jini (http://www.jini.org).

7.5 TV-Anytime Forum

The development of personal video recorders will bring about a significant change in the way consumers access TV, as PVRs act as home servers for audiovisual content and therefore free people from linear broadcasts, allowing them to access their personalised content on demand. There are already a number of 2nd generation PVR products on the market today but they are all proprietary designs, have no interoperability, and in some cases are tied to particular service providers.

The TV-Anytime Forum (TVAF) [14] is an association of organisations that aims to define the platform for future generations of personal digital recorder (PDR) products, applications and services that give consumers access to what they want, when they want it, and how they want it. TV-Anytime will provide this by developing open standards and specifications that enable interoperable audiovisual and other services based on mass-market high-volume digital storage on consumer platforms — referred to as local storage. This forum has approximately 150 members from around the globe representing many industry sectors, e.g. content creation, consumer electronics manufacturers, IT industries, professional equipment manufacturers, component manufacturers, software vendors, broadcasters, Internet broadcasters, telcos and service providers. The TVAF was formed in 1999 with the emergence of mass-storage technologies (hard disk drives) that met the demanding needs of the consumer mass markets in terms of reliability, noise, power consumption, data transfer rate, and capacity at an appropriate cost.

There are five main working groups, each with their own requirements and specifications.

- Business Models

  The Business Models working group develops the functional and non-functional commercial requirements to meet the needs of a range of cross-industry commercial models and scenarios. These are passed to the relevant technical working groups to test and push forward the specification series.

- Content Referencing and Location Resolution

  The Content Referencing Specification provides a location-independent content reference identifier (CRID) and the process by which this CRID is resolved into times and locations where the content can be found.

- Metadata

  This working group is developing a standardised format and dictionary for describing content.

- Rights Management and Protection

  The aim of this working group is to maintain rightful use of content by developing a standardised API to address existing and new content protection schemes.

- System and Transport

  The System and Transport working group covers the end-to-end system design ensuring that all the TVA specifications work together and are compatible with other standards bodies.

The TV-Anytime Forum maintains close links with many other standards and industry forums to define and agree scope.

Widespread availability of standards-based PDR devices in the home will enable BT and other service providers to exploit and complement their capabilities to deliver richer and more engaging multimedia broadband services.

7.6 Other Industry Forums

There are many other international standards-based forums who are looking at the requirements for CPE in the home. These include ANSI, DSLForum, OpenDSL, ISO/IEC, Cenelec, IETF, ATSC and ATVEF, who are involved in setting standards for xDSL CPE, in-home networking and TV-related standards. The DVB and FS-VDSL forums are building on elements from these established international bodies.

The standards being produced for terminal compliance with the DVB-IPI specifications should allow for these devices to connect to any IP access network, e.g. xDSL and cable. Therefore it is planned that the DVB-IPI standards for home network end devices should also work with the emerging standards from the cable industry, who set standards through an organisation in the USA called CableLabs [15]. The cable industry has common standards for digital cable based on DOCSIS. Cable standards are being produced for packet services over IP (including voice and video) known as PacketCable and a home network architecture called CableHome. The cable industry has also specified standards for OpenCable, which include hardware specifications and middleware specifications for interactive TV. The middleware specification is called the OpenCable Application Platform (OCAP) software specification and is largely based on the DVB MHP middleware specification.

7.7 Summary

This chapter has presented an outline of the major standards areas pertaining to the customer premises equipment that will be connected to future broadband multiservice networks.

The DVB Forum is a key driving force for the standardisation of consumer equipment, and is currently tackling some critical issues for service providers such as:

- interactive TV content interoperability;
- standards for delivery of DVB services over an IP network;
- interface and protocol standards;
- home networking;
- security and digital rights management.

It is essential that the DVB standards for consumer equipment and content are compatible with future standards for broadband multiservice access networks such as those being defined in FS-VDSL. Any service provider committed to delivering multiple services (video, voice and data) over broadband needs to be aware of these standards and to influence them to ensure they are driven in the right direction.

The developments in the CPE standards forums are key to bringing down the cost of consumer equipment and enabling interoperability. BT should continue to play an active role within these organisations to achieve a vision of a rich, media-enabled broadband network, capable of delivering services to multiple devices in the broadband home.

References

1 DVB Forum — http://www.dvb.org/

2 FS-VDSL — http://www.fs-vdsl.net/

3 MHP — http://www.mhp.org/

4 DAVIC — http://www.davic.org/

5 Luetteke, G.: '*MHP Multimedia Home Platform*', presentation to DVB Seminar (February 2001).

6 Vogt, C.: '*A guided tour of MHP*', World Broadcast Engineering, White Paper (March 2000).

7 Perrot, P.: '*DVB-HTML — an optional declarative language within MHP1.1*', EBU Technical Review, White Paper (September 2001).

8 W3C — http://w3c.org/

9 ETSI TS101812: '*MHP Specification*' — http://www.etsi.org/

10 Consumer Electronics Shows, Las Vegas — http://www.cesweb.org/

11 DVB TM2520r1 DVB-IPI: '*Architectural Framework for the Delivery of DVB Services over IP-based Networks*', (September 2001).

12 DVB TM2549 DVB-CPT: '*Call for Proposals for Content Protection and Copy Management Technologies*', (September 2001).

13 FS-VDSL Committee, FS0060R0.6: '*VDSL Customer Premises Equipment Specification*', (Draft)(December 2001).

14 TV-Anytime Forum — http://www.tv-anytime.org/

15 CableLabs — http://www.cablelabs.com/

Part Three

LIVING IN THE HOME

S G E Garrett

So far this book has considered the technology for bringing broadband into and around the home — but what use will be made of this enhanced method of communication? Before looking at the application technology (which follows in the next part), user needs have to be considered. Perhaps the simplest way of finding this out is to go and ask — unfortunately this is not easy for new technology. Asking people how they would use technology and products they have never experienced can give misleading results. Equally important is gaining an understanding of not only what people do in their homes but also how they behave there. Social, economic and cultural trends are important — the home today is not static.

In this third part, the four chapters discuss the usage of new telecommunications technologies, drawing conclusions about the implications for future communications applications in the home.

Chapter 8
Domesticating Broadband Access — What Really Matters to Consumers

This chapter provides an analysis of a range of survey, interview and usage-log data on the level of use of broadband Internet by households in the UK and Europe. It discusses the (few) discernible differences between PSTN and broadband Internet users in mid-2001. The analysis suggests that, while there are few socio-economic and demographic differences, broadband users tend to use a wider range of applications, access them more frequently and for longer. However, because most broadband users have been, on average, users of the Internet for longer than PSTN users, these effects may be to do with their Internet competencies rather than the nature of broadband Internet itself. The chapter also suggests that the interaction of speed, flat rate and 'always-on' is a key feature of broadband and a key value model to the user. It concludes by discussing the implications of the usage patterns of these 'early adopters' for current and future portal, application and service investment strategy.

Chapter 9
Is the Future Really Always-on?

For many, the advantages of 'always-on' (AO) are still not clear and the concept can be difficult to understand and distinguish from traditional network connections. How important is the ability to be AO compared to other network characteristics? What will be the quality of the service (speed, availability and reliability) and which services will take full advantage of AO? Is it important to be able to control the information flow? What will be the business model for AO services — would it be cheaper or more expensive than other networks? What are the security and privacy concerns, and are there any social consequences such as less face-to-face contact and user isolation?

This chapter starts by describing the work done to capture the user's attitudes and expected behaviour towards AO. This includes a review of the initial testing of the concept through a number of focus groups and the development of usage scenarios to illustrate the possibilities for services. The results from a number of user trials are also presented. These were carried out in five different European countries and provide a snapshot of different AO services across Europe. The trials identified a number of technical and operational issues and the chapter makes recommendations for the operation of AO services, and the implications for AO terminals, devices and services. The user trials identified a significant difference between being '*always-on to a network*' or '*always-on to a service*' — which seems to be a much more compelling prosposition for users. This is further explored with a description of some ongoing research work investigating this issue through the development of a device-unifying service (DUS) which provides users with a single virtual terminal for all of their devices and access to persistent network-based services, which are 'always-on'.

This work was undertaken within EURESCOM, the European Institute for Research and Strategic Studies in Telecommunications, a leading institute for collaborative R&D in telecommunications. It was completed as two separate projects (P1003 'Exploiting the always-on concept', and P1101 'The device-unifying service') with different collaborative partners in each.

Chapter 10
Digital Living — People-Centred Innovation and Strategy

This chapter provides a summary of a research programme at BTexact Technologies that is aimed at helping a technology innovation company to consolidate its innovations, to see opportunities for the exploitation of its technologies, and to create socio-technical visions that can help to drive technological innovation itself. As a by-product, the programme has also created strategic knowledge that is of critical importance to public and private policy/decision makers alike. This research is a key part of the BTexact Technologies approach to the creation of, and response to, disruptive technologies. Understanding 'usage by people' is absolutely critical to figuring out what is disruptive about technologies, why this is so, and therefore how to make money out of them. Since this is critical to many companies' core competencies, the value of the research reported here is self-evident both to BTexact and to its customers. Without it, they will only ever make money by accident, a strategy that shareholders do not seem to find amusing.

Chapter 11
Digital Homes — For Richer For Poorer, Who Are They For?

Networking homes to support and use high technology services is usually seen as the domain of top-of-the-range, executive properties. However, there is growing evidence not only that this is not necessarily the case, but also that technology can be used to bring cost-effective benefits to a wide range of people. This chapter looks at the digital home from a 'customer' viewpoint, where the customers are house builders or developers, housing associations or local authorities, home owners or tenants. Their individual requirements and expectations are explored, along with a framework for identifying and delivering the advantages of the digital home.

8

DOMESTICATING BROAD-BAND — WHAT REALLY MATTERS TO CONSUMERS

B Anderson, C Gale, M L R Jones and A McWilliam

8.1 Introduction

The 'broadband revolution', or access to the Internet via high-speed connections, has been hailed as the future of the Internet. But what is the real evidence? What do 'ordinary' people do with it once they have it and how can these behaviours be turned to profitable advantage? This chapter draws on a range of material in an effort to answer some of these questions, especially the investigations into the domestication of broadband in Europe and, in particular, in the UK. We say 'domestication' for a specific reason. Not only does it remind us that there is a complex process of assimilation and working out 'what this thing is for' going on [1], but it also reminds us that broadband Internet is a new and strange beast as far as the general population is concerned.

The material we use to examine these processes comes from the following three sources:

- relevant background literature;
- primary research conducted with BT broadband customers by BTexact Technologies' Customer Behaviour Laboratory under the umbrella of the 'Digital Living' research programme [2] — this research involved semi-structured interviews with 19 broadband households together with automatic logging of their Internet applications usage over a period of 12 months from May 2000 to May 2001; ten of the households were married couples, five worked from home, most were of high socio-economic status and all had been early adopters of the Internet (the households were part of a commercial, i.e. paid subscription, trial of ADSL Internet in West London, and during this trial they were not encouraged to make any extra or particular use of the Internet connection);

- data collected as part of a European information and communication technology (ICT) users survey conducted from 30 October to 11 December, 2000 as part of the EURESCOM P903 'Cross-cultural attitudes to ICT in everyday life' project [3, 4] — this survey collected data on some 211 broadband[1] Internet users as a by-product of surveying a representative sample of more than 9000 households across Czech Republic, Denmark, France, Germany, Italy, the Netherlands, Norway, Spain, and the UK (where used below this data set is referred to as 'P903 data').

According to *The Economist* [5], in mid-2001 there was a huge variation in the rate of broadband adoption between different countries. South Korea came out top with 9.2 broadband connections per 100 inhabitants, compared with 2.25 in the USA, and a mere 0.08 in Britain — and this is despite the popularity of the Internet in all three countries. However, this data is perhaps somewhat misleading because it does not indicate how many current Internet users have broadband access nor whether these 'connections' are actually in use. As Fig 8.1 shows, the percentage of Internet users with broadband access varies across Europe. Of the nine countries surveyed the average is 8% although it is rather higher in the Netherlands (17%) and France (10%) and, interestingly, this survey suggested that the UK also ranks quite highly (10%).

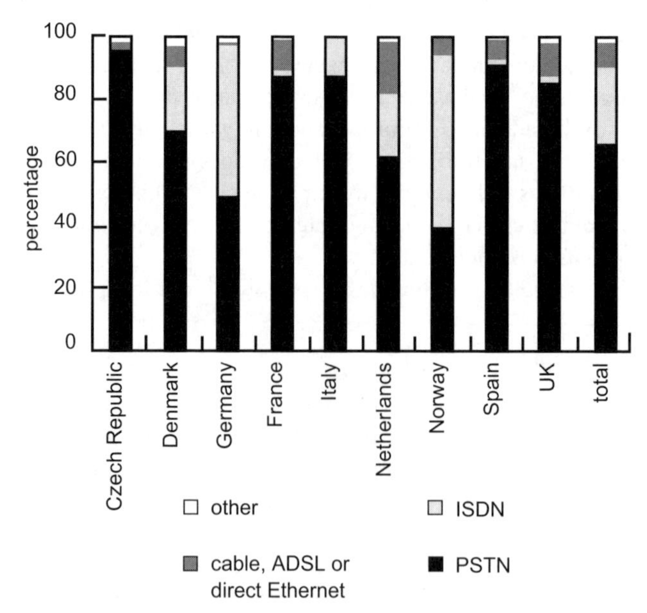

Fig 8.1 Penetration of domestic broadband Internet access in a range of European countries in late 2000 (source = P903 data, base = all households who had Internet access, number = 2412).

[1] Defined as all those with an ADSL, cable modem or direct Ethernet connection in their home.

The Economist [5] attributes the low adoption rates to a number of overlapping political, regulatory and technical factors. Meeting demand can be a huge problem because of the massive amount of resource and investment needed to build the infrastructure. In the UK the picture is patchy, with DSL and cable modems only available in specific areas where investment in infrastructure is thought likely to see a reasonable revenue return and where the physical connectivity supports the service[2]. As a result the diffusion of broadband Internet [6] is constrained by geography as well as price and utility (see Fig 8.2). This is, of course, changing as broadband availability becomes more widespread.

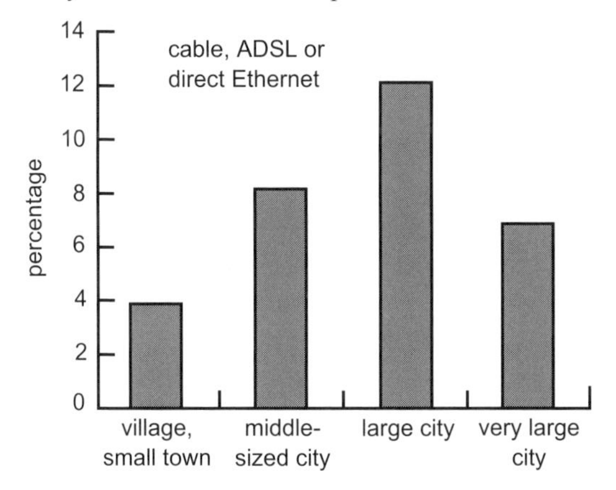

Fig 8.2 Percentage of Internet user households in each settlement type who have broadband Internet access (source = P903 data).

8.2 Broadband Socio-Demographics

So who are the broadband users and are they any different from PSTN Internet users? They are, by definition, an early adopter segment [4] because they are the early adopters of broadband Internet. When comparing their usage with that of 'normal PSTN Internet users', we must always be aware that their behaviour might well be that of a particular group of technophiles rather than some sort of early picture of future broadband customer behaviour.

This section uses the P903 data to explore the current socio-demographic nature of households with broadband Internet access and compares them with households who have PSTN Internet access. In doing so, it lumps all the 'broadband' users together as one sample although we acknowledge, of course, that broadband households in one country may be structurally different (due to differing public policy or market conditions) from those in another. Unfortunately the sample sizes preclude meaningful national level analysis.

---

[2] DSL throughput decreases with the distance from the exchange.

The P903 data suggests that in late 2000 there were differences between the life-stage distributions of broadband and PSTN Internet users in Europe. As Fig 8.3 shows, a higher proportion of PSTN users are in the 'aged 30-60 couple with children' stage than broadband users, while the reverse is true of single households and especially those without children. These differences in distributions are statistically significant[3].

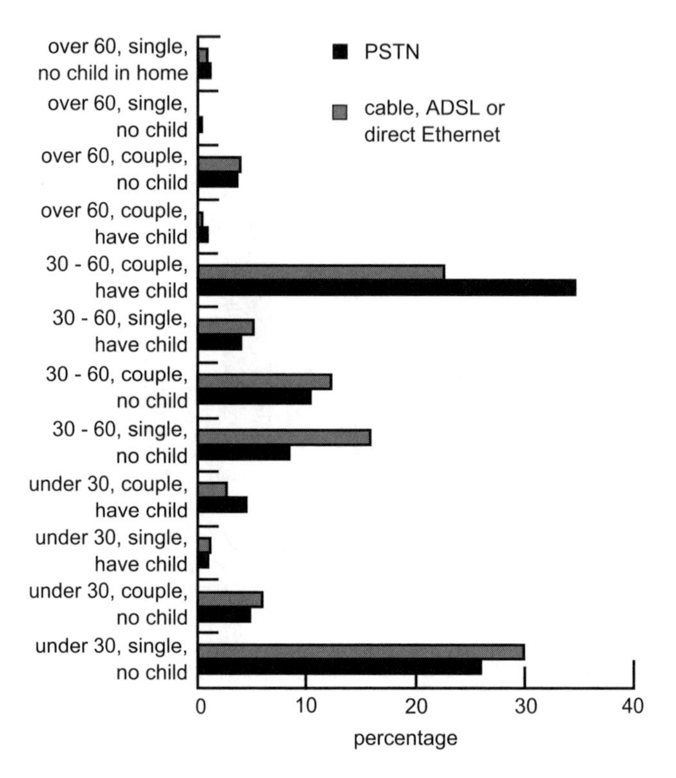

Fig 8.3 Percentage of broadband or PSTN Internet users in each life stage (source = P903 survey data, base = all randomly surveyed households who had PSTN or broadband Internet access).

The P903 data shows no significant differences between PSTN and broadband Internet users in terms of age (Internet access rates fall with age), gross household income (Internet access rates increase as income increases), educational attainment of the household-head (those with secondary or tertiary education are more likely to have access) or their current employment status (economically active households are far more likely to have Internet access). These results confirm Hoag's similar results [7] in an early study of cable modem adopters in the USA.

[3] chi sq = 27.67, df = 11, p = 0.004

In addition, households with broadband access did not have more TVs, set-top boxes, video recorders, games consoles, mobile telephones or PCs than households with PSTN Internet access. However, they did have on average more CD ROMS connected to the PCs in the household[4] and more PCs with Internet access[5]. The latter suggests that either acquiring broadband has prompted them to increase the number of PCs they have or that the kinds of people who have signed up for broadband access have tended to have multiple PCs already. Our interviews also suggested that many broadband customers were maintaining PSTN access 'just in case' of technical problems with their broadband access. Indeed in some cases the most technologically able had devised an in-home network to enable several PCs to share a printer, scanner and broadband connection.

Individuals who had broadband Internet access in their households, and who were Internet users, had been using the Internet for, on average, 3.48 years, while comparable PSTN users had been using it for 2.95 years. The differences between these two are statistically significant[6].

As well as collecting socio-demographic data, the P903 survey also collected attitudinal data on Internet use.

This data (summarised in Table 8.1) shows that broadband users agreed more strongly than did PSTN users with the assertion that using the Internet is fun and that it is easy to use. Broadband users disagreed more strongly with the assertion that it is too expensive and that it is easy to get lost.

In terms of socio-demographics, there seems little to differentiate households with broadband Internet access from those with PSTN access except that some life stages (especially singles) appeared more likely to have acquired broadband access. A confounding factor in this analysis may be that the geographical availability of broadband might be greater where there are more singles, such as urban areas. However, the length of time since an individual started using the Internet does appear to be a factor in broadband take-up and, when combined with their possibly greater technical competence (see Table 8.1), this has significant consequences, as will be discussed below.

8.3 Broadband Usage

Hoag [7] suggests that broadband users tend to make more use of FTP and the Web than PSTN users and also to spend more time on-line. She also suggests that they make more use of a wider range of applications and are more satisfied with their Internet experience [8]. We take these two issues in turn.

---

[4] $t = -2.05$, $df = 202.32$, $p = 0.02$, 1 tailed

[5] $t = 1.75$, $df = 240.81$, $p = 0.04$, 1 tailed

[6] $t = -2.83$, $df = 227.777$, $p = 0.0025$, 1 tailed

Table 8.1 Scaled responses to a selection of the P903 attitude questions. Respondents were asked to record their response on the standard Likert scale of 1-5, where 1 is 'agree' and 5 is 'disagree'.

	Mean score (PSTN)	Mean score (broadband)	Sig
Using the Internet is fun	1.81	1.48	$t = 4.93$, $df = 275.80$, $p < 0.001$
Using the Internet is too expensive	2.95	3.39	$t = -4.16$, $df = 261.34$, $p < 0.001$
The Internet is easy to use	1.78	1.53	$t = 3.57$, $df = 1758$, $p < 0.001$
It is easy to get lost when using the Internet	2.96	3.28	$t = -2.77$, $df = 255.75$, $p < 0.001$

8.3.1 Time On-line

According to the P903 data, individuals who were currently active Internet users, and who had broadband access at home, spent, on average, 3.67 hours using the Internet in the week leading up to the interview while comparable individuals with PSTN access spent, on average, 2.16 hours using the Internet[7]. It should be noted that this is not the order of magnitude difference in usage which is sometimes quoted in the media (for example, as reported in *Wired* [9]) or in other studies such as MediaOne Labs' trial of PC-based and wireless broadband Internet usage where they found an average daily usage of just under 2 hours for the PC-based access and just under 2.5 hours for the wireless Webpad [10]. It should also be noted that asking survey interviewees to make *post hoc* estimates of time spent on particular activities is a notoriously inaccurate method and this may explain a possible underestimation by the P903 survey respondents (see Robinson and Godbey [11] for a detailed discussion), although the degree of difference in these results suggests that some other factor is at work. For example it is possible that McClard and Somers' households [10] were selected for the trial precisely because they were high Internet users already.

As Fig 8.4 indicates, usage rates for broadband users tends to increase with Internet experience[8] although this does not appear to be the case with PSTN users, whose usage rates appear to be relatively flat with respect to Internet experience. This suggests that the heavier usage of broadband customers is not simply because,

---

[7] This difference is statistically significant: $t = -4.803$, $df = 204.81$, $p < 0.001$

[8] Although not necessarily with the length of time they have been using broadband — the P903 data does not tell us.

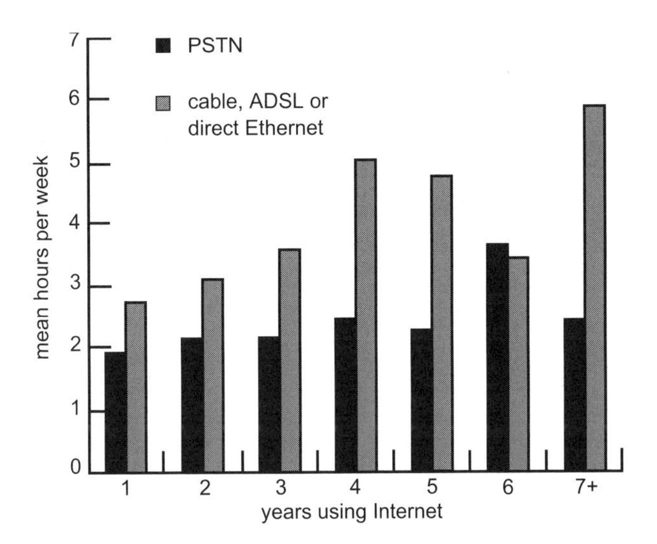

Fig 8.4 Mean hours spent using the Internet 'last week' by broadband and PSTN Internet users against years using the Internet (source = P903 data, base = all households who had PSTN or broadband Internet access).

on average, they tend to have been using the Internet for longer than have PSTN customers.

However, when interpreting these results, it must be remembered that this data is cross-sectional. Thus each cohort of users represented in Fig 8.4 may have had specific usage patterns which may or may not have changed over time as their experience of using the Internet increased. For example, the heavy broadband users who had been using the Internet for 4 years may not have been heavy users when they first started using the Internet but on the other hand they may have been much heavier users than they are now. Unfortunately cross-sectional data such as that furnished by the P903 survey cannot help us to answer this question.

8.3.2 Internet Applications Usage

The P903 data shows that broadband Internet users make use of a greater number of Internet applications/services than do PSTN users[9]. The data also shows that they make more frequent use[10] of the services they use ($t = 6.293$, $df = 1745$, $p < 0.001$, 1 tailed).

---

[9] Respondents were asked if they had used a range of 40 services in the last three months. Broadband users had used an average of 12.16 while PSTN users had used an average of 9.42 ($t = 05.13$, $df = 244.966$, $p < 0.001$).

[10] Respondents were asked whether they used it every day (1), a few times a week (2), once a week (3), a few times a month (4) and less than once a month (5). Their 'frequency of use' score was calculated by taking the mean of this interval data. Clearly this figure has no real everyday meaning but it is at least an indicator of frequency of use.

The 'Digital Living' broadband usage logs suggest that daily usage flattens significantly due to the lack of usage (and temporal) tariff structures (see Fig 8.5). However, usage is still fundamentally constrained by life-style — much of the UK population is 'at work' or 'at school' between 09:00 and 16:00. As a result, in the future we should expect to see the greatest usage increases by those who are 'at home' during the day such as carers (including mothers), young families, the unemployed, retired or housebound.

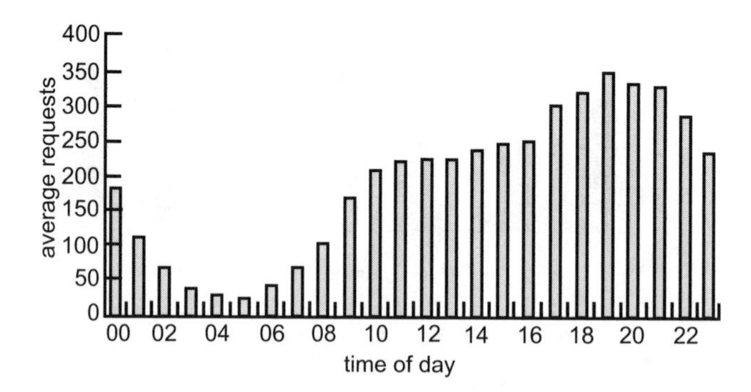

Fig 8.5 Average number of http (Web) requests per hour by time of day for the months of May to July 2000 (source = Internet usage logs for 19 BT broadband households).

The 'Digital Living' interviews showed that broadband Internet usage starts to fit more closely into patterns of daily life — into spare slivers of time. For example, users checked traffic[11] or weather before leaving for work, they looked up restaurant telephone numbers and then booked a table rather than looking in the telephone book and they watch the 18:00 BBC news at 19:35 because that's when they have a spare moment.

In addition, it appeared that broadband users used the Internet as a primary source of information whereas previously they might have used other forms of media (such as books or newspapers). As one person remarked:

> ".... instead of 'I've got to dial up, I'll get an engaged tone, and it's going to take me X amount of time to get connected'... if I've got to turn my computer on to get a recipe and I'm on a 56k modem, those recipe books are going to win. Now ADSL wins ..... if I have a question that needs answering that's where I go .... so you just end up relying on it, using it more."

Furthermore, broadband appeared to encourage people to try different types of content. For example, they would use it for real-time share prices whereas previously they would have consulted the paper-based *Financial Times*. However, it

[11] Indeed this is confirmed by the P903 data which shows that broadband users were more likely to have looked for traffic information (44% said they had versus 30% for PSTN users), although there was no significant difference for weather information.

was interesting to note that the respondent who said this still took the FT at the weekend in order to catch up on more in-depth analysis.

The breadth of Web usage of the BT broadband customers was quite remarkable. During the period of May to September 2000, the 19 households made http requests to at least 1523 different Web site domains; the mean number of requests per domain was 24.6 while the mode was 1 (34% of sites were 'visited' only once). The most frequently visited site (real.com) received 13% of all hits, the second (bbc.co.uk) received 9% but the 10th most visited (egroups.com) received only 2% (see Fig 8.6). In addition, the 'top 10' most visited sites were different for each household. The households whose usage was logged all had very different Web usage patterns, no doubt reflected by very different interests and needs. This re-emphasises that the 'mass market' is really no such thing — it is in fact an aggregation of an enormous number of niche markets.

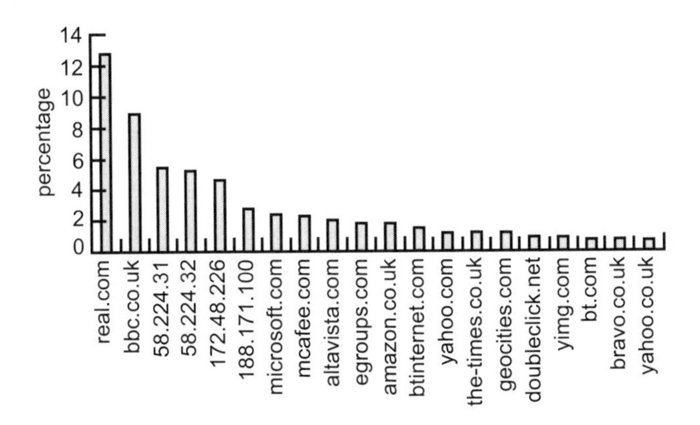

Fig 8.6 Percentage of total http requests made to the top 20 Web site domains[12] for the months of May to July 2000 (source = Internet usage logs for 19 BT broadband households).

Some of this usage is enabled by a change of location of the broadband equipped PC:

> "... one of the things that I did do is to move the ADSL down from the spare bedroom, ... into the living room. Because having a permanent connection to the Internet becomes part of your life-style, part of your life ... I can't do without it now."

---

[12] Note that Web site addresses were aggregated by removing the leading '*.' where * could be any sequence of characters. Thus the hits on www1.bt.com and www2.bt.com were aggregated into a sum for bt.com. This is not a foolproof algorithm for summing hits on particular domain names but we believe that it gives a better indicator than no aggregation at all.

Families in particular often moved the PC into household spaces where it became a shared appliance, albeit with a spaghetti of wires and poor product aesthetics. Once there it becomes a central focus of family activity.

8.3.2.1 Streaming-Media Usage

It might be expected that broadband Internet users would be more likely to use streamed media than PSTN users, and to use it more intensively. The P903 data shows that broadband users are more likely than PSTN users to use on-line radio (24% of broadband users had used it against 16.5% of PSTN users), on-line TV (20% versus 9%) and music downloads (54% versus 29%). Broadband users also make more frequent use of on-line radio[13] and music downloads[14] but not of on-line TV.

For the 19 BT broadband households who had their Internet usage logged, two sites[15] '209.191.132.13' and bbc.co.uk dominated their usage of streamed media, receiving between them over 45% of Realpayer/Netshow requests (see Fig 8.7). Times of usage appear to match well-known TV viewing patterns — busy periods are weekday early evenings, and especially on Wednesday and Thursday evenings. There is some evidence of use on Saturday and Sunday early afternoon although this is skewed by the high usage of particular households on particular days. It is likely that different content types will be more popular at different times of day (due to life-style factors). Time-use diary data collected during other digital living projects shows that at the population level people do not tend to watch that much TV on Friday and Saturday evenings because they tend to be involved in other social activities often taking place outside the home although there is considerable life-style/life-stage variation.

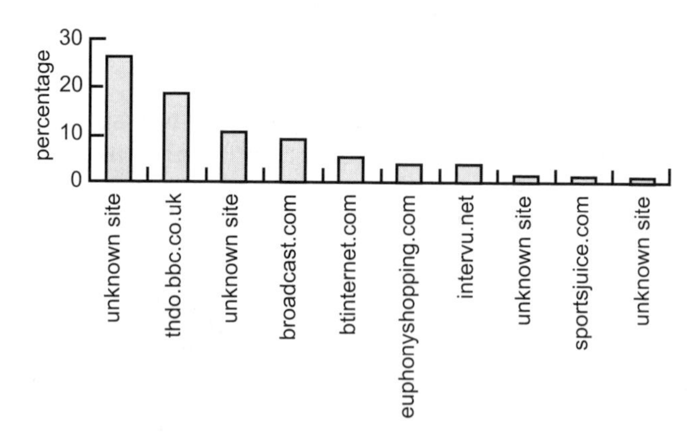

Fig 8.7 Percentage of Realplayer/Netshow requests made to the top 10 sites between May and July 2000 (source = Internet usage logs for 19 BT broadband households).

[13] $t = 2.589$, $df = 292$, $p = 0.005$, 1 tailed

[14] $t = 4.686$, $df = 539$, $p > 0.001$

[15] Domain names have been aggregated as previously discussed.

However, the usage of streamed media was relatively low overall, with only an average of 1.92 transactions per household per day. This low usage, and also that reported by the P903 data with respect to on-line TV may well be due in part to the relatively little broadband-ready content available at the time. However, it should also be noted that 'watching TV' on a PC is a very different experience to 'watching TV' on a TV with respect to the usual location, angle and social context of use.

8.3.2.2 E-mail Usage

According to the P903 data, broadband users are no more likely to use e-mail than PSTN users, although their frequency of use score was significantly higher[16].

The BT broadband customers used PoP e-mail applications and/or Web-based e-mail. Figure 8.8 suggests that most e-mail is sent in the early morning, gradually tailing off to late evening when there is another peak. However, given that several of these households were also running businesses from home, this figure is probably showing a typical work-day pattern overlaid by their (and other household's) domestic patterns of use. Also of note is that with broadband (always-on and flat-rate) many people tend to leave their connection open and their e-mail application 'logged in' so that it continuously checks for new e-mail.

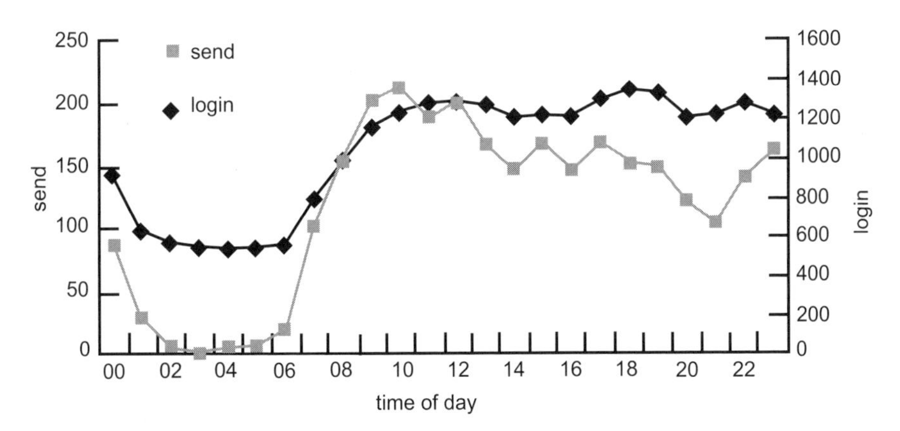

Fig 8.8 Number of PoP/SMTP e-mail send and log-in transactions per hour between May and July 2000 (source = Internet usage logs for 19 broadband households, note that the value (y) axis is cumulative over time to emphasise the usage patterns).

As Fig 8.9 shows, while their usage of Web sites may have been extremely wide ranging, the 'portal' that the BT broadband households did go to was their e-mail service. While they used 25 different e-mail servers, the top one (mail.btinternet.com) captured 56% of the traffic. As we will discuss below, this has implications for what a 'useful portal' might be.

---

[16] $t = 3.995$, $df = 244.35$, $p < 0.001$, 1 tailed.

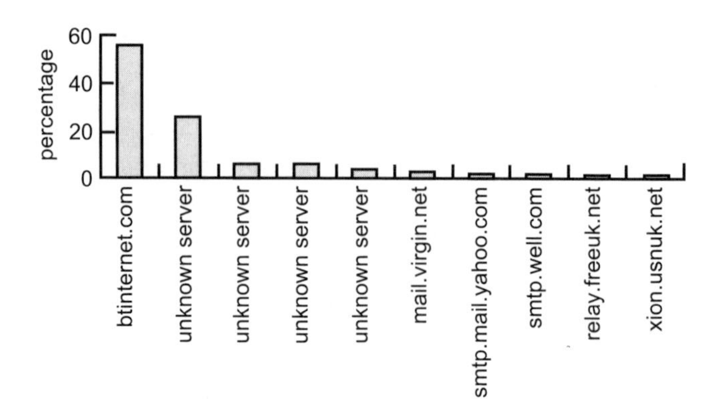

Fig 8.9 Percentage of e-mails sent by the 19 BT broadband households via the most heavily used e-mail servers.

8.3.3 Other Applications and Services

The P903 survey also suggested that broadband users were more likely than PSTN users to create their own Web pages (22% had done so against 15% of PSTN users), to use text (IRC) chat (43% versus 27%), usenet newsgroups (36% versus 24%), multi-user games (28% versus 15%) and single user games (37% versus 22%). Broadband users were also more likely to use their Internet connection to telework[17] (11% had done so against 5% of PSTN users). As noted elsewhere these results do not mean that broadband Internet leads to changes in behaviour. It may simply be that these early adopters of broadband Internet are more likely to do these things because of the kind of technophiles they appear to be.

Few broadband or PSTN users had made use of Internet-based video-conferencing (7% and 4% respectively) and few intended to (76% of broadband users and 78% of PSTN users said they had no intention of doing so). In contrast, 15% of broadband users and 11% of PSTN users had made use of Internet-based telephony and roughly a further 20% (of each) said they intended to. It should be clear from even this relatively coarse data that there are grave doubts over the usefulness of video-based communications applications in the minds of current PSTN and broadband users. This does not necessarily mean that they will never be of any use, but it does mean that in their current incarnation[18] and for this group of people, there is little opportunity.

---

[17] i.e. work from home rather than travelling to their office.

[18] For example, considered as an adjunct to telephony — in fact, research on the use of video in the collaborative work setting has demonstrated that the power of video is less as an add-on to telephony than as a way to 'see at a distance' to judge if a colleague is available for a telephone call, to show objects or models, or simply for a feeling of awareness and co-presence [12]. These same needs are, of course, true of the domestic context.

8.3.4 Changes in Internet Usage Behaviour

The 19 BT broadband households were asked how they had used the Internet prior to broadband, and what, if anything, had changed since getting the service. While these sorts of retrospective questions do not necessarily provide reliable data, all participants claimed to use the Internet far more than they had prior to broadband. The reasons given for this were some combination of the speed, permanent connection and fixed-rate pricing. Participants were then asked what they were using the Internet for and to account for any extra time they now spent on-line. They were specifically asked about what kind of sites they looked at, whether they shopped and banked on-line, and whether they downloaded and uploaded files. The following findings emerged.

- Finding information

  The Internet is more often the medium of choice for finding information. The speed, permanent connection and fixed rate mean that participants are more likely to use the Internet as the first port of call when trying to find information, where they might previously have used other media (e.g. books or newspapers). The nature of the information varies according to the needs of individual participants, but stated items of interest included local restaurants, cinema details, financial information, recipes, bus and train timetables and antiques.

- Exploring content

  For the same reasons as detailed above, participants are more likely to use broadband to explore different types of content. Again the type of content varies between individuals, but the finding is consistent across the sample.

- Buying on-line

  People are more inventive and explorative in their Internet use. As people successfully use the Internet to explore different types of content and find information, they become more willing to explore the different things for which it can be used. For example, many of the sample cited previous fears about on-line shopping because of the potential of credit card fraud, and of downloading software because of the possibility of viruses. However, having the speed and convenience of broadband makes them more likely to attempt such novel things, and having successful transactions creates a sense of trust and thus a willingness to explore further. This behaviour is sometimes seen in other Internet users, but it appears accelerated and accentuated with broadband. Indeed, the P903 data suggested that, while broadband users were no more likely to have conducted eCommerce transactions than PSTN Internet users (25% of broadband users had done so compared to 22% of PSTN users), they were more likely to have done it more frequently ($t = 1.73$, $df = 62.10$, $p = 0.044$, 1 tailed). Similarly broadband users were as likely as PSTN users to have made banking transactions via the

Internet (27% versus 23%) but they did so more frequently ($t = 2.26$, $df = 394$, $p = 0.012$, 1 tailed).

- Ease of access

  Broadband better supports people's life-styles by allowing them to quickly connect to the Internet when they want, for as long as they want. It does not, however, necessarily change people's life-styles, nor turn them into 'mouse potatoes'. Instead, it allows people to use the Internet as much or as little as they choose, without any of the associated problems of dialling up.

While the kind of self-report recollections mentioned above are useful, and the cross-sectional P903 data provides some indicators, as far as we are aware there has been no systematic quantitative attempt to study the changes in behaviour of Internet users as they move from PSTN to broadband access. Only a 'tracking study' which follows a group of individuals over time, measuring their behaviour continuously if possible, would offer significant new insights into what behaviour changes occur in order to provide a basis for broadband service providers' potential revenue models.

8.4 Perceptions of Broadband

Hoag [7] found that users with high-speed access felt the Internet to be more interactive, more compatible with their life-styles and needs, and less complex compared to PSTN users. We have already seen some of the same results from the P903 data, and the dominant theme to emerge from the BT broadband interview data was that all participants were positive about broadband and wanted to keep it. Some comments were only marginally positive. For example, one participant said:

> "What ADSL means to me is the difference between something that's tediously slow and something that's bearable and I'd like it to be much better".

The majority, however, were extremely positive, for example:

> "If you took it away from me tomorrow I'd just have to put up with it wouldn't I ... god forbid, no, no, I wouldn't want to go back to how things were ... it's a marvel"

> "I'm hooked, I can't go back to the way it was".

> "There isn't a single person that I know that I haven't told about ADSL and said get on the waiting list; it has to be done."

Given the overall enthusiastic attitude to the service, participants were then asked to identify the features of broadband that made it such a positive experience. Three features were mentioned by all participants, and will be detailed individually.

8.4.1 Fast

When asked to identify the key features of broadband that made them unwilling to give it up, all of the BT broadband customers first mentioned the connection speed. Participants' reasons for valuing a high-speed Internet connection varied, but included:

- less frustration when surfing because page content loads almost instantly;
- less frustration when downloading files, especially to the home-worker — as one home worker said:

  "I can download so much faster, I downloaded something that would have taken me four hours before, in eight minutes ..."

  this aspect of broadband also being perceived as making downloads more reliable, because the less time a download takes, the less likely it is to crash during the process — in addition, the shorter the download time, the more efficiently the participant can work, because long downloads prevent the Internet and PC being used for other things;
- a fast Internet connection means that less effort and investment is needed on the part of the customer, and they are more prepared to use the Internet to try out new things — one participant gave the following example:

  "Let's face it if you have to spend half an hour waiting for something to be downloaded then you lose interest. If it's in minutes, it's fine, you make a cup of tea, come back and it's there. But half an hour, honestly, I mean... you have to be really dedicated to something to wait that long. And then it may turn out that it's not what you want. So you just give up and look it up in a book instead, and look through pages, and it's much faster ..."

- as is often the case with updated technology, participants quickly became used to the 'new and improved' way of accessing the Internet, and were very reluctant to give it up — one participant compared broadband to hot water: non-essential but highly desirable and highly inconvenient to live without.

8.4.2 Always-on

With broadband solutions, a connection to the Internet is established as soon as the PC is switched on. This 'always-on' (AO) feature consistently rates as one of broadband's most attractive benefits. The BT broadband customers suggested that the elimination of the frustrations associated with the 'boot-up dial-up' barrier was the reason for this. Dial-up connections could be 'psychologically off-putting' if people wanted to go on to the Internet for a short time because they were so cumbersome, with the result that they would often prefer to use an alternative source of information. As one interviewee remarked:

> "One of the reasons why I love ADSL is that it makes the Internet so much more accessible. With modems, all that kind of dialling up crap, all the slowness of the connections, it's an inhibitor."

Being able to use the fixed line telephone at the same time was also a key advantage. Many described the frustration of being cut off while downloading files or surfing. In addition, the immediacy of e-mail was seen as a major benefit — being able to see when e-mails arrive and to access them instantly was an important reason for wanting AO:

> "And what I hated ... was that you'd have to dial up to see whether you'd got mail and if ... there isn't anything there it's frustrating, it's costing you money for nothing. And then if you don't dial up and then you find you've got thirty e-mails ... it's just very, very irritating."

AO appeared to make the Internet supportive of people's life-styles in a way that dial-up connections could not:

> "It's just like having a TV or a hi-fi, it's just there and when it's not there you think, damn, it's really convenient. It also means that ... banking or whatever, you can do it as soon as you think about it. It means you have to remember less."

As discussed in the opening section, the current broadband revolution among the consumer population of the UK is still in its infancy. This means that key features such as high speed and permanent connections are sufficiently attractive to suggest they will continue to be important triggers for broadband adoption for existing PSTN Internet users for some time to come. Nevertheless, as access becomes more widespread and consumers begin to have more choice in terms of their broadband providers, it is likely that high-speed AO Internet access will increasingly be viewed as the norm. Speeds offered by dial-up connections will be seen as antiquated, in the same way as we currently view mainframe computers. Indeed, one person from the BT broadband interviews, when asked whether they had a modem prior to broadband, commented: "I can't remember, it's pre-history that."

High-speed permanent connections will effectively become 'hygiene factors' to the whole Internet user population — they have got to be there, consumers will view them as their right, and only when they are absent will they be an issue. This lends weight to the idea that access providers should cultivate relationships with PSTN consumers in order to focus on services and applications to meet their needs in the future.

8.4.3 Flat Rate

Most broadband solutions have fixed flat-rate pricing via, for example, monthly subscriptions. Despite the fact that the perceived expense can be a barrier to initial adoption, the Digital Living participants found the guarantee of no extra call costs to be a definite benefit because it provided absolute control over their Internet budgets.

The combination of pay-per-minute pricing and slow dial-up modem connections could be extremely frustrating, as one person commented:

> "If the network's really busy then everything just grinds to a halt and [with normal Internet access] you are paying for it to grind to a halt."

Indeed, as was shown in Table 8.1, the P903 data suggests that broadband users appear to be less worried about the costs of Internet use.

So which is the most attractive broadband benefit? The BT broadband interviews suggested there was no consistent view among interviewees as to which of the three main benefits of broadband (high-speed, always-on and flat-rate charging) was the most attractive. Thus, although individuals may prefer one benefit over another, it was concluded that all three are mutually dependent as we will discuss below.

8.5 Implications

The results reported in previous sections have a series of implications for any organisation (public or private) which is seeking to benefit from the 'broadband revolution'. These range from possible applications and services to strategic investment options. We discuss some of these below.

8.5.1 Commercial Strategies — to Portal or not to Portal?

Current and near-future broadband users who are, in general, already very heavy Internet users do not use broadband portals — they already know what they want and they know where to get it on the Net. As a result they spend very little time using their ISP's Web server (i.e. broadband portal) — instead they go to their usual sites such as news.bbc.co.uk, chillout.real.com and so forth, as these respondents remarked:

> "I don't feel any particular interest at all in any of the offerings that are available really .... I don't really care what it looks like, it doesn't really

> matter, ... most of the time I'm just going somewhere else so I just click on favourites to get where I want to go, or I type in an address."

> "Frankly the idea of portals is utterly useless to my way of thinking. I'd use the existing brand names, I don't think the companies like BT, or any of the other portal providers, are in any way capable of competing with the ... BBCs of this world."

This implies that portal advertising revenue that depends on 'eyeballs' and hits is unlikely to generate much revenue from this early adopter segment, most of whom are already experienced Internet users. In addition, our analysis of Web usage shows that these early adopters have such differing information needs that attempting to cover them all in one portal would be impossible.

Thus if predicted advertising revenue is being used to offset the subscription price then unless a mass market of new Internet users can be attracted to (and equipped with) a broadband service[19], the subscription price will have to rise in the near future to make up for the lower advertising revenue.

However, there are alternatives to 'content-based portals'. Rather than creating a place to 'watch stuff', 'applications portals' could be created through analysing the things people value, then applying products and services to those values to create a site where people can 'do stuff' to achieve their needs. If their ISP's portal is the only place they can do this, it will lock people in to the ISP's applications, make the place addictive so that the advertising revenue model works and so that ISPs can charge for a package of valued applications (rather than content).

8.5.2 Ways of Thinking

8.5.2.1 The Broadband Virtuous Loop

Section 8.4 discussed the perceptions of broadband in terms of speed, flat rate and always-on. How do these features interrelate given that, when asked which they valued above all others, participants varied in their responses. We postulate the following balance between the three elements, highlighting the key psychological features of each that combine to make broadband so attractive to participants.

The model in Fig 8.10 is not merely theoretical, but is derived from what broadband triallists said about the service during interviews. Fundamentally, we should regard a flat-rate tariff, always-on, broadband service as a single proposition, rather than simply regarding it as a technology which is inherently desirable. The model proposes that each element of the proposition helps to positively reinforce every other element in a mutually reinforcing way.

---

[19] Until broadband access moves beyond 'PC-based' technologies, this is highly unlikely.

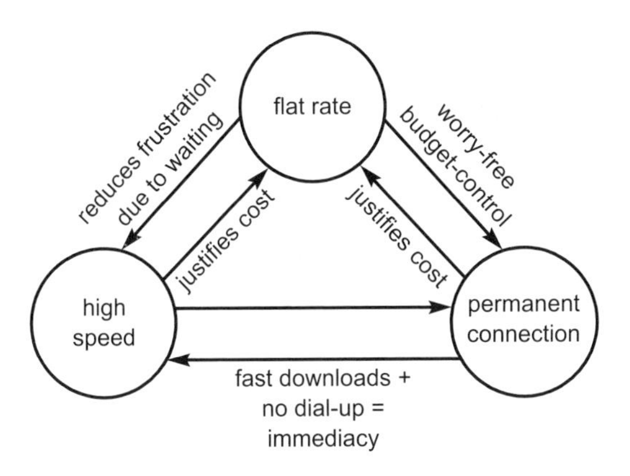

Fig 8.10 The broadband virtuous loop.

- Relationship of speed to tariff

The increased speed of broadband (relative to the standard 56 kbit/s and to ISDN) is seen as justifying the additional cost:

"But if anyone had to ask me should I go for ISDN at £30 a month or ADSL at £50 a month I'd say no question, ... go for ADSL."

The flat-rate element is also a significant factor in people's minds:

"What I like about it, and it is purely psychological, is that I pay a flat rate and I don't have to worry how long I am on the Internet. ... So I just don't care, when I want to go on and surf, I surf. And it could be an hour, two hours, three hours, it doesn't matter."

"I think they're [speed and flat-rate tariffing] completely locked together ... But if somebody does ... eventually offer unlimited usage all the time, then no, not for me because it's not fast enough."

There is also a relationship in the opposite direction. The frustration of having to wait when using a 56 kbit/s link is compounded when people know they are paying by the minute[20]. In that context, people appreciated that if they did ever have to wait while using a broadband service, at least it would not cost them any more.

"It is frustrating when you wait. It's a damn sight more frustrating when you're on your telephone line and you're wondering how much it's all costing you! ADSL has gone some way to overcoming those issues."

---

[20] These interviews were carried out prior to the introduction of flat-rate packages for narrowband access.

This will become a stronger consideration as broadband takes off and customers start to experience the inevitable effects of contention, and as non-broadband services continue to deliver service at narrowband rates.

- Relationship of tariff to always-on

  Having a permanent connection is greatly valued and is what alters the way people use the Internet — this serves to justify the extra cost (compared to dial-up ISP costs):

  "I think it is over-priced ... but then the convenience of having it means that I don't want to do without it."

  In return, the flat-rate tariffing means that people are no longer worried about how long they spend on-line — and so the full benefits of the permanent connection can be realised. Thus the technical possibility of a permanent connection only adds benefit if the tariffing does not inhibit its use:

  "I wouldn't be using anything like £50 a month in call charges I don't think, but nonetheless knowing what you are going to pay is an advantage."

- Relationship of speed to always-on

  The higher speed results in significantly faster downloads (appearing to be effectively instant for things like e-mail and Web pages); and the permanent connection means no more waiting for the PC to turn on and the dial-up connection to be made:

  "[With 56 kbit/s dial-up] dialling in and waiting for the connection and all that is kind of a bit cumbersome. I feel if I have to do it I just think 'oh forget it', very often if you don't have much time you just don't want to do it."

  "It's a toss up for me which is the greater advantage ... I think they go hand in hand ... people who are permanently connected on a 56 kbit/s modem are not going to see a huge amount of benefit. It's still a bore to download pages."

  Together, these add up to a level of immediacy that dial-up narrowband services cannot provide. It was this type of immediacy which users noticed and started to utilise in order to use their own time more efficiently:

  "It also means that ... banking or whatever, you can do it as soon as you think about it. It means you have to remember less."

From this model, we can see that tampering with any one of these elements could potentially undermine the entire proposition from a customer viewpoint. On the positive side, we can also see that as the technology comes under increasing pressure when more customers come on-line, and as service providers start to provide services which properly use the available bandwidth (resulting in longer downloads), elements such as tariffing and service quality (especially reliability of the permanent connection) can play a role as service differentiators.

8.5.2.2 Changing Modes of Provision

How should we understand the sorts of change described in section 8.3.4? It may help to consider Gershuny's analysis of modes of provision [13]. Take food, for example. All human beings need to eat to survive, we always have and always will. That doesn't change. What changes is the way in which that need is satisfied — the mode of provision changes. The social and industrial processes which enable human beings to eat can be viewed as a structure of costs whether that cost is time, money or an integration of the two.

What is critical to understand is that the relative 'costs' of various parts of the chain that meets the need to eat have changed and are still changing as a result of (and as a driver for) technological innovation. Thus far less of the cost of food production (in terms of time) is spent in basic agriculture and far more in processing and retailing than was the case 250 years ago (or even 50 years ago). Even though we still eat, the way our food is created has changed considerably.

As with food, so too with other human needs. Leisure, relaxation and entertainment are now provided in ways not possible 25 years ago. So too social communication, clothing, heat, personal care. We still shop, but the way we do it (and where) has changed. This becomes clear from the transcripts of the interviews with the 19 BT broadband households — people are still doing the same old things (communicating with each other, finding stuff out, buying stuff, having fun, running their lives) but they are achieving them in new ways (e-mail/chat/instant messaging and the telephone, on-line weather and traffic at my convenience, my share prices on-screen and sometimes the FT on paper, watching a celebrity chef on TV, and getting the recipe off the Web site, sending e-mail home/to work as reminders).

Understanding that people still do the same things they have always done, but achieve them in new ways is key not only to understanding what is changing and what is not, but also to understanding the significance of various technological innovations and market opportunities. If we can start to understand what core needs human beings have, we can start to see whether or not our favoured technological innovations will really be revolutionary by asking the following questions:

- Do they change the mode of provision of any human need?
- How will that mode of provision change?

and we can then ask:

- Is that a market opportunity?
- For whom?

and the critical question then becomes:

- Can we tell in advance?

As we have seen, thinking about broadband Internet access in this way can help us to start to understand what changes in modes of provision it enables and thus where and how money can be made from it.

8.6 Summary

This chapter has provided a picture of the current 'state of play' in the domestication of broadband Internet access in Europe. The quantitative data is from 2001, but the qualitative picture is largely unchanged at the time of writing (summer 2002). It has done so by bringing together several different data sources, each of which provides a distinct but complementary view of what people do with broadband Internet. We have seen that broadband Internet users differ only marginally from PSTN Internet users but that the differences may be commercially significant. These differences can be summarised as:

- at present singles without children are a more significant adoption segment for broadband than they are for PSTN;
- broadband users tend to be more PC centric — they have more PCs connected to the Internet and they appear more likely to have new hardware;
- broadband customers tend to have been using the Internet for longer and to spend more time using it;
- broadband customers make more frequent use of a wider range of applications — this may be due to their higher on-line time and flat-rate charging encouraging experimentation, but it could also be a reflection of their greater Internet experience;
- broadband customers are more frequent users of e-mail, but again this could be related to their greater Internet experience;
- broadband users are more likely to have used 'multimedia' applications such as streamed audio but they do not necessarily make a habit of it — there is some evidence for a preference for streamed audio, which is to be expected given that it is a 'background' media and suited to current broadband bandwidth limitations.

This summary highlights the finding that broadband Internet users tend to be more experienced and technologically competent. They also have diverse and particular needs, which have implications for their (non-) use of broadband portals and other content services.

We have shown how studying broadband users, and indeed any ICT users, in this way can provide actionable insights for any organisation that seeks to benefit from broadband Internet whether that benefit is driven by a commercial or public policy agenda. However, we have also noted that to date there is a large gap in our knowledge. While we can make informed guesses based on cross-sectional data, we simply do not know what actually changes in the lives of individuals and households who acquire and use broadband Internet because — as far as we are aware — no-one has conducted a before and after tracking study to find out. Only then will we be able to prove that getting broadband Internet leads to less (or more) TV watching, less (or more) listening to the radio, less (or more) use of other traditional media content, less (or more) use of broadband communication services to support local or distant social networks (see Chapter 10 for more details of such studies).

References

1 Silverstone, R., Hirsch, E. and Morley, D.: '*Information and Communication Technologies and the Moral Economy of the Household*', in Silverstone, R. and Hirsch, E. (Eds): '*Consuming Technologies*', Routledge, London (1992).

2 Anderson, B. and Tracey, K.: '*Digital Living: The Impact (or otherwise) of the Internet on Everyday Life*', in Wellman, B. and Haythornwaite, C. (Eds): '*Special issue on The Internet in Everyday Life*', American Behavioral Scientist, **45**(3), pp 456-475 (November 2001).

3 Eurescom — http://www.eurescom.de/public/projects/projecttables/

4 Mante-Meijer, E., Haddon, L., Concejero, P., Klamer, L., Heres, J., Ling, R., Thomas, F., Smoreda, Z. and Vrieling, I.: '*ICT Uses in Everyday Life: Checking it out with the people*', ICT markets and users in Europe, Confidential EURESCOM P903 Project Report, EDIN 0161-0903 (2001).

5 '*Broadband Blues*', The Economist, p 62 (23 June 2001).

6 Rogers, E.: '*The Diffusion of Innovations*', The Free Press, New York (1995).

7 Hoag, A.: '*Speed and the Internet: The effects of high speed access on household usage*', Presented at the 25th Annual Telecommunications Policy Research Conference, Alexandria, Virginia (September 1997).

8 Hoag, A.: '*Measuring usage and satisfaction: cable modems and the Internet*', D-Lib Magazine (March 1998).

9 Wired (June 1999) — http://www.wired.com/wired/archive/

10 McClard, A. and Somers, P.: '*Unleashed: Web tablet integration into the home*', Proc of the CHI2000 conference on Human Factors in Computing Systems, pp 1-8, ACM Press, New York (2000).

11 Robinson, J. and Godbey, G.: '*Time for Life: The Surprising Ways Americans Use their Time*', Pennsylvania State Press, University Park (1997).

12 Finn, K., Sellen, A. and Wilbur, S.: '*Video-Mediated Communication (Computers, Cognition, and Work)*', Lawrence Erlbaum Assoc (1997).

13 Gershuny, J.: '*Social Innovation and the Division of Labour*', OU Press, Oxford, UK (1983).

9

IS THE FUTURE REALLY ALWAYS-ON?

M R Gardner, C Gibbs and T van Do

9.1 Introduction

The development and deployment of always-on (AO) services will make it feasible to offer uninterrupted connection to both fixed and mobile nets, without the cost or time penalties previously incurred when using the Internet. Furthermore, it is expected that information and communications technology (ICT) will become an everyday part of life. This poses a challenge to most telecommunications operators, and may well lead to fundamentally different network and service characteristics involving a range of different network and device types.

Always-on may change the way people use telecommunications; it is a challenging new concept for service and network providers, as well as customers. Furthermore, it will be a decisive competitive characteristic of an operator's services.

What is always-on? The AO concept can sometimes be difficult to distinguish from traditional network connections, as highlighted by the following questions.

- How important is the ability to be AO compared to other network characteristics?
- What will be the quality of the service (speed, availability and reliability) and which services will take full advantage of AO?
- Is it important for users to be able to control the information flow?
- What will the business model be for AO services — would it be cheaper or more expensive than other networks?
- What are the security and privacy concerns?
- Are there any social consequences such as less face-to-face contact and user isolation?

From a network-centric point of view, 'always-on' is the network functionality of providing sustained access for the user (or their representatives, such as automatic household devices) with no or minimal use of network capacity when there is no traffic. For Ethernet users the office local area network (LAN) provides a good example of the concept.

However, the user perception of AO can vary and there can be confusion between 'always-on' and 'accessing on request'.

A number of focus groups were run by each project partner to capture and identify the users' understanding and their attitudes towards AO. Focus groups are a common method, used especially in market research, for ascertaining consumer opinions, attitudes and needs towards new or existing products. Generally the method involves collecting together several (usually 6-9) people representing a desired population to discuss informally, for 1-3 hours, a selected theme. One or more moderators guide the session, steer the discussion and focus it on the predefined themes. They may also present items to be evaluated and take records of the session, etc. In addition, one or several observers support the work of the moderator, especially in taking records of the session. Many variations of this general schema exist, using different methods to present the items being discussed, recording participant responses, eliciting discussion or structuring and focusing the discussion.

The focus groups for this project also generated a number of different scenarios for AO usage. This helped to clarify the user needs for AO services across a range of different market segments. These user needs were then tested in a number of user trials within the partner countries.

The tools and techniques used by the team were based on an in-house developed methodology called SUNA (scenario-based user needs analysis) which has now been applied to a number of projects to help capture and define user needs for new services and technologies. The SUNA method was developed from the idea that telling stories is a good way to get a group of people to think 'out of the box' (i.e. beyond preconceived constraints), and to synthesise their inspiration and experience into a logical whole. The resulting stories convey creative and innovative solutions that can be readily grasped by people outside the project group. These can be sanity checked and/or used to buy in support from within the participating organisation(s) more effectively than traditional reports. The subsequent processes then carry the group through the necessary scoping, clarifying and refining exercises to a point that provides a logical link to the beginning of a formal development methodology such as the unified process [1].

9.2 Focus Group Findings

The general advantages of focus group methods are as follows:

- it is a comparatively cheap way to gather user opinions compared to individual interviews as it involves several people at a time;
- compared to questionnaires the situation is more interactive, so richer and more accurate data can be gathered;

- when creativity and imagination are needed, small groups are better than individuals as people often get new ideas from each other;
- focus groups are better than direct observation methods for getting an overview of activities that span whole days.

For the purpose of exploring initial user reactions, preferences and attitudes towards always-on features, focus groups were found suitable for various reasons, in particular because it was known that field trials would be carried out after the focus groups.

These would verify findings with some of the most promising service concepts and user groups, and give more precise accounts on actual behaviour with the product. The focus groups were also designed to help select user groups and services for the field trials.

The following market/user segments were studied by the project:

- rich residential segment (Spain);
- remote work teams 1 — working on physical objects or with people (Finland);
- remote work teams 2 — working on information (Finland, Sweden);
- eCommerce sellers — small to medium enterprises (SMEs) (Italy);
- eCommerce buyers (Italy);
- students/tele-learners (Great Britain, Finland, Spain);
- stock market participants (France, Italy);
- independent professionals doing information work (France, Italy);
- heavy users of technology (Italy).

The focus groups explored the following dimensions across each of these target customer segments:

- current user behaviour;
- understanding of the AO concept;
- relating AO to consumer life-styles;
- brainstorming of possible AO scenarios;
- focus on selected AO services;
- assessing the user readiness for AO.

Based on this structure, a number of focus group sessions, targeted at different market segments, were run by the project team. The following general findings were derived from these sessions.

- Definition of 'always'

  In general, users like to 'turn off' their device/terminal. For connections to the network users expect to be able to switch off their devices at some point in time.

This leads to an important factor that influences greatly the technical implications of AO services — this being that 'always' is not necessarily the case, but rather 'immediately as the user turns it on'. In this case, 'AO' is actually understood as 'immediate access on-line on request'.

- Access

  Fast and cheap access to the network, combined with a service available on-line, can be more attractive than fixed-line AO access in many cases. Always-on access to services, in the sense of pushed information without user-initiated request, is attractive only in rare cases. Furthermore, the quality of the information is often more important than information push or alerts.

- Terminal type

  The participants of the focus groups mainly saw services as related to only one terminal type (such as either a fixed line or mobile, PC or mobile/WAP telephone) and converged services using alternative terminal equipment were seldom suggested.

- Concept of continuous contact

  It seems that the general concept of continuous contact to network services can have negative connotations for many user groups. On the other hand, many examples of real AO services, not too far away from the currently used services, are attractive for several user groups.

- Perceived benefits

  The focus groups reaffirmed a number of perceived benefits of being always on:

  — instant access to services and information without having to actively establish a connection each time they are used;

  — the ability to immediately receive new or updated information;

  — space always being available for communication;

  — the means of remotely controlling and monitoring equipment and appliances.

- Network versus service

  The focus groups also re-confirmed the important separation between 'always-on to the network' and 'always-on to a service'. Most current AO services are specific to particular networks (e.g. ADSL). As users own more device types (e.g. PC, mobile), they are also using different networks (e.g. PSTN, GSM, LAN). The ability to be always-on to their services is perceived to be more important than AO network connections, as an AO service could be accessed by a variety of different networks and devices.

9.3 Usage Scenarios

The focus groups provided information on users' attitudes and behaviour to current (or currently perceived) AO services. This was used to define a number of usage scenarios which illustrated how future AO services could influence everyday life of users within different market segments. Some of the devices and services described in the scenarios are already available. They have helped to identify some imaginative uses of AO for the future.

From the focus groups a list of possible always-on services were identified. Table 9.1 illustrates the services identified for the UK-defined market segment 'distributed family group'.

Table 9.1 Sample services identified for the 'distributed family group' segment.

Voice mail On-line banking, accessed through: — PC, SMS alert/mobile phone Intelligent house: — remote appliance control — external temperature assessment tool Web site hosting URL push E-mail diary service File sharing eMessaging Videoconferencing Personal search facility Hyperlinks Data sharing Web page design tools On-line education information On-line travel information, accessed through: — PC, in-car navigation tool, and palm-top device Local information service, accessed through: — palm-top device — SMS/mobile phone E-mail (with multimedia attachments and remote access)

From the service identification it was possible to define a number of scenarios that illustrated the usage of these services for a given user and market segment. The following is an extract from a 'distributed family group' scenario:

> '......... School term is just about to finish; Oliver and Katie have two weeks left at Brimstone Park Middle School before the summer break. Jacob finished his GCSEs five weeks ago and has been working in a newsagent's ever since. He has decided to go to Westminster FE College in September where he can take a Textiles and Design course. He will be moving to London to live with Laura and Nathan for the duration of his course. After work one day he logs on to his PC and e-mails Laura to ask her about her role as a buyer. When she receives the e-mail alert via her mobile telephone from Jacob she is working in a store on Oxford Street. She listens to the e-mail using her mobile telephone and the text-to-speech facility. She has this on her hands-free kit and a colleague overhears Jacob's request. He suggests that Jacob comes up to London and shadows him at work to find out more about their work. She sets a reminder in her personal digital assistant (PDA) to discuss this with Nathan.
>
> Later that afternoon Nathan calls Laura to say he is going to be late home due to a work deadline ......'

The scenarios can then be used to extract and identify the key user needs for each of the market segments. For example, the key user needs for the UK 'distributed family group' segment were as follows:

- UK distributed family groups (children away studying)

  — need live conferencing facilities;

  — need on-line banking;

  — like on-line reservation facilities;

  — need e-mail and remote access to e-mail;

  — need travel information;

- UK distributed family groups (divorced)

  — need conferencing facilities, for different branches of the family;

  — need remote access to e-mails;

  — like voice-to-text translation tool;

  — need on-line banking;

  — like on-line messaging services;

  — need WAP mobile telephones for all family members;

- UK further and higher education

  — need access to courseware outside lecture hours;

  — like on-line student services;

  — need on-line banking facility;

  — like access to alternative information sources;

  — need access to library outside library hours;

  — like aids to family communication, e.g. e-mail message alerts, remote access to e-mail;

  — need WAP phones with personal alarm and satellite locator.

The outputs from the focus groups and scenarios, particularly the user needs, were useful as guides both for planning the user trials and as input to modelling the business issues for future always-on services. They also captured and demonstrated some of the requirements for future AO services across the five European countries taking part in the project.

9.4 Evaluating User Behaviour — User Trials

The user trials were designed to evaluate the user behaviour when confronted with using AO network services in a number of different scenarios. The types of scenario differed greatly from country to country. Some tested current adoption to AO, others a more futuristic use of AO services (such as the Spanish partner's 'intelligent house' concept). Some of the services were based on off-the-shelf services (e.g. Italy, Sweden and UK), while others were based on new integrated services (e.g. Finland and Spain). The trial user groups were based on a subset of the market segments that were studied in the focus groups and scenarios.

9.5 Trial Findings

This section gives details of the main findings (per country) from the user trials. Tables 9.2-9.6 provide an overview of the trials in each country.

- Italy — ADSL early adopters:

  — the most important ADSL benefits were perceived to be the real-time connection, faster connection speeds, the always-on connection, the availability of a second telephone line and wideband connection;

  — it was clear that all the users considered this kind of connection useful, but the cost was still considered a strong limit to future subscriptions;

— ADSL was seen to be useful overall for surfing, downloading and searching information quickly, but some specific ADSL services were missing, e.g. push services (for continuous updating of news information) and real-time interactive TV, full-screen movies, and high-quality audio services;

— all the users complained about the necessity to install some antivirus and firewall software to protect their PC.

Table 9.2 Overview of user trial — Italy.

Italy	
Platform available	ADSL (fixed connection)
Services available	• point-to-point videoconference • multi-point videoconference • news on demand • audio-streaming • video-streaming • e-mail • Internet browsing • (multimedia) instant messaging • audio chat • video chat • game on line
Target segment	Early-adopters
User sample	9 residential users — AO sample and control group. Expert Internet users — selected on the basis of previous use of desktop videoconference, instant messaging, e-mail.

- Sweden — mobile/local distributed workteams:

— problems occurred with portability and the ability to keep mobile terminals on for long periods;

— current terminals are bulky, generally delicate and the battery life short;

— services require adaptation to the AO concept, and to the limited screen size and input capabilities of mobile terminals;

— services need to become more network aware, e.g. to inform users if the service becomes disconnected and to automatically reconnect when the network connection is established;

— services most frequently used were e-mail, presence, calendar and Web surfing, and these were mostly being used passively, i.e. to view information;

— users enjoyed being connected and in touch (there was a strong sense of 'addiction' to being connected).

Table 9.3 Overview of user trial — Sweden.

Sweden	
Platform available	Wireless LAN, hand-held (Compaq Ipaq)
Services available	• intranet • Internet • e-mail • presence service/instant messaging • positioning
Target segment	Mobile/local distributed workteams
User sample	14 local internal users at Telia Research AB

- Spain — rich residential segment:

  — system reliability was the main concern for the participants;

  — in this simulated 'real-life' situation the AO concept was more easily understood, with a high percentage of the trial users feeling that the intelligent house would not be possible without permanent access;

  — users valued communications services most highly;

  — one of the main advantages was the flexibility and comfort AO gave when using the intelligent house.

Table 9.4 Overview of user trial — Spain.

Spain	
Platform available	Intelligent House (fixed and mobile connections)
Services available	Domestic control (including remote control) through: • PC and laptop • TV • mobile telephone: SMS and keyboard • Webphone • alarms (fire, water, gas surveillance) with: — entryphone through fixed telephone — cameras — light control — automatic blinds
Target segment	Rich residential segment
User sample	30 users, divided into two groups, depending on their technology knowledge and use

- Finland — distance learners:

— users had little need for the mobile Internet in their work;

— the mobile Internet did not have a great effect on the users' communication patterns;

— the usability of the device combination (consisting of Ipaq and Nokia 7110) was not felt to be good;

— the participants were downloading and reading e-mail more than sending it;

— the connections to the Internet were often unreliable;

— the introduction of an SMS service that notified of incoming e-mail increased the downloading of e-mail, although the participants themselves did not feel that it affected their downloading;

— the two most active users were calling the server frequently (on average 4.7 times a day), with calls to the server generally increasing in the evenings after work.

Table 9.5 Overview of user trial — Finland.

Finland	
Platform available	Compaq Ipaq + Nokia 7110 telephone (data connection over infra-red) + normal WWW/Internet/e-mail connections from home and/or work PC
Services available	• e-mail — also AO mobile • WWW • pushed notifications (SMS) of incoming e-mail • other Internet services
Target segment	Distant learners and/or virtual team members.
User sample	5—10 users with mobile AO connections and a comparison group in the same activity, e.g. distance-learning course, with their normal connections

- UK — ADSL residential users:

— ADSL users downloaded more material and used more entertainment and information sites than those with PSTN;

— it could not be confirmed that they would spend more money or time on-line, or on e-mail.

Table 9.6 Overview of user trial — UK.

UK	
Platform available	ADSL (fixed connection)
Services available	• News on demand/other on-line information services • audio-streaming • video-streaming • e-mail • Internet browsing
Target segment	ADSL and PSTN residential users
User sample	20 residential customers (split between ADSL and PSTN) — this was an 'independent measures' design

Both fixed and mobile connections were tested and general distinctions between these identified. The users found more problems with mobile devices than with the service, in contrast to the fixed device users who found more service problems. Always-on is a concept with great potential for mobile terminals, but current mobile devices are generally not well adapted for it.

9.6 Technical Issues

The trials have shown that AO connection could be used with benefit on a range of different services, including communication, information, eCommerce, entertainment and monitoring. The trials also highlighted that AO users may use multiple communication tools and network connections. A number of technical recommendations for AO network services were identified from the trials.

9.6.1 General Recommendations

These are divided into two sections — network service providers and application service providers.

- Network service providers should:

  — provide a permanent and transparent IP connection with sufficient bandwidth to enable seamless connection to the network;

  — provide access to any application service provider to enable a greater range of services to make use of the AO network capability;

  — provide support for QoS controls when needed to help maintain consistent levels of service;

  — ensure secure transmission so that traffic cannot be intercepted (nor eavesdropped) by third parties;

— provide roaming functionality to allow services to operate across different networks;

— provide user information such as geographical position (where applicable) or whether a user's terminal is switched on (this can be used by service providers).

- Application service providers should:

  — provide network and terminal adapted services;

  — ensure that access to services is granted only to authorised users,

  — ensure that user information is stored and handled securely.

Network service providers supply both fixed and mobile network access. However, users will expect to be able to use the service provided by a network service provider (fixed or mobile access) to reach any IP application, irrespective of provider.

9.6.2 Network Operation Recommendations

The eight recommendations in this section relate to the facilities required to ensure reliable network availability.

- Investigation of roaming and multi-access issues

  Future mobile IP networks must function in a similar way to present mobile telephony networks (such as GSM) with regard to the roaming capability between different network service providers.

- Mobile devices

  These will most likely be able to use two or more access networks, e.g. use a wireless LAN in local hotspots and GPRS for wide-area coverage.

- Always-on

  This requires devices to be always reachable. Networks must be designed to allow sessions to be initiated by network servers or other users. This implies that devices must maintain a public IP address (or other means of interoperable addressing) while they are switched on.

- IP addressing/migrating to IPv6

  Mobile devices will require a unique IP address in order to be addressable for sessions initiated from the network or other users. This will greatly increase the demand for public IP addresses. Due to the limitations of IPv4, this will require migration to IPv6.

- Provision of transparent IP connections

  Network service providers should be careful when deploying network equipment (such as network address translation and firewalls) that can cause problems for many application layer protocols, thereby blocking many always-on services.

- Support for quality of service (QoS)

  Network service providers should provide QoS standards for services such as voice over IP (VoIP) and streaming media.

- Security

  Network service providers must ensure that networks are designed to prevent eavesdropping and interception.

- Availability and network coverage

  As more user information is stored on network servers, networks used to access these servers must have a high availability and coverage.

9.6.3 Terminals and Devices Recommendations

These recommendations cover the capabilities of the mobile device/terminal.

- Fixed terminals

  Fixed terminals must be capable of being always-on — in order to be able to access always-on services and to be available to IP communications services.

- Mobile always-on devices

  These need to be more portable and easier to use. Important factors are size and weight, robustness, durability and battery life.

- Security

  This must be built into mobile devices. Authentication, encryption and other security measures, e.g. virtual private network (VPN) client software, must be built into mobile terminals to provide secure acess to services and data.

- Open terminals that support third party software

  Basing (mobile) terminals on open platforms should allow application service providers (ASPs) to make their services available on a wide range of terminals.

- Personalised mobile devices

  Users should be able to configure their mobile devices and service preferences. This enables many more personalised capabilities which adapt to each user's profile, and the device and bandwidth characteristics.

9.6.4 Services and Applications Recommendations

The following recommendations relate to service availability.

- Always-on providing instant access and availability

  Open standards provide network and terminal independence. In order to achieve AO, services should be based on open standards [1]. This enables standardised information exchange and storage. One example is XML which can be used to enable data to be stored and transferred in a standardised way [2].

- Secure data storage in the network

  By storing data, e.g. address lists, on secure network servers as opposed to locally in terminals, information can be made available from any terminal and network.

- Access

  Access to IP services will not be restricted just to PCs. Services will be accessed from various terminals with different computational power, screen sizes, and input capabilities.

- Login

  Users should have a single login for all services.

- User accessibility

  Services must provide facilities for the users to make themselves unavailable, as well as available.

9.6.5 General Conclusions from the Trials

In addition to the above specific recommendations, there are more general inferences that can be drawn from the trials.

- Time on-line

  In all the trials an AO connection generally meant more time spent on-line (by the user).

- Information gathering

  Information services were very popular. However, it was important that the users have control over where and when they receive information. The quality of connection is valued more than being always-on. Control of information is therefore an essential requirement for information services, especially when information is 'pushed' to the user.

- Mobiles

  Mobile devices are still technically immature:

  — limited working time, dependent on battery capacity, limits the AO connection;

  — usability;

  — limited display capability;

  — lack of compatible and useful services;

  — new services should be developed which take into account the network and terminal characteristics;

  — connection reliability.

- Quality

  In general, irrespective of the service involved, quality of connection is an essential issue. Fast and reliable access is one of the main requirements of users.

- Security

  Security and privacy are of concern to users. This raises the question as to whether always-on increases the security and privacy risks? Further work is needed here.

- Impact on life-style

  Users are concerned about the impact on their life-style, particularly the merging of different aspects such as work, study, family, hobby time, etc. Their main fear was that better opportunities for networking activities would reduce time spent with family or on other social activities.

Also, it is clear that new user behaviours are developing, for example, more distributed use of the Internet, more 'playing' (both accessing entertainment and game sites, and playing on-line games using a mobile device) not related to work. These behaviours indicate exciting market opportunities for telcos. Most of the increase in the use of services was reflected in 'passive' user behaviour (i.e. users 'consuming' information rather than producing and sharing their own information). This is probably due to the asynchronous nature of current mass-market AO network services such as ADSL. It is also possible that some terminal types are simply not suitable for the current services accessible to them.

Current technology shows great promise. With the increase in broadband AO network coverage the benefits of mobile always-on services will become more apparent.

AO technology and services and the AO concept itself are both still in their early phases of development. From the results of these trials it is clear that AO could cause changes in the way people use network services. It should lead to the development and deployment of new services, which will better exploit the characteristics of these networks.

9.7 Always-on Services — the Device-Unifying Service

The trials also highlighted the distinction between being always-on to a network connection and being always-on to a service. With the scenario of users having a range of different devices (e.g. PC, laptop, telephone, mobile, PDA) the ability to be always-on to services becomes much more important, such that a user can seamlessly switch between networks and devices while maintaining the same always-on service session (e.g. their e-mail service). This has led to the concept of developing a device-unifying service (DUS).

Always-on service is not a new concept. For many years we used mainframe-based services which ran centrally and were independent of the user terminal. They were based on terminal standards such as the DEC VT100 and IBM 3270 terminals. The World Wide Web is just an extension of this simple client/server model with the Web browser acting as a simple 'terminal' on to the centralised Web services. A new breed of Web portal services (see Clickmarks [3]) are now offering multichannel access to centralised Web services, whereby users can access the same services from a range of different terminal types (e.g. WAP telephone, PDA, PC). The key difference with DUS is to allow always-on service sessions to be dynamically switched between different devices and networks, and to do this in a completely seamless way. Thus data sessions, like Web and e-mail, will be treated in the same way as multimedia sessions, such as voice and videoconferencing.

The device-unifying service is aiming specifically at extending the always-on concept to support multiple heterogeneous devices. The user is now 'always-on' not only from a unique device but also from a variety of devices. The goal is to ensure that the user's services and applications can be presented and moved smoothly and naturally between their devices. This is extending the concept of the virtual home environment [4] which enables the user to receive customised and personalised services, regardless of location, network or terminal type.

9.8 The 'Multiple-Device' Scenario

Nowadays the user is confronted with several different devices including a fixed or cordless telephone, a mobile telephone, and a PC or a workstation. Even though the future is not clear, there will no doubt be a vast number of communications, computing and other electronic devices around us. The number of different devices available to users might even be expected to increase dramatically both in type and in number. These devices might very well be autonomous in that they are able to function individually and independently of each other. However, the quality of the services they offer may be enhanced if they co-operate. Almost all of these devices are not only able to function independently of each other and without any co-ordination, but they are most often not even aware of the presence of other devices. As the owner, the user is required to handle them all and does not always succeed since often they cannot perform many tasks at the same time. For example, both the fixed telephone and the mobile ring simultaneously; the user has to alternate between calls or terminate one of them.

Another inconvenience is the repetition of similar tasks for each device. For example, the user has to enable a voice-answering service both on their cellular telephone and fixed telephone when they are busy and do not want to be disturbed.

The user may also want to use several devices with different functions such as a mobile telephone, laptop, PDA, digital camera, printer, an electronic paper, etc. These devices should be integrated in such a way that they can work together. For example, it should be possible for the user to talk on the mobile telephone with another person while exchanging pictures taken by the digital camera and sketches from the electronic paper, etc.

When the user is confronted with several different devices (as illustrated in Fig 9.1), the management of all the different devices might be complex and time consuming. It can be difficult for a non-technical user to master them all. For

Fig 9.1 The multi-device problem.

example, there will be several terminal profiles and several user profiles to which the user has to relate. The user has to define their profile and preferences on every device. From the user's point of view, it would certainly be desirable to have a single user profile for all devices.

9.9 Ownership and Use of Devices

Another important issue is the combination of mobile and stationary devices. The user might visit places away from their home domain where there are stationary devices offering different types of services, e.g. printers, screen displays. As the situation is today, it is difficult for the user to make use of these devices at the remote sites. Most often they cannot utilise the services being offered and have to manage with the limited capabilities offered by their own mobile device.

With the device-unifying service, the user's different devices can act together as one virtual terminal (as represented in Fig 9.2) with multiple output and input options and offer the user an optimal multimedia service. There can be several communications devices such as fixed telephone, cellular telephone, PC or a workstation all offering human-to-human communications in a slightly different way, and several devices with completely different functionality and capabilities, such as a laptop, a digital camera, a TV screen or a printer, working together to make up the virtual terminal.

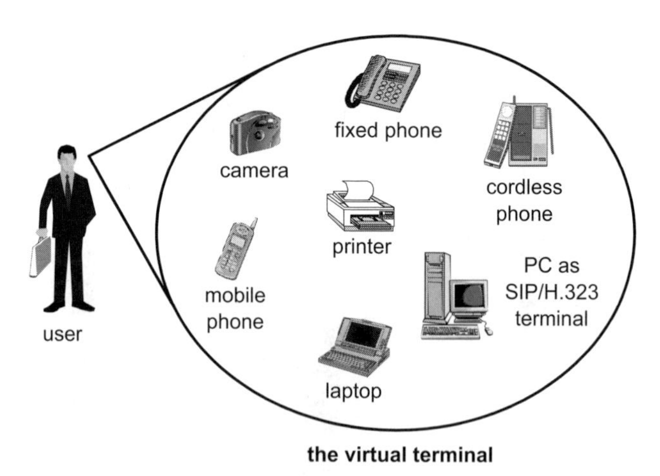

Fig 9.2 The DUS virtual terminal.

In the case where several devices act together as one there is also need for a user-friendly and straightforward management and co-ordination of the different devices within the virtual terminal.

To make this possible a control mechanism is needed that supports the collaboration and communication between the devices which might be connected to different access networks. The DUS will provide this control mechanism. It should help the user to exploit such unification of devices in an easy and straightforward way.

9.10 DUS Requirements

So far the project has defined a number of requirements for the DUS.

- Unifying different devices

  DUS helps the user combine all their computing and communications devices so that they can behave as one device with multiple input and output capabilities. This can provide the user with, for example, an optimal multimedia session using only a number of simple single-media devices. These devices should be co-ordinated and have the possibility to be used together in the same session, even if they are connected to different types of network.

- Unique user profile

  The DUS will handle the management of the different devices and the different terminal profiles. For example, if a user updates their address-book on one device, it will automatically be updated on all other devices to which they have access. The DUS will likewise allow the user to set up and modify their preferences for all devices at one place — once and for all.

- Communications service customisation

  This will allow the end users to customise their communications service (such as when they want to be called, on what device, under what condition, by whom, etc). DUS also allows both parties to have control over how the service will be supplied.

With customisable redirection of incoming services, a user may for example specify that all calls during daytime should be redirected to their office telephone and thus be handled by a secretary. A specific conference call could be redirected to their desktop PC beside their communications device based on user preference profiles. When the user is asleep, they may wish to receive all calls as voice-mail — except for urgent calls from their boss. Or in the evening, when they are at home, all incoming calls should be received at their home telephone.

Again, a particular conference call that includes video should be received at their home PC (for example), while all voice services received while travelling in their car should be delivered at their cellular device.

- Static, dynamic or automatic configuration and reconfiguration

  There are equal alternatives as to how the user could configure their virtual terminal:

— statically such that all devices that are to be used in all future sessions have to be predefined and thereby known by the DUS (this static configuration can also include a time-table or location-table that decides which devices should be used, where and when);

— dynamically, allowing new, unknown devices to be added to the DUS as they are needed;

— automatically, but this requires some kind of service discovery that is able to recognise nearby services and devices and report this to the DUS.

- User centric

  An important feature of DUS is that it provides a user-centric service. The situation today is that a device will be used to communicate with another person. For example, one might try to reach a person by successively dialling their home telephone, office telephone, cellular telephone or whatever other telephones might be around. If one is successfully able to contact the desired person or not is completely dependent on whether or not the right telephone is dialled. Very often one will unintentionally disturb other people before the desired person is reached.

  DUS should offer a service that is truly user centric in that one will be able to address a person directly.

- Stationary devices at visiting sites

  DUS will make it easier and more straightforward to borrow stationary devices at visited sites that may offer better input and output capabilities than the mobile devices which the user is carrying, provided that the owner of the stationary devices allows such use.

- Total mobility

  DUS should support personal and session mobility in addition to the terminal mobility, which is supported by the underlying network like GSM, UMTS, etc. By combining these three types of mobility one could provide more flexible communication services for the user.

  With personal mobility, communications services treat people rather than devices as communication end-points. Personal mobility should give the user the total control over how they can be reached.

  Session mobility allows devices to be switched in the middle of a service session (for example, seamless switching from a cell-phone to an IP-phone in the middle of a conversation).

  With total mobility the user will have more flexibility using communications services, they can make use of their mobile devices and also stationary devices at visited sites. The user could also move the input and output of a session from a set of devices to another device. For example, visual output from a mobile device with small display could move to a larger and better fixed screen. In this way the user's session remains always-on while the device and network access may change. In other words the user may manage their own and available devices in a more flexible way than the existing networks allow.

- Dynamic multi-party communication

  Both parties may add or remove one or more new devices/parties to the ongoing service session independent of what kind of access network they are using.

- Using services on multiple devices successively

  DUS provides the possibility for the user to redirect streams and hence use multiple devices during one session, by successively switching from device to device dynamically. This could happen repeatedly.

- Using services on multiple devices simultaneously

  DUS also provides the opportunity for the user to split and multiply streams and hence use multiple devices at the same time. New devices should always be able to be added to or released from the service session dynamically.

- User-activity-driven service

  DUS could support a new kind of communications service based on user activity. This type of service generalises the location-based services that have appeared in many other systems.

  Instead of customising the communications service based on the current user location alone, DUS should allow the current user behaviour (such as 'I am talking to an important person') to be tracked and used for customisation. The users control what behaviours are tracked as a way to control privacy. DUS should allow users to control privacy policies, such as which information is tracked and to whom the information can be released.

9.11 Summary

This chapter has described the work done by the EURESCOM P1003 project [5] to understand the user needs for future always-on network services. It has been shown that although AO services are starting to appear there is still some way to go before the network technology and on-line services fully exploit the characteristics of being always-on.

An important issue is the explosive growth in mobile devices, which is leading to the scenario of heterogeneous networks and devices. The user is now 'always-on' not only from a unique device but also from a variety of devices such as a mobile or fixed telephone, a stationary PC, a laptop computer, etc. The goal is to ensure that the user's services and applications can be presented and moved smoothly and naturally between their devices. This has led to the development of the device-unifying service on the EURESCOM P1101 project.

The key difference with DUS is to allow always-on service sessions to be dynamically switched between different devices and networks, and to do this in a completely seamless way, so that data sessions such as Web and e-mail are treated in the same way as multimedia sessions such as voice and videoconferencing.

This chapter has described the scope and requirements for the DUS. Currently the project is developing the core DUS platform and a number of usage trials by the project partners are planned to fully evaluate the concept in use.

References

1 Fowler, M. and Scott, K.: '*UML Distilled*', Second Edition Addison-Wesley (2000).

2 Selkirk, A.: '*Using XML security mechanisms*', in Temple, R. and Regnault, J. (Eds): '*Internet and wireless security*', The Institution of Electrical Engineers, London, pp 45-62 (2002).

3 Clickmarks — http://www.clickmarks.com

4 3GPP (3rd Generation Partnership Project): '*Technical specification group services and system aspects service aspects: The Virtual Home Environment*', 3G TS 22.121 (1999).

5 EURESCOM — http://www.eurescom.de/public/projects/

10

DIGITAL LIVING — PEOPLE-CENTRED INNOVATION AND STRATEGY

B Anderson, C Gale, A P Gower, E F France, M L R Jones, H V Lacohee, A McWilliam, K Tracey and M Trimby

10.1 Introduction

It is perhaps sobering to note that the telephone was originally conceived as a means to deliver musical entertainment to the home. It was only as its social use for interpersonal communication emerged that telephone companies realised just how this 'disruptive' technology was going to make them very rich [1]. The history of the telecommunications industry is littered with similar examples of technologies, propositions or business cases that turned out either to make money in unexpected ways or not to make money at all. We might name SMS (an unpredicted[1] phenomenon), ISDN (reborn as an Internet access technology in mainland Europe), xDSL (also reborn as an Internet access technology), video-on-demand (reborn as 'TV when you want it'), and the videotelephone (not reborn at all).

How could analysts have apparently got them so wrong? How is it that they continue to get it wrong? Why is it that telecommunications companies seem to make money (if they do) largely by accident? After all, such a strategy is not likely to win the confidence of many shareholders. Is it because the uses to which people put telecommunications technologies are unpredictable or, worse, unknowable? Or is it simply that we have not been looking at the world in the right way and thus not doing technological innovation that has a high chance of intentionally creating useful things?

Being optimists we believe the latter to be the case and by way of a response we have built a programme of socio-technical innovation which closely integrates social science research with technological design innovation in order to:

---

[1] But perhaps not unpredict**able**. If we had noticed that making telephone calls is a *public performance* and that SMS takes the public out of the performance, perhaps we might have seen SMS coming.

- generate commercially valuable strategic knowledge on social trends, changes, technology 'impacts' and likely futures;
- spot opportunities for the development of new technologies based on an understanding of what people do, what they might want to do (but often cannot articulate), or what they want to do but cannot;
- develop and refine technologies (and their associated intellectual property) to reduce the risks of investing in the development of non-starters.

We have called this programme 'Digital Living' and results have been published in Anderson et al [2], Gower et al [3], Lacohee and Anderson [4], and Anderson and Tracey [5] (see also Chapter 8). As a companion to Gower et al [3], which sets out the form and function of disruptive design, this chapter sets out the key research methods (and methodologies) that underpin the social scientific part of this research programme together with a broad overview of some of the results to date.

Section 10.2 describes the social science foundations of this research and section 10.3 gives details of other related studies. Section 10.4 focuses on an analysis of the 'impact' of acquiring household Internet access, looks at the implications of young people's use of mobile devices for the digital home, reviews the ICT usage of urban dwelling gay men, and explores the Internet usage of early-adopting broadband households. In each case we outline the strategic implications of these studies and, in the case of the sections on youth, gay men and the broadband households, we present some of the results of our grounded design innovation. We cannot here go into more than surface detail on each activity, but we make reference where possible to other published material.

We conclude with an overview of the main issues that these studies bring to the fore and, in particular, their implications for viable business cases and revenue models that can be specified in advance of product or service launch.

10.2 The Social Science Foundations

Given our interest in the impact or effects of information and communications technologies (ICTs) on individuals in the home, the core of our social science research programme is a longitudinal panel study. Around this we have constructed a series of 'triangulation' projects using other methods and, in some cases, other groups of people who are of particular interest. In tandem with this social science (and welded to it), we use a number of creative processes to engage the insights gained 'from the field' with the development of new technologies through 'grounded design innovation' (see Gower et al [3]).

10.2.1 The Digital Living Household Panel

The panel was initiated in July 1998 through a university research contract with the Institute for Social and Economic Research (ISER) at the University of Essex, UK.

The selection of the panellists and the survey fieldwork was carried out on behalf of ISER by NOP Research Group and resulted in the establishment of a longitudinal panel, and completion of the first wave of survey fieldwork by March 1999. The second wave of quantitative panel fieldwork was completed in April 2000 and the third in April 2001. While there will be no further data collection waves using this particular sample, we have recently launched an EU 5th Framework IST Programme funded project ('e-Living') to expand this research model to five further countries in Europe until at least 2003[2].

10.2.2 Recruitment Strategy

The households in the sample were recruited in 1998 by using a standard postcode reference and random selection procedure to generate a panel of individuals that are demographically representative of the UK population. Interviews were then arranged with every household resident aged 16 or over. Two screening questions were applied to eligible households.

- Do you have a telephone?

  Only households with a fixed line telephone were selected. Since some 6% of UK households [6] did (and do) not have fixed line telephones, this means that any population level statistics, such as average bill size or time spent on the telephone, based on the panel are marginally larger than similar figures for the UK population which will include the 'unphoned' in their statistics.

- Do you have a computer at home?

  The households were selected so that 50% had a PC of some sort. Thus, in effect, the initial panel of households constituted two samples — a representative panel of non-PC owners and a representative panel of PC owners.

The requirement for face-to-face recruitment meant that, while the panel was spread across Britain (no households were recruited in Northern Ireland nor in northern Scotland), there were geographical clusters to reduce interview costs.

10.2.3 Quantitative Data Sources

10.2.3.1 Survey Data

This data collection took the form of both a survey and a time-use diary for completion by all individuals aged over 16 which was repeated on a 12-month cycle. A second time-use diary, explicitly for younger household members, was designed for completion by all individuals aged 9-15.

---

[2] See http://www.living-digital.com

- Household questionnaire

  This questionnaire was completed at wave 1 by an NOP fieldworker during a face-to-face interview (CAPI) with the nominated household response person, while waves 2 and 3 used telephone interviews (CATI). In brief, the questionnaire covered ownership of the household's goods and services, ownership of ICT, and socio-economic data.

- Individual questionnaire

  This questionnaire was completed by an NOP fieldworker during a face-to-face interview with every member of the household aged 16 or over. In brief, the questionnaire covered personal usage of ICT, personal consumption and communication behaviour, social networks, attitudes and socio-economic data. As above, the wave 2 and wave 3 data collection was via telephone interviews (CATI).

- Time-use diary

  All household members aged 16 or over were asked to complete a week-long time-use diary. This diary splits each day into 96 distinct 15-minute segments and invites panellists to record which of a range of predefined activities (see Table 10.1) they are doing during each 15-minute segment. Not only does this capture the main or 'primary' activity but respondents are also asked to record any other activities in which they are also involved simultaneously. Younger members of the household (9-15) were asked to complete a similar but differently presented time-use diary[3].

- Following rules

  Naturally it is not possible to follow a household over time since the constituent members may change in a variety of ways. It is only possible to follow individuals over time and apply a set of rules for who should be interviewed in subsequent waves. In common with the BHPS[4], this panel adopted the following rules:

  — an attempt was made wherever possible to interview an individual who had been interviewed in a previous wave;

  — any new members of the household in which those individuals lived were also interviewed. This had the effect that where a wave 1 individual had left a household and formed another, all individuals in those new households were then interviewed.

---

[3] Both diaries are available from http://www.iser.essex.ac.uk/mtus/studies/uk-1998-2001.php

[4] See http://www.iser.essex.ac.uk/bhps/

Table 10.1 Time-use categories provided for the Digital Living panel.

1	Sleeping, resting	19	Sports participation, keeping fit
2	Washing, dressing	20	Hobbies, games, musical instruments
3	Eating at home	21	Watching TV/Cable/Satellite TV
4	Cooking and food preparation	22	Watching videos/laser disks
5	Care of own children or other adults in the home	23	Listening to radio, CD, cassette
6	Cleaning house, tidying, clothes washing, ironing and sewing	24	Reading newspapers, books, magazines
7	Maintenance, odd jobs, DIY, gardening, pet care	25	Being visited by friends, relatives in own home
8	Travel (to and from work, shops, school, cinema, station, etc)	26	Receiving telephone calls
9	Paid work at work place	27	Making telephone calls
10	Paid work at home (not using PC)	28	Playing PC games/games console
11	Study at home (not using PC)	29	Reading/writing e-mail
12	Courses and education outside the home	30	Browsing Web, or other Internet use
13	Voluntary work, church, helping people (not in own home)	31	Study at home (using PC)
14	Shopping, appointment (hairdressers/ doctors, etc)	32	Paid work at home (using PC)
15	Going to concerts, theatre, cinema, clubs, sporting events	33	Other PC use
16	Walks, outings, etc	34	Doing nothing (including illness)
17	Eating out, drinking (pubs, restaurants)	35	Other (please write in)
18	Visiting or meeting friends or relatives		

- Response rates

  Wave 1 recruited some 2400 individuals in 999 households across Britain. As Table 10.2 shows, some 1740 individuals completed the wave 1 questionnaire although only 1101 of these also completed the time-use diary. Similar data is shown for waves 2 and 3.

  Table 10.2 also shows that we have longitudinal time-use data on 301 individuals and survey data on 836 individuals in all three waves with larger numbers responding at any one wave or in at least two waves. The wave 2 sample was boosted to allow for non-response, although 1101 of the 1740 wave 1 individuals completed the survey in wave 2 — a retention rate of 58.6% which is within expectations for this kind of repeated survey. Similarly the retention rate for the time-use diary is within expectations given that wave 2 diaries were sent to individuals by post rather than passed over during a face-to-face interview.

Table 10.2 Number of respondents in waves 1-3 of the Digital Living panel.

	Wave 1	Wave 2	Wave 3	Longitudinal sample
Diary completed	1101	689	733	
Survey completed	1740	1566	1568	
Diary completed in all 3 waves				301
Diary completed in at least 2 waves				689
Survey completed in all 3 waves				836
Survey completed in at least 2 waves				1587

10.2.3.2 Call Record Collection

All households who were BT customers were asked to give permission for BT to collect their incoming and outgoing call records for the duration of the panel and to match them to the detailed socio-demographic data collected via the surveys. Roughly 400 households consented and their call records were collected on a continuous basis for the duration of the panel study. For each incoming or outgoing call these records provided:

- origin and destination;
- date and time of call;
- call duration;
- call cost;
- call type (national, local, international);
- line type (PSTN, ISDN, etc).

It should be noted that records could not be collected for calls that were:

- non-chargeable (i.e. 0800);
- from mobile telephones;
- from a licensed operator other than BT.

In practice, this meant that while the study had records of all chargeable outgoing calls from those households that had consented, it had records of a lower proportion of their incoming calls.

10.2.3.3 Internet Usage Logging

A selection of households who had a Windows 95 (or later) PC and an Internet connection were asked to install software which collected and analysed data on their Internet usage. This data included:

- when they connected to the Internet, for how long and how often;
- which Web sites they went to and when;
- when and how many e-mails they sent and received;
- which streamed content sites they used, when, and for how long;
- similar data on a range of other Internet applications including NetMeeting, secure http (i.e. eCommerce), telnet, ftp.

Of the 15 households from within the panel where the software was installed, usable data is available from 12. In addition, some 20 households from a market trial of broadband Internet access also installed the software on their PCs (see Chaper 8).

10.2.4 Qualitative Fieldwork

Following the first wave of quantitative survey fieldwork, a selection of households were approached for qualitative study. These studies included both structured and unstructured interviews, photo records, prompt-based discussions and repeat visits in what became a 'long conversation' between the qualitative researchers and the selected households [7].

Altogether the qualitative data was drawn from 104 individual interviews in 70 separate households carried out between December 1998 and October 2001. Ages of participants ranged from 13 to 67, with 55 being male and 49 being female. Forty-three interviews were carried out with individuals from the longitudinal panel selected according to their life-stage, the technology they owned, and the technology that they reported they were likely to purchase in the near future. Of these, 16 focused specifically on the role of the social network in Internet adoption and usage and 27 were more general interviews.

During the course of the interview a communications network was drawn up showing all the people with whom that person had regular social contact (see Fig 10.1). The interview covered:

- whether each individual within their social network lived locally or remotely;
- whether they were a relative or a friend;
- the strength of the relationships;
- the duration and direction (whether or not they were proactive or reactive in communication acts with each person) of communication with each individual;

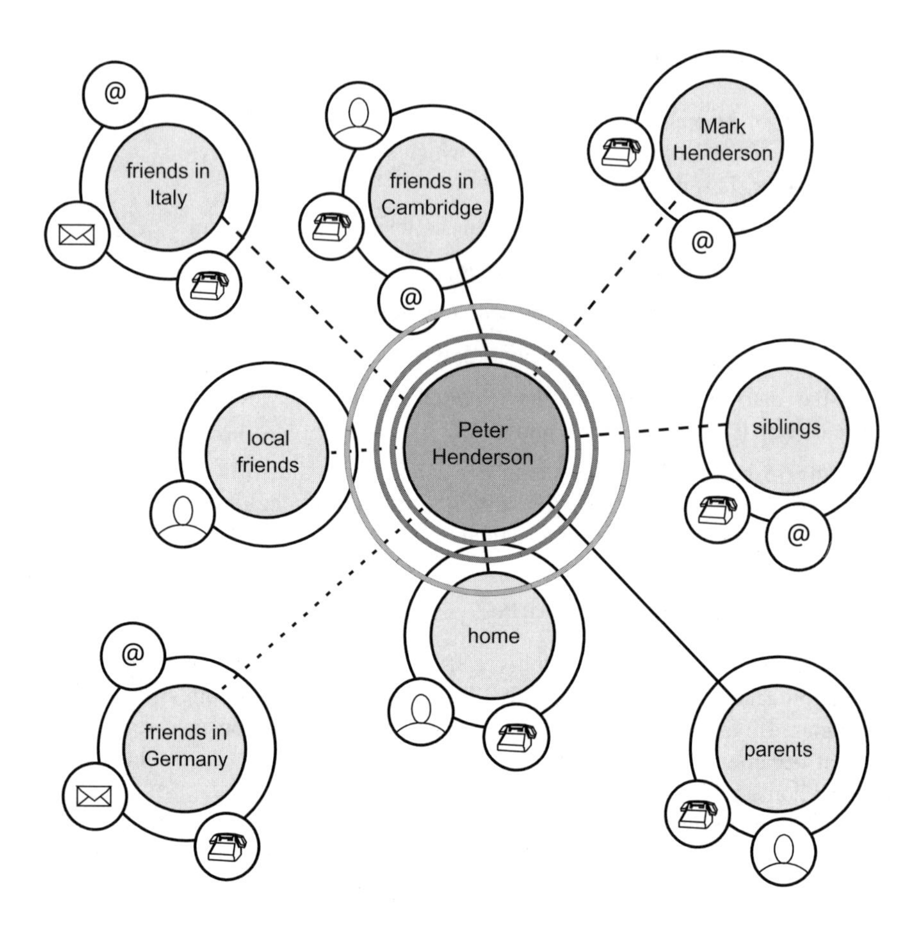

Fig 10.1 Schematic social network diagram.

- how often they communicated with them, how often they saw them, what forms of technology they used, what they communicated about, what time of day they communicated.

In principle, the interviews were designed to elicit details of the individual's communication habits, but, in practice, they often became wide-ranging conversations about technology-related behaviours of all kinds.

10.3 Other Studies

The same (or derived) research methods have been applied to a number of other groups of individuals or households who have been studied under the umbrella of Digital Living:

- teleworkers in a number of European countries [8];
- rural Cornish teleworkers, SOHOs and consumer households;
- students' use of mobile telephones in the UK [9];
- families and individuals in St. Petersburg, Russia;
- homosexual urban-dwelling men in the UK (see section 10.4.3);
- UK children (aged 11-16) through a research contract with LSE [10];
- households who took part in the BT Interactive Broadband Internet Access trial in West London, UK in 1999-2000 (see section 10.4.4 and also Chapter 8).

Similar data to that described above for the longitudinal panel members exists on these groups and several of these studies are summarised in section 10.4.

10.4 A Brief Tour of the Results

The following sections give a brief overview of results that can now be placed in the public domain. In most cases the sections summarise material that has been or is about to be published elsewhere. For the sake of brevity we note here only some of the main findings and their significance for technology innovation companies.

10.4.1 The Impact (or Otherwise) of the Internet on Everyday Life

There is an ongoing debate about the extent to which the Internet is changing people's lives. One way of addressing this question is to look at how or if people's use of time changes when they acquire access to, and start to use, the Internet. In the qualitative interviews with Internet users, one topic of discussion was the extent to which the Internet had an impact on the way interviewees spent their time and to what extent Internet use displaced other activities.

Although this might appear a relatively straightforward question, informants found it extremely difficult to pin down any clear or explicit changes:

"It's difficult to say if it [the Internet] displaces one activity or another."

The range of activities which were reported as possibly being displaced included watching television, spending time in the garden, reading newspapers, magazines and books, going to the supermarket, making telephone calls, going to the pub, doing nothing, writing letters, sleeping, playing computer games and typing on a typewriter. However, no one activity was mentioned by more than a handful of informants and even the heaviest of users felt that any displacement was marginal at best. One possible reason for this may have been the relatively low level of daily or weekly usage in the UK (between 1 and 3 hours per week on average) compared to the USA, although even those who spent as much as six hours per week using the Internet in the evenings could not pinpoint any major displacements. As the informants' time use appeared to evolve and change continuously, it became apparent that, rather than a straight substitution effect, a range of activities were adjusted or multi-tasked to enable Internet use to fit in. In addition, other factors have a significant influence on patterns of time use. For example, during the summer months, one respondent's television viewing, game playing and Internet usage were all displaced by spending time in the garden when the weather was good.

Therefore it might be expected that changes in time use would not be significantly associated with a simple transition such as acquiring Internet access. It might also be expected that an analysis of patterns of changing time use would show that the acquisition of Internet access is having relatively little immediate impact on people's lives. If so, we can conclude that conceptualising the relationship between technological change and social change, in terms of 'impact' or time-use 'substitution', may be over simplistic.

The quantitative time-use diary data from the Digital Living panel can be used to explore this issue in a relatively straightforward manner (see Anderson and Tracey [5]). The only time-use changes that could be associated with gaining Internet access were a decrease in time spent on hobbies and an increase in time spent studying at home using a PC, eating at home, watching TV and e-mailing/Web surfing. This last is, of course, a staggeringly obvious result. The only changes that could be associated with losing Internet access were less time eating at home, and more time watching videos. Interestingly, there was no evidence of a decrease in the amount of telephone communication received or initiated by new Internet users as a primary activity, even though Internet use in these households at this point in time would have used the fixed telephone line and thus prevented simultaneous voice calls.

In general, what was noticeable from these results was not what turned out to be significant, but what did not. There was no evidence that individuals who had acquired Internet access in their household, and who used it, were spending less time watching television, reading books, listening to the radio or engaged in social activities within or outside the household in comparison with individuals who had not.

Indeed in some cases they appear to be doing more of some of these. These results, based on longitudinal data, refute those of Nie and Erbring [11] which were based solely on cross-sectional data and therefore could not measure true change before and after a transition.

These findings suggest that changes in an individual's time use cannot be attributed solely to the change in access to the Internet. Of course, significant results can only associate changes in time use with changes in Internet access because a great deal of other significant events could have taken place in the lives of these individuals between waves 1 and 2. Further analysis is needed to unravel these effects using suitable regression models. It should also be noted that major changes in most people's time use are very unlikely to occur over these sorts of time period unless they undergo a significant life transition. In itself getting or losing access to the Internet does not appear to be such a transition, partly because technologies do not have measurable 'impact' overnight — there is a period of domestication [12].

The results imply that the simple impact model of Internet access and usage is not a useful explanatory tool. Not only have few significant effects been found but a range of confounding processes and triggers may make this kind of analysis over-simplistic. As a result, the impact model does not enable much purchase on the problem of how to understand and explain the place of the Internet in everyday life. This can only come from a deeper understanding of the triggers for and processes of its domestication, and a more detailed examination of how individuals and households are making sense of, and integrating, its applications and services into their lives.

As Anderson and Tracey show [5], qualitative data from the Digital Living study has started to draw out the complex relationship between Internet take-up and usage and an individual's changing personal circumstances in just this way. At the micro level, certain conditions and transitions in an individual's life may be significant triggers for Internet take-up or usage and, simultaneously, become causes of change on patterns of time use.

Work-related transitions, such as shifting from home-based to office-based work, or the reverse in the case of new teleworkers or new self-employment, seem to affect both access to, and the style of, Internet usage (see also Akselsen et al [8]). Changes within employment can also trigger Internet adoption through an explicitly or implicitly recognised need to improve work skills or competencies. Retirement also appears to be a significant trigger for household Internet adoption, particularly for those with distributed social networks and for those who have computer or Internet skills that they have learned in their workplace. Other household-related transitions that need to be considered are the departure of household members, perhaps to distant universities, to start employment or to set up independent households, because they have an impact both on the communication needs of the leaver and on those left behind, which Internet applications can meet. There is also some evidence from our qualitative data that household formation transitions, such as couples forming cohabiting partnerships or the birth of a child, also trigger Internet acquisition (or loss of access) and changes in the style of use.

It should be clear that these effects are extremely important in any attempt to understand the role and place of Internet applications and services in people's lives. In particular, it shows that analysis of these sorts of transition needs to be taken into account when conducting any kind of study into the change in people's patterns of activities that may be associated with the adoption of a particular technology. With the qualitative findings taken into account, a plausible explanation emerges and an important conclusion can be drawn — acquisition of the Internet and usage of its different applications is not necessarily changing individuals' lives but may be embedded within the normal social change of everyday life. As a result it seems clear that simple replacement effects are unlikely because other significant events are ongoing in individuals' lives.

By conducting analysis in this integrated and iterative manner, the research has started to tease out some of the motivations and triggers for Internet acquisition and usage, such as the role of life-style and life-stage related transitions, which to date have largely been ignored.

To give a concrete example of the commercial value of this approach, imagine you are an Internet service provider and you are about to invest in creating an entertainment portal where users can 'watch', 'listen' and 'play with' content supplied by a host of content producers. Critical to your business model is an assessment of exactly how much eyeball time your portal can steal from the television because you can then levy appropriate advertising charges. So you need to know how the dynamics of TV watching and Internet surfing work. Does one minute more on the Web lead to one minute less on TV? Or what? Only longitudinal analysis can tell you. As this section has shown, our ICT panellists filled in at each wave a time-use diary that was sufficiently fine-grained to collect this data and to provide answers to these questions (which can be found in Gershuny [13]).

10.4.2 Young People and Location-Based Services

The research described in this section is based upon the social scientific findings of a EURESCOM report from a strategic study of youth and mobile location-based services (LBSs)[5] which was carried out between November 2000 and May 2001[6]. It focuses on young people as a possible group of customers for location-based services to be provided on future mobile devices which can (and will be) used inside the home as well as out and about. One reason for this choice of age group is that current mobile technologies are highly popular among young people. In late 2000, 85% of 15 to 24 years olds in Europe owned mobile telephones[7].

What first became apparent is that the research group of interest — young people — is diverse, covering a huge range of life-styles, ages, cultures and so on; this presented us with the dilemma of how to identify commonalities among youth. Our approach was to differentiate between them in terms of their current life-stage, since those in the same life-stage tend to have some similar needs and desires. Life-stages can be defined in a variety of ways and for this age group one useful way of doing this is to refer to their level of physical (and cognitive) maturity, in particular their stage of adolescence, a concept which emerged as a stage between childhood and adulthood at the end of the 19th century [14]. Sociological youth research further differentiates between early-, mid- and late-adolescence with the approximate age ranges for these stages being 10 to15 years, 15 to17 and 17 to 19 years respectively.

[5] See http://www.eurescom.de/public/projects/projecttables/

[6] This section of the paper summarises selected material published as EURESCOM confidential reports accessible **only** to EURESCOM shareholders. The final report is found at http://www.eurescom.de/secure/projectresults/P1000-series/1045d1.asp and a more detailed review of social science research is found at http://www.erurescom.de/secure/projectresults/P1000-series/1045ti1.asp.

[7] Source = EURESCOM P903 data; for more information see http://www.eurescom.de/public/projects/projecttables/

In this section we examine some of the commonly found behaviours associated with youth, in particular focusing on the importance of relationships with their peers. We differentiate between the stages of adolescence at which the behaviour most commonly occurs, where necessary. In adolescence and particularly in early to mid-adolescence, peer culture is hugely important and young people start to have greater independence from their parents, identifying more strongly with peers than their family [15][8]. Within the peer group there will be a system of nicknames, jokes, styles of clothes, songs and artefacts. Peers influence one another in their choice of such 'cultural items', hence knowing the latest trends and information to keep up with one's peers is of concern to this age group. Older adolescents, however, are also keen to express their individuality. Part of belonging to a group of peers involves not only emotional and social support but also teasing, gossip and infighting.

Younger peers tend to interact within the home or in some known place, perhaps the home of a known peer-group member, as a result of the increasing 'privatisation' of young people's leisure activities [17]. Activities with peers outside the home are popular with older teenagers, such as clubbing, dancing, youth clubs, drinking, dating, going out on the town and 'hanging around' with friends, visiting friends, pop music, pub-going and cinema [18]. In order to have fast updates of social events, to arrange their social lives and gossip, young people want to be more or less constantly accessible and available to their peers — indeed, most adolescent communication occurs between peers. Much of this peer group behaviour is reflected in young people's current use of mobile telephones and has no physical boundaries — mobiles are used both inside and outside the home because they are a personal communications device not just a mobile device. Indeed a recent UK study of final year students' use of SMS (short messaging service) found that 61% of all text messages were sent or received while at home [19]. We would expect this also to be the case for others who spend considerable amounts of time at home.

The mobile telephone plays a vital role in flexibly arranging and then rearranging schedules, especially *en route*. The majority of 15 to 24 year olds use SMS to co-ordinate social arrangements; and the number and frequency of text messages increases when teenagers have just or are about to spend time with their peers, i.e. when making arrangements to go out or having just arrived home [20]. SMS, for instance, is used to spread the word about parties and places to go in real time and young people emphasise the importance of knowing this type of information quickly [21].

An LBS could support scheduling by offering services such as tracking the whereabouts of friends to aid meeting up, as the service concepts below describe:

- a tracking service allows groups of individuals to co-ordinate social outings — before they leave they can send an electronic invitation to multiple friends'

---

[8] It is worth noting that the age of greater independence from one's parents and identification with peer culture differs across cultures. For example, for Finnish children this is seen from an early age (around 9-10), but starts at a later age in Switzerland (age 12-13) and Spain (age 15-16) [16].

mobile devices to arrange to go out, while the acceptance of this invitation by the recipient triggers access to their location information so that the other invitees can track his or her progress until arrival at the agreed destination;

- to find out in which of your favourite places your friends are hanging out on Saturday night, you can subscribe to a service that allows you to access information about your favourite pubs, bars and clubs — this not only tells you whether any of your friends is currently present but also informs you how busy it is, and indicates any special offers and events;
- so that your friends can join you at a party or other social event that is in progress, you can send an electronic message allowing the recipient to track your location details.

The private and personal nature of the mobile is of particular importance for teenagers, especially those in early and mid-adolescence who use mobile telephones for 'hyper-coordination'. This is the expressive rather than task-oriented use of mobiles, e.g. the exchange of text messages expressing interest in potential boy/girl friends, jokes, chain messages and other fun-related content. While there is a concrete content, there is also a 'meta-content', i.e. the receiver is in the thoughts of the sender and the next time they meet they will be able to base some of their interaction on the messages. This behaviour helps them to maintain the social and emotional bonds with their peer group and often occurs shortly before or after some sort of social activity - such as just after getting home from school [21]. There could be an opportunity to support this peer group emotional and social communication or 'hyper-coordination' behaviour through new (location-based) services:

- leave a location-sensitive message or a photo in a specific place for your friend — just to remind them about the last time you were both here or to recommend a good shop nearby from which to buy that CD you both like, and then, when they are next in that location, they receive your message on their mobile device and know you were thinking of them.

Constant availability to peers is important to the younger teenagers, especially those in the early and mid-adolescent periods; the mobile telephone is ideal for this purpose, especially when used inside the parental home for peer interaction while avoiding potential 'regulation'. One advantage of always being reachable is that one's availability via a mobile telephone indicates popularity among peers [22]. The number of text messages sent and received is displayed to peers as a sign or quantification of a person's 'popularity' within their peer group, as is the fullness of the mobile telephone's automatic dialling register. The majority of the 14 to 18 year olds interviewed by Ling and Yttri [21] had between 100 and 150 names in their dialling registers; it was important to them that their register was full even though many numbers were used infrequently. As such, the mobile telephone is used to enhance both self-identity and impression management with regard to peers or self-expression. Older adolescents (aged 19 plus), however, are at an age where they often want to limit their availability to friends [9]:

- the number of people whose location you have permission to track can be used as another sign of your popularity and also supports accessibility and availability among peers;
- the number of location-specific messages you receive can be shown to peers to indicate your popularity in the same way as standard text messages.

In summary, this section has outlined just a few of the service concepts arising from an investigation of the habits and behaviour of youth. Young people appear to be a natural market for future mobile services, including LBSs, since mobile telephones are already an intrinsic part of most adolescents' life-styles. The mobile telephone offers immediate and private communication with peers as well as the ability to get fast updates on the latest happenings and to make arrangements while on the move — all of which are common/central behaviours and habits. To best fit into the life-styles of young people, future services should be focused around applications that promote their links with their peers such as allowing new forms of 'hyper-coordination', new ways of scheduling (e.g. finding out the locations of friends), and even faster ways to find out the latest gossip or places to go. Of course, many of the habits and interests of young people overlap with those of other groups of people, as will be seen below (see section 10.4.3), and therefore many of the services suggested would also appeal to a wider range of consumers. Finally it should be apparent that mobile devices are personal as well as location independent. It is critical to understand that personal communications and scheduling services (for example) will be used both inside and outside the digital home. Customers will expect seamless service wherever they are. Thus digital mobility must not be contrasted with the digital home — people are inherently wireless and their services need to be too.

10.4.3 The Pink Report

This study focused on technologies used by gay men, why they use them, and how this differs from other market segments. While full results of this study can be found in Green et al [23], this section provides a brief overview and an indication of the sorts of technology that are being developed as a result of the grounded design process outlined in Gower et al [3].

In-depth interviews were conducted with twelve gay men aged between 22 and 45 years. Respondents were recruited through Usenet newsgroups and the authors' personal contacts using the snowball technique. Six were living together as couples and six were single. All interviewees used the Internet, ten had access at home, one had access at work and one used cyber cafes. The interviews focused on life-style and use of technology.

10.4.3.1 Identity, Image and the Internet

These men define themselves first and foremost as 'gay'. Whereas a heterosexual man might define himself by profession, all of the respondents focused first on their sexual identity, and what it meant to be gay. Life revolves around 'being gay', and the Internet and gay portals are a very important way of building and maintaining a sense of identity for this group. Part of being gay is about creating the right image; it is very important to gay men to look and feel attractive. Respondents often described themselves as very vain, the driving force behind which is to secure sexual partners. Respondents also reported that image is very important within their community, and therefore maintaining themselves physically was also a way of maintaining status within the community, and conforming to the social standards of the peer group.

Respondents also describe many examples of prejudice and situations where they felt they had to conceal the fact that they are gay. Several interviewees had, for example, come out to their heterosexual friends, but not at work. Having to conceal the fact that they are gay leads to feelings of isolation. The Internet is a compelling medium for the gay community because it provides a sense of safety and anonymity where they can meet others with whom they share an affinity. It is important to note that at this time there is no other medium that supports this kind of anonymity. The Internet is therefore not merely a substitute, additional or alternative communications forum (as it may be for other market segments), it is, in many situations, the ideal and preferred medium of communication.

The Internet, and chat rooms in particular, enable this group to be part of a community that is mediated by various communications technologies. Chat rooms in particular act as a bridge between an on-line and a face-to-face community in that they provide the opportunity for face-to-face meetings.

Not only does the Internet support this kind of 'macro' level community but it also facilitates the sharing or exchange of digital 'gifts' [24]. A common behaviour among social networks is the circulation of jokes, cool sites, URLs or 'content'. In this, gay men are no different. They often used e-mail to exchange pornographic pictures gathered from the Web, and it was not unusual to find men with collections of 40 pictures or more sometimes gathered over extremely short periods of time. Given the impending roll-out of broadband and the decreasing cost of digital still and movie cameras it is reasonable to expect that this particular segment will increase the scale and size of these 'multimedia exchanges'. Indeed it is likely that the asymmetry of ADSL will become a key constraint to these patterns of electronic 'gifting' for a segment who seem highly likely to use the technology for content creation (pornographic or otherwise) for both social/fun reasons and economic gain.

10.4.3.2 Mobiles and Text Messaging

For this group, creating the right image includes technologies that are 'worn' in public. For example, owning a mobile telephone is seen as vital by the men interviewed — in response to the question: 'Which piece of technology that you own could you not live without?' the sample invariably replied: 'My mobile'. For this group it is the primary mode of communication and also a fashion accessory. Owning the 'right' mobile telephone is very important[9]. The 'right' mobile telephone is invariably described as 'slim', 'sexy' and something that ' ... would not make an unsightly bulge in the pocket.'

The ability to send text messages is seen as vital to this group; a mobile telephone without this capability is viewed as worthless, a view shared by many teenagers. Text messaging is so popular with this group that they are prepared to spend substantial amounts of money on it (e.g. one interviewee spent £54 in a month), even though they describe many messages as being 'about nothing at all'. 'Nothing at all' often referred to flirting, and text messaging was seen as an ideal way to indulge in this activity. Some use text messaging because it is cheaper than making a call and some use it to ask the recipient to ring them or to see if they are free to visit at home or to go out. In both cases the nature of the medium supports this kind of unobtrusive or 'background' communication (ignoring SMS messages is not considered 'rude').

10.4.3.3 Application, Product and Service Implications

There is scope for the development of devices and services that support new styles and methods of location-independent communication. Given the importance of fashion it seems obvious that such devices need to be 'personalisable', not just at the level of the application/service but at the level of the device itself. Taken together with the importance of image it seems clear that there is a gap in the market for the 'Louis Vuitton of mobile telephones'. These men regularly change and update their mobiles even though they are fully functional. It seems that what they are looking for is a range of life-style products (for example, special edition mobile telephones) that meet their aspirations. Given the dynamic nature of fashion it would also seem important to enable owners to change the form of their device over time in much the same way as the current configurable mobile face-plates and badges.

There are many more unmet needs and unarticulated aspirations that could be explored within this and other contexts, examples of which are listed below.

---

[9] See also Nafus and Tracey [25].

- Products

  — A range of life-style telephones (Fig 10.2) with limited functionality place extra emphasis on style and fashion rather than functionality. They include 'social' mobiles that are used only when 'going out' (i.e. clubs/bars) and limited functions that cater better for those specific social environments.

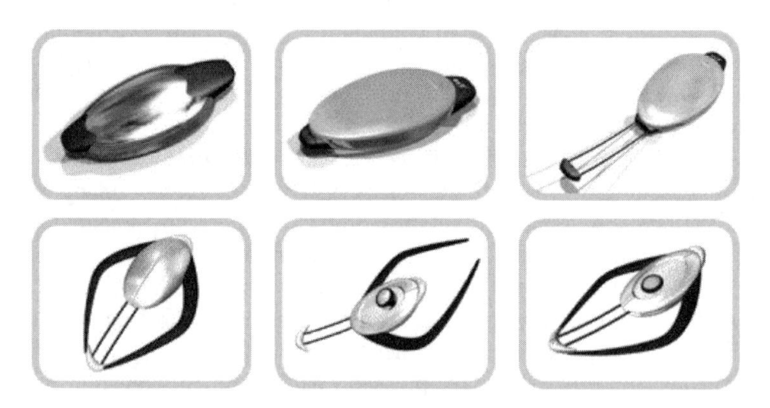

Fig 10.2 Life-style mobiles.

— Devices with a large area of screen (Fig 10.3) (or, in time, LEP[10] 'skins') allow users to download the latest cover designs, screen fonts and animated graphics, a choice of which can be updated weekly to support greater personalising of an otherwise standard design.

Fig 10.3 Surface display devices.

— In terms of gay etiquette giving someone your mobile number is seen as *de rigueur*, even if you hardly know them. As a result, on examination of their mobile telephone address books, it was clear that respondents had enormous calling circles — this led to the proposal of a SIM card back-up device (see Fig 10.4), a separate product that invisibly keeps an updated back-up of your telephone's call-list and therefore both ensures that the loss of a telephone does not mean the loss of the majority of your social network, and makes a change of

[10] Light emitting polymer — essentially thin, flexible plastic displays.

telephone seamless, as your new mobile can be instantly updated. The back-up device could attach as part of the charging device, so that back-up becomes thoughtless on a regular basis, or could simply be 'in the network'.

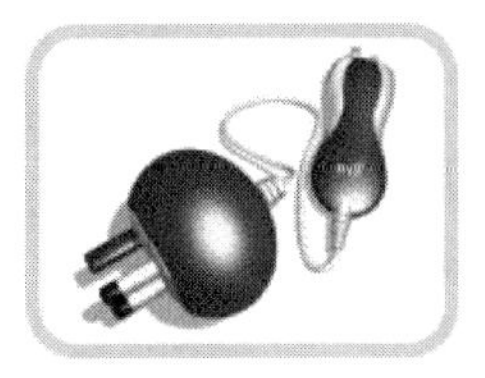
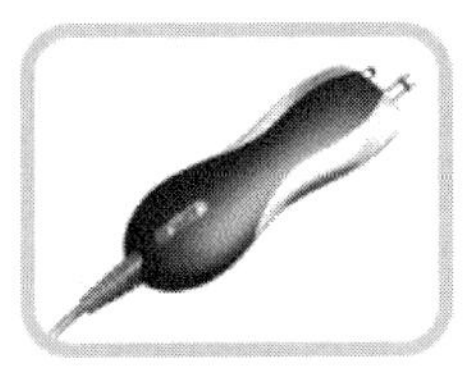

Fig 10.4 SIM card back-up device.

- Services

  — A range of 'social presence' services and functions cater for some of the attributes that are common to this market sector, such as extensive call lists, regular number exchanges, and the fear of isolation (or absence) from their social network.

  — Quick number exchange services make it 'unsexy' to have to type a number in. This type of interchange could transfer additional data, apart from just a telephone number, such as extra information on the user's hobbies, likes and dislikes, URLs, photos or movies.

  — Mobile chat groups take the Internet chat room concept into the mobile world, allowing instant messaging via mobile device access. Individuals could start private groups and invite friends to join through SMS, which would include log-on and password details. Individuals could be members of several groups, for differing sets of friends. Users could be notified via SMS when there is activity on a related account, allowing them to connect and join in, or log on later and view the text transcripts to catch up on the gossip or arrangements. This would also work seamlessly with on-line (PC-based) chat rooms so that the access method becomes irrelevant — you can be in the chat rooms/IM groups you want to be, anywhere on the planet, using any device.

It should be clear that these products and services are not of interest only to gay male technophiles, since many of them are also applicable to a number of other potential market groups. It should also be clear that gay men are not the only group that uses the relative anonymity of some communications technologies to support their social networks. Other groups, who may suffer similar 'marginalisation' or social pressure, also do so, such as geographically and socially isolated Muslim women in North America [26]. Finally, as depicted in section 10.4.2, it is clear that any organisation seeking to deliver service to the home has to think about how existing mobile services will relate.

10.4.4 Broadband Living — HomeLAN

Our final results section describes the concepts and services developed from the analysis reported in Chapter 8.

Not only does that research analysis have implications for strategy but, as with the studies of gay men, it also has obvious implications for appropriate products and services.

We have used insights from these studies of broadband users to generate concepts for products and services. For example, the advent of residential local area networks (LANs) provides an opportunity for a centralised Internet gateway, network hub and file server. Centralising the storage and processing of digital information and access to the Internet also enhances the viability of low-cost thin-client devices such as Web pads. The HomeHub server (Fig 10.5) is designed to visually complement a living space. It also provides a PowerPad for recharging eBooks, Web pads and other portable thin-client cordless devices (for more details see Gower et al [3]).

Fig 10.5 HomeHub server.

The wet area Web pad (Fig 10.6) provides access to Internet news and traffic reports via a local wireless communications capability. It incorporates a detachable tentacle base that allows the pad to be supported by being stuck to, or wrapped around objects such as bath taps or shower pipes.

The BedBug Web pad (Fig 10.7) is designed for children aged 7-12 years and provides Internet access, videophone (via a camera and microphone antenna), games station and alarm clock functions in addition to streamed TV broadcasts. The Web pad includes a floor-based multi-positional support arm that allows it to be cantilevered over the bed. It can also be detachable from the stand for use in other areas of the house that are supported by a wireless LAN.

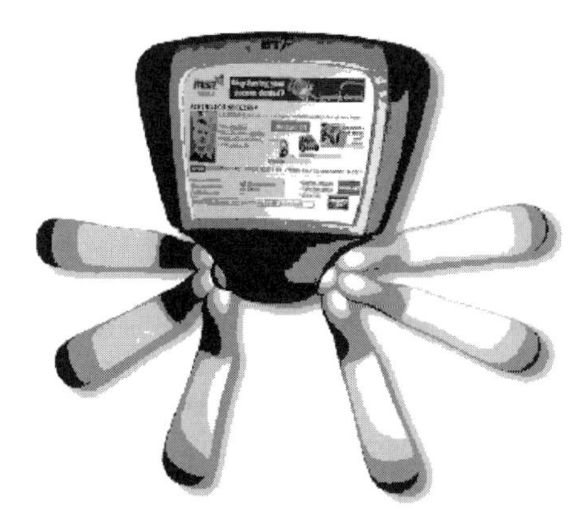

Fig 10.6 Wet area Web pad.

Fig 10.7 BedBug Web pad.

10.5 Synergies and Reviews

10.5.1 Gifting and Exchange

Throughout society we find widespread evidence of 'gifting'. For example, people send URLs, images (of themselves or others), jokes and so forth. Knowledge is also 'gifted' where 'gurus' offer technical help to their social network members and SMS messages/mobile telephone calls are used to 'gift' emotion through non-functional 'thinking about you' calls or messages. Information and communications technologies are also gifted as is seen in shared use of mobiles by households and children, although, with the increasing penetration, this is perhaps becoming less frequent.

In all of these cases the gifting is part of a tacit, informal exchange system where the giver may expect (at some point in the future) the 'gift' to be reciprocated. Such gifting systems (as opposed to formal monetary exchange) have long been the focus of study in cultural anthropology and it is likely that conceptual models of reciprocity, such as that developed by Mauss [27], will be of some use in explaining why, and in what circumstances, people exchange what kinds of things with each other.

This may seem of purely academic interest but it is not. One of the main reasons people use information and communication technologies is to exchange 'stuff'. Not only can this be greetings, URLs, e-mail messages and so forth (as discussed in section 10.4.3), but it can also be presents (eCommerce) or content (e.g. 'booking' a movie for your girlfriend's birthday). Understanding these patterns and using them to define novel 'gifting' services is therefore vital [24] because broadband connectivity (whether in the home or elsewhere) could support the emergence of new exchange services such as:

- gifting of DIY media objects (photos, videos, audio recordings, games, Web pages, etc);
- gifting of commercial content — booking a VoD (video-on-demand) for your friends when you go round for dinner, sending your Mum that episode of Eastenders that she missed;
- gifting personalised media collections such as 'mixed' CDs or personal albums.

It is important to realise that goods and services are a communication system, and offering help in figuring out the language of gift exchange and the meanings of gifts is an important task for some retailers, especially those with no off-line presence. Given the explosive growth in eCommerce in the UK in the past year, understanding such basic, pan-human needs can provide an important competitive advantage to eRetailers.

10.5.2 Identity, Fashion and Symbolism

It is apparent that how a device, a product or a service looks is becoming as important for some groups as what it actually does. We can illustrate this point with respect to gay men and young people. In particular, they emphasise the importance of the look of mobiles. That this should become an issue for mobile telephones is hardly surprising — they are items which are 'worn' in public like clothes or a haircut. Whether they are utilised to 'give off' signals or simply do so unintentionally, there is, however, no doubt that other people 'read' their significance in the specific cultural context [28]. This is also true of the services accessed via these devices. For example, the number of SMS messages young people receive is sometimes portrayed as a 'popularity' indicator.

With the likely emergence of new mobile information and communications devices (e.g. UMTS/3G), this focus on style, fashion and symbolism will grow in importance since all these devices will, by definition, be used and carried into public spaces and 'public life'. Those products and services which are likely to be successful will be those that either do as little 'damage' as possible to people's public presentation of self and/or offer people new ways to present that image of themselves.

However, we should also note that fashion and 'presentation' matter within the home. People use their homes, whether explicitly or implicitly, to create images of themselves for the perception (and approval) of others [29, 30]. As a result, there are many aspects of the current 'digital home' technologies which people simply do not want 'on show'. Designers of domestic technologies will need to take heed of this phenomenon and join those who are going some way towards creating products, which either have symbolic/aesthetic appeal[11] or which are designed to 'disappear into the woodwork', if they are to achieve the kind of penetration they anticipate.

10.5.3 Transitions

We are starting to see the significance of transitions in people's lives and the interrelationship between these transitions and their usage of information and communications technologies.

The first examples involve personal transitions such as 'coming out' for gay men, stages of adolescence such as primary to secondary school, secondary to university or work, and leaving home. In all these cases the technologies people were using were wrapped up in their experiences of these transitions. In contrast, household transitions such as children leaving home, the arrival of a new child or changes in work status can all lead to changes in the access to (i.e. influence purchasing decisions) and usage of different kinds of technology.

It should be obvious that the prime motivation behind implementing a panel study of the kind described here is to be able to follow people over time to understand what kind of transitions they experience, and how this influences their access to and usage of various information and communications technologies. Clearly the results reported in Anderson and Tracey [5] are a first pass at this kind of analysis using the quantitative panel data for just two waves to try to analyse the impact in reverse, i.e. what difference does going on-line make to people's lives.

As we have seen, the answer, in terms of time use, is not a lot. However, this is perhaps the wrong way to approach the issue of transitions — more fruitful would be to start from individuals who have undergone some sort of transition and analyse the changes in their lives.

---

[11] Such as Apple.

10.5.4 Unanticipated Uses (and Unanticipated Revenues)

Technologies have always generated unanticipated uses. The use of the telephone for social chat was a use not anticipated by the Bell Telephone company and indeed was one that they tried to stop through advertising guidelines for 'proper use' [1]. This tradition of finding uses for technologies that their designers and marketers did not perceive still continues. It is part of current mythology that SMS messages were originally for use by engineers, but they have been recruited into the repertoire of everyday communications methods to spectacular effect, both in terms of usage and revenue generated.

Indeed 'technologies' such as SMS also lead to unanticipated social outcomes. For example, we have reported the use of an SMS message to see if 'anyone is there', when it would have been inappropriate to make a telephone call or stop by a friend's house. Such 'malleable' technologies thus allow new ways to navigate complex social mores because they are largely value free — in other words it is up to the user to decide what the 'technology' is for and the answer is that it is different for different people in different contexts.

This issue has been well known in the field of computer supported co-operative work (CSCW) for some years[12]. Here it is common place for users of work-related systems (workflow, document and process management, etc) to adapt them to their own purposes in ways that were not foreseen by the designers of the systems. Rather than viewing these new uses as 'improper' or 'user error', CSCW sees in them an opportunity to capture users' creativity and fold these uses back into the product or system. In other words, the users become the co-designers of the system and the result is usually a workplace system that is far better suited to the work practices of those users and therefore far more likely to be used.

Extrapolating to the domestic context, the Digital Living research has already shown that the diversity of stakeholders and usage contexts for household technologies means that both of these may be unknowable at the time of conception and design. Thus perhaps the most fundamental challenge facing the design of products and services for the domestic context is understanding what a thing is for, and who its users will be, and this cannot be defined in advance. Thus any approach to the design of these artefacts which assumes that the definition of user, task and goal is possible will be of limited value[13].

This suggests that we should throw the idea of 'formal requirements capture to drive development' out of the window. Instead we suggest a different model of product development, one that involves the potential users/customers in a rapid cycle of design, use and redesign, rather similar to the alpha and beta-test cycles common to many Internet software applications. During this cycle of 'design-tests',

---

[12] See Robinson [31] and also, more recently, a special issue of Computer supported Cooperative Work on Adaptive Workflow Systems: Volume 9, Issue 3/4, November 2000 (http://www.wkap.nl/issuetoc.htm/0925-9724+9+3/4+2000).

[13] For a number of examples of this problem in practice, see Lacohee and Anderson [4].

user feedback, comments and suggestions can be built into the product or service and, perhaps most importantly of all, the likely revenue streams (and hence the business case for full launch) can emerge, evolve and be refined.

Perhaps the most radical end-result of this approach is the free-to-use launch of products and services and the very rapid secondary development and enhancement (of both the product and the revenue model) of those that appear to be successful. The extent to which this model is new is perhaps moot, but it is most certainly radically different from the 'innovate, design, build, launch, market, sell, wait for revenue' pipeline model because it allows the business case/model to evolve during development, and thus respond to unanticipated use, rather than (usually) being incorrectly specified in advance.

Of course, it may be that such adaptive revenue models require a highly adaptive organisation, or are one part of a response to the problem of reacting to disruptive technologies. If uses, and thus revenues, cannot be predicted in advance, then at least we must put in place adaptive organisational mechanisms so that emerging uses and revenues can be exploited rapidly.

This then is our partial answer to the problem posed in the introduction. Rather than persisting in attempting to predict usage and revenue models and thus form coherent business cases that turn out to be wrong, we suggest that grounded innovation and adaptive revenue models can lead us to a more effective, flexible and responsive innovation process.

10.6 Summary

This chapter has provided a summary of a research programme at BTexact Technologies which is aimed at helping a technology innovation company to ground its innovations, to see opportunities for the exploitation of its technologies, and to create socio-technical visions which can help to drive technological innovation itself.

As a by-product, the programme has also created strategic knowledge that is of critical importance to public and private policy/decision makers alike. Perhaps most importantly of all, this research is a key part of BTexact's approach to the creation of and response to disruptive technologies, because, through thinking about changes in modes of provision, it provides exactly the kinds of insights that are required to answer such questions as: "... does (the technology) enable a broader population of less skilled or less wealthy customers to do things for themselves that previously only experts or wealthy individuals could do?" "... does this product or service help customers get done more easily and effectively what they are already trying to get done?" [32] As Cosier and Hughes note: "... it is the usage by people that creates social, economic and business disruption." [33]

It is therefore self-evident that understanding 'usage by people' is absolutely critical to figuring out what is disruptive about technologies, why this is so and therefore how to make money out of them. Since this is critical to several of BTexact's core competencies (and to those of its customers), the value of the research reported here to BTexact and its customers is also self-evident. Without it, they will only ever make money by accident and as we noted at the start, such a strategy is not likely to amuse their shareholders.

References

1 de Sola Pool, I.: '*The Social Impact of the Telephone*', MIT Press, Cambridge, Mass (1977).

2 Anderson, B., McWilliam, A., Lacohee, H., Clucas, E. and Gershuny, J.: '*Family life in the digital home — domestic telecommunication at the end of the 20th Century*', BT Technol J, **17**(1), pp 85-97 (January 1999).

3 Gower, A., Lacohee, H., Jones, M., Tracey, K., Trimby, M. and Lewis, A.: '*Designing for disruption*', BT Technol J, **19**(4), pp 52-59 (October 2001).

4 Lacohee, H. and Anderson, B.: '*Interacting with the telephone*', in Kraut, R. and Monk, A. (Eds), Special Issue of the International Journal of Human-Computer Studies entitled: '*Home Use of Information and Communications Technology*', **54**(5), pp 665-699 (May 2001).

5 Anderson, B. and Tracey, K.: '*The impact (or otherwise) of the Internet on everyday life*', American Behavioral Scientist, **45**(3), pp 456-475 (November 2001).

6 Oftel: '*Homes without a fixed line phone — who are they*?' (1999) — www.oftel.gov.uk/cmu/research/unphoned.htm

7 Silverstone, R., Hirsch, E. and Morley, D.: '*Listening to a long conversation: an ethnographic approach to the study of information and communication technologies in the home*', Cultural Studies, **5**(2), pp 204-227 (1991).

8 Akselsen, S., Gunnarsdóttir, S., Jones, M., Julsrud, T., Marion, R., Martins, M. P. and Yttri, B.: '*The impacts of telework on quality of life — preliminary results from the EURESCOM Project 904*', paper presented at the conference: '*2000 and Beyond: Teleworking and the Future of (Tele)Work*', Stockholm, Sweden (2000).

9 Standen, N.: '*Social places, personal spaces: how, where and why UK students use, display and conceptualise mobile phones in a university environment*', paper presented to the New Technologies and Visions of Society stream at the European Sociological Association Conference, Helsinki (2001).

10 Livingstone, S.: '*Children on-line: emerging uses of the Internet at home*', Journal of the Institution of British Telecommunications Engineers, **2**(1) (Jan-Mar 2001).

11 Nie, N. and Erbring, L.: '*Internet and society: a preliminary report*', Stanford Institute for the Quantitative Study of Society (2000), available at — http://www.stanford.edu/group/siqss/Press_Release/Preliminary_Report.pdf

12 Silverstone, R., Hirsch, E. and Morley, D.: '*Information and communication technologies and the moral economy of the household*', in Silverstone R and Hirsch E (Eds): '*Consuming Technologies*', Routledge, London (1992).

13 Gershuny, J.: '*Web-use and Net-nerds: A Neo-Functionalist Analysis of the Impact of Information Technology in the Home*', Working Papers of the Institute for Social and Economic Research, Paper 2001-1, University of Essex, Colchester (2001).

14 Gillis, J.: '*Youth and history: tradition and change in age relations, 1770-present*', Academic Press, London (1981).

15 Pulakos, J.: '*Young Adult Relationships: Siblings and Friends*', The Journal of Psychology, **123**(3), pp 237-244 (1989).

16 Suess, D., Suoninen, A., Garitaonandia, C., Patxi, J., Koikkalainen, R. and Oleaga, J. A.: '*Media Use and the Relationships of Children and Teenagers with their Peer Groups*', European Journal of Communication, **13**(4), pp 521-538 (September 1998).

17 Livingstone, S. and Bovill, M. (Eds): '*Children and their Changing Media Environment: a European Comparative Study*', Lawrence Erlbaum Associates, Hillsdale, NJ (2001).

18 Hendry, L.: '*The influence of adults and peers on adolescents' life-styles and leisure styles*', in Hurrelmann, K. and Engel, U. (Eds): '*The Social World of Adolescents: International Perspectives*', Walter de Gruyter, Berlin (1989).

19 Krefta, M.: '*The silent revolution: a study into the implications of SMS on society*', unpublished report, School of Geography, University of Bristol (2001).

20 Ling, R. and Helmersen, P.: '*It must be necessary, it has to cover a need. The adoption of mobile telephony among pre-adolescents and adolescents*', presented at the Conference on the Social Consequences of Mobile Telephony, Oslo, Norway (June 2000).

21 Ling, R. and Yttri, B.: '*Nobody sits at home and waits for the telephone to ring: micro and hyper-coordination through the use of the mobile telephone*', in Katz, J. and Aarhus, M. (Eds): '*Perpetual Contact*', Cambridge University Press (2002).

22 Taylor, A. S. and Harper, R.: '*The gift of the gab?: A design-oriented sociology of young people's use of mobilZe*', submitted to the European Conference on Computer Supported Co-operative Work (CSCW) (2001).

23 Green, N., Lacohee, H. and Wakeford, N.: '*Someone I should have deleted: The integration of new technologies into the lives of gay men in the UK*', Sexualities, to appear (2002).

24 Zafiroglu, A.: '*Gifting and eCommerce*', Internal BTexact Technologies Report (2001).

25 Nafus, D. and Tracey, K.: '*The more things change: mobile phone consumption and concepts of personhood*', in Katz, J. and Aarhus, M. (Eds): '*Perpetual Contact*', Cambridge University Press (2002).

26 Bastani, S.: '*Muslim women online*', Arab World Geographer, **3**(1), pp 40-59 (2000).

27 Mauss, M.: '*The gift: the form and reason for exchange in archaic societies*', Routledge, London (1950).

28 Goffman, E.: '*The presentation of self in everyday life*', Doubleday, Garden City, New York (1959).

29 Miller, D. (Ed): '*Material Cultures: why some things matter*', UCL Press, London (1998).

30 Madigan, R. and Munro, M.: '*House beautiful: style and consumption in the home*', Sociology, **30**(1), pp 41-57 (1996).

31 Robinson, M.: '*Design for unanticipated use...*', in De Michelis, G., Simone, C. and Schmidt, K. (Eds): '*Proceedings of the Third European Conference on Computer Supported Co-operative Work*', pp 187-202 (1993).

32 Christensen, C. M. and Sundahl, D. L.: '*Foreword*' to special issue on '*Disruption*', BT Technol J, **19**(4), pp 5-6 (October 2001).

33 Cosier, G. and Hughes, P.: '*Editorial*' to special issue on '*Disruption*', BT Technol J, **19**(4), pp 7-8 (October 2001).

11

DIGITAL HOMES — FOR RICHER FOR POORER, WHO ARE THEY FOR?

P A Rout

11.1 Introduction

The smart home, future home, connected home, intelligent home, technology home, digital home — whatever you want to call them, they are only for the seriously rich. They are the top-of-the-range, prestige, executive properties that would be the flagship of the house builder or property developer. It is where bespoke entertainment, climate control, security and home office systems are controlled at the mouse click of a personal computer. Where technology is king and the latest fads are the current 'must haves'.

But is this really the case? No!

The underlying technology currently available or under development in this area is not necessarily expensive, nor is it the sole domain of the 'executive' customer. There is growing evidence, and indeed demand, that introducing technology and services into the home can provide cost-effective benefits and not just for the rich, but throughout the community, including the poorest and digitally excluded. The importance is not in the technology itself, but in the benefits it can bring and the scale of its capability. These benefits will depend on who the customer is, what they want to achieve, the services they value and those for which they are prepared to pay.

This chapter looks at the digital home from the customer perspective, based on practical experience gained from a range of current and recent projects. It identifies who the customers are, the types of service they are looking for and the underlying issues that can hold back or accelerate developments in this area. It also describes a framework for engagement with the customer which is being successfully employed by BTexact Technologies.

11.2 From 'Home Sweet Home' to 'Home Smart Home'

Compared to other areas, such as the development of the motor car, homes have changed very little over the past 50 to 100 years. In the 1960s optional extras for cars still included windscreen washers, heaters and simple radios. Today standard features on a good family saloon can include air conditioning or climate control, radio and CD entertainment systems, remote central locking, electrically operated sun roof and windows, information console and trip computer, auto dim interior lights, security alarm and engine immobiliser.

Modern houses rarely have comparable features, even when they cost upwards of ten times the price of a standard family car.

The introduction of electricity transformed how our homes were illuminated, with individual lighting controlled at the turn of a switch. Electricity also enabled the widespread use of technology for heating, cooking and entertainment. However, these have typically been stand-alone systems that can only be used or enjoyed within designated rooms or parts of a room depending on where the electrical sockets are installed. Families congregate around the television, radio or CD player for their entertainment (Fig 11.1). It does not seem so long ago that a key selling point of a house featured how many electrical sockets you had! Other advances such as central heating are now commonplace, but again these are typically controlled from a single point (the thermostat on the wall or a mechanical valve on each radiator) with little or no scope for dynamically setting the temperature in different rooms.

Fig 11.1 'Home sweet home' entertainment.

Another transformation took place with the development of the telephone. For many years this was a single instrument hardwired into the wall and typically located in the hallway. If the telephone rang you had to run and answer it, and out of this grew a strong market for telephone tables and sitting on the stairs for a chat. We have now progressed from wiring multiple telephones into multiple walls, to using various types of roaming, cordless telephones, including the ubiquitous mobile. In addition, the telephone line is no longer being used just for voice, but is our connection to the world of data, linking our PCs to the Internet and beyond.

Home security has traditionally meant strong locks, keeping your windows closed and telling your neighbour when you are going on holiday. This is often followed by the debate about whether the curtains should be open, closed or half-and-half when you are not there. Of course you could always buy a timer switch for the table lamp or living room light to fool any potential burglars into believing that you are at home. Exterior security lighting has come into play, with passive infra-red sensors being used to detect when someone or something is moving nearby. Some homes have more comprehensive security systems installed, often with remote monitoring by a third party, but again this is not the norm.

Each of the services outlined above are typically installed on a stand-alone basis, with little interaction or common infrastructure connecting them together. The digital home looks to support these and other services, but in a more effective, integrated and flexible manner.

11.2.1 The Killer Application

So as we move from 'home sweet home' to 'home smart home', what is the killer application that will make this happen in the mass market? While there probably is not one in isolation, there are a number services, and resulting benefits, that, when added up, start to make a compelling case for the digital home. In order to address some of these services, they are divided into three categories — home working, home entertainment, and home control.

- Home working

  Growing numbers of people are now working from home on either a full-time or part-time basis. In these situations they would normally expect to have access to the full range of information technology and associated services they would find in a normal office environment. These could include:

  — secure remote working;

  — access to remote office information systems;

  — access to remote file storage systems;

  — access to third party services, such as remote banking;

  — high-speed Internet access;

  — internal and external e-mail;

— simultaneous multiple data connections;

— fixed and mobile telephone services;

— use of peripherals such as printer, facsimile and scanner.

Also, an increasing number of homes have more than one PC, separate machines for home use and work use, and older machines which have been replaced by a more up-to-date model. In these situations it is very useful to be able to connect and share data, network services and peripherals between machines in different parts of the home without needing an information technology degree or computer networking qualification to achieve it. It is interesting to note that many spare bedrooms are being converted to a 'home office', but how many will be converted back when it comes to selling the house? What will look better on the estate agent's advertisement — a four bedroom house or a three bedroom house with an office?

- Home entertainment

  If there is an area closest to being a killer application, then in the short-to-medium term it is likely to be home entertainment (Fig 11.2). Rather than having to go to a dedicated piece of equipment or room to be entertained, for example to view a television programme or listen to a CD, a networked property could distribute and control services throughout the home. These services could include:

  — analogue, digital or satellite TV,

  — video and DVD,

  — multiple PC gaming,

  — audio (CD, radio, Internet radio).

  Live and recorded TV programmes could be simultaneously distributed to a number of televisions and PCs, audio zones could be established so your favourite music could follow you round the home, and interactive multiplayer PC games could be played across your home network.

- Home control

  Home control covers a wide range of potential digital home services, including:

  — home security;

  — energy management;

  — local and remote control of home functions;

  — caring services for the less able.

Fig 11.2 'Home smart home' services.

Security within the digital home can be addressed on two levels:

— maintaining the physical security of the property, its contents and its occupants;

— maintaining the security of the home's IT and network services.

Existing physical security services, such as stand-alone and remotely monitored intruder detection and video surveillance systems, can be integrated and enhanced within the digital home. Since these and other services are likely to be controlled within the home network, it is essential that access to specific services, information and the network itself is strictly limited to authorised users.

Prime drivers for implementing home control in a digital home could be to save money, save time or to reduce the drudgery of a task. One way of saving money is to monitor and make best use of the utility services that are used in the home, for example electricity, gas, heating or water. Energy management and metering systems can be used to monitor consumption and where appropriate used to select the most cost-effective source of supply.

The control of home functions, such as lighting, central heating, hot water, laundry facilities and security, can be used to help set your ideal environment as well as save you time and money. When combined with other features, like automated curtain tracks, the security of your property can be enhanced, for example when you are away on holiday. Secure remote access to these functions also adds to the flexibility of this control — so all the services can be ready for you when you arrive home at whatever time of day.

The final area of home control which is creating a great deal of interest is in the use of technology to provide support and peace of mind to those less able to look after themselves — the caring home or Telecare. Individuals who are old, frail or are recently out of hospital may be able to look after themselves, but need some reassurance that help is near to hand and their welfare is being monitored. Projects are currently under way to investigate how technology can be used to help maintain their independence in a safe environment, including the 'Millennium Homes' initiative by Professor Heinz Wolff that is being supported by BTexact [1].

Taking any one of these services in isolation may not in itself prompt many people to invest in a networked home. However, when a number of these are combined, particularly with a home working and home entertainment capability, the benefits and attractions begin to add up. Currently no players in the market appear to have a viable business case for home control or telemetry without combining this with other services. This points to a business model where local government and other third parties (finance, utilities) might have a considerable interest in bringing down the cost to the customer and thereby stimulating demand.

11.2.2 The Underlying Technologies

From the user's viewpoint the detail of the underlying technology needed to support the digital home is largely irrelevant. Their main concerns generally focus on what the benefits are, what the costs are, and whether it is future proof, sustainable and easy to install and operate. The enabling technology is key to delivering the benefits of the digital home and BT is taking a lead in developing this technology:

- structured cabling [2];
- home area networks (see Chapters 3 and 6);
- residential gateway (see Chapter 5);
- security;
- broadband (see Chapter 4);
- home control systems;
- home network standards (see Chapter 7);
- smartcards [3].

An additional factor that comes into play when accessing the underlying technology is how to integrate the requirements for the digital home design into new buildings and existing properties. This is currently being addressed by BT and the equipment suppliers, and solutions will evolve along with the technology.

The key challenge, and also the key opportunity, is to provide an integrated, multi-channel approach to delivering digital home services to the customer.

11.3 Who are the Customers?

So who are the customers for the digital home? Ultimately it is you and I — the home owners, lease holders and tenants. But there are other key players involved, other 'customers' for digital home services and solutions. This section sets out who these customers are, and then, from practical experience, addresses their individual needs.

- Home builders

  These are the national and international construction companies who are building new properties on green and brown field sites.

- Developers

  These are companies working in conjunction with home builders on new property developments or who are refurbishing existing properties, such as old factories, warehouses or groups of properties, for residential use.

- Facilities management companies

  These are the companies who manage the property developments once the construction or refurbishment is complete and the residents have moved in.

- Housing associations

  These are non-profit-making organisations who provide social housing for tenants. Their involvement may also extend to providing additional tenant community services and taking the lead in urban renewal programmes.

- Local authorities

  These are local government organisations responsible for urban regeneration and development programmes, including housing departments of local councils.

- Home owners

  These are freehold and leasehold residents of private housing.

- Social housing tenants

  These are rental tenants of social housing managed by local authorities or housing associations.

The relationship between the different 'customers' is complex, particularly when it comes down to identifying who is responsible for the different elements of the digital home, what they want to invest in it, what they expect to get out of it and who owns the ultimate customer — the resident of the property; and this is before we even consider the involvement of the different service and end-user equipment providers. In practice, as development consortia often involve the builder, developer, facilities management company and housing association, the specific drivers and responsibilities for specifying and delivering IT services into the home can therefore become blurred.

The following sections take a look at the needs of each key player in turn.

11.3.1 Home Builders

Home builders are increasingly looking to include elements of the digital home to give themselves a business edge by enhancing the specification of their properties, making their homes more attractive to potential buyers and charging a premium for the extra services.

Innovation can also be a key driver or contractual requirement for a particular development, such as the architectural design, the use of novel or sustainable materials, energy conservation and the use of information technology. This raises the question as to what aspects of the digital home builders are interested in and what they will 'pay for' in relation to the overall cost of the property. In general, the builder, in conjunction with the developer and possibly the facilities management company, may only be interested in providing the basic infrastructure to support the digital home, but not the actual services or applications which may or may not be taken up by the residents.

This infrastructure would typically include the basic trunking and voice/data cabling across the site and within the individual properties, for example, by the installation of structured cabling to each room. Exceptions to this are where common services are provided to all properties, like lobby security or concierge services. However, as the driver for such services could come from the developer or facilities management company, this is likely to be a consortium issue rather than just a home builder initiative.

A major benefit of the new-build market is the ease with which the IT infrastructure can be installed, since the trunking, cabling and associated equipment can be installed at the same time as other services during the building process. This limits the amount of disruption to the property and to the occupier who will not move in before the property is complete. The disadvantage is that the supplier of the IT services has to be ready to install in line with the other building work, be it the pouring of concrete or the installation of the mains electricity cabling. An alternative would be to educate the tradesmen into installing ducting, cable and associated equipment to an acceptable standard.

Opportunities will exist to demonstrate the value, benefits, features and flexibility of the digital home. By making good use of a show house or show flat owned by the builder or developer, the benefits and features can be demonstrated in order to stimulate sales. It is important for all involved, including the end-service and customer-equipment providers, to make best use of these facilities to showcase the digital home and its capabilities. There is little point, for example, in installing structured cabling within a property if:

- no one knows that it is there;
- no one knows how to use it;
- its value cannot be demonstrated to a potential buyer.

11.3.2 Developers

While developers are obviously involved in the construction of new-build projects, as outlined above, another key area of activity is their re-development and refurbishment of existing properties for the residential market. One area that has seen increasing activity in recent years is the conversion of redundant industrial buildings to residential and live/work units. Examples include factories, warehouses, schools and wharf buildings that may have stood unused and neglected for many years but are now prime candidates for development into prestige living accommodation. In such developments, the facade or shell of the building is usually retained and renovated, while the interior is developed to varying degrees depending on the options being offered to potential buyers. These can range from supplying an empty shell with basic services installed for the owner to complete, through professionally designed fitted apartments, to top-of-the-range penthouse accommodation. A key selling point for this type of property is the inherent space, light and flexibility they offer, particularly with the open-plan loft apartments that are becoming increasingly popular. The developers are also selling an image, a contemporary life-style, and will often look to create a sense of community and belonging within the property.

The infrastructure and services offered by the digital home are well suited to support this type of development. The inherent flexibility of how services can be accessed and distributed within the apartment can complement a changing, open-plan living space. It can also be used to support inter-apartment and shared services within the property as a whole, such as local information systems, community Internet sites, voice and data communications and distributed entertainment services (including digital TV and shared libraries for DVDs, videos and music CDs). In conjunction with a property management or concierge service, the residents may also benefit from integrated security services, energy management savings and reduced-cost data communications.

11.3.3 Facilities Management Companies

The facilities management company is responsible for managing the property or site on a day-to-day basis and supporting the needs of the residents. This will typically include the provision and maintenance of common services, such as concierge facilities, security, energy management, distributed entertainment services, fault reporting, community information and Web site, and communications with residents. These services will require controlled IT communications to and from individual homes and, on larger development sites, within each accommodation block.

Central security services (for example CCTV and electronic locks) can be provided to monitor and control access to building, courtyard and car park entrances, and linked to individual homes if appropriate. The facilities management company may wish to act as a focal point for obtaining remote home-metering information (electricity, heating and water) so that beneficial rates can be obtained from a preferred utility supplier, or, in the case of a development running its own combined heat and power plant, directly bill residents for their energy use. Entertainment services such as satellite, terrestrial or digital TV may be fed to individual homes in a controlled and consistent way, and also reduce the need for multiple dishes or antennas that could spoil the look of the development.

Effective methods of direct communication between residents and the facilities management company are key to the smooth running of the development. Ease of access to key contacts and systems is essential for reporting specific service faults or concerns, keeping residents informed of community issues and notices, and addressing individual account matters. Where appropriate, a range of communications channels could be made available to suit the residents' particular needs or preferences (e-mail, Web, mobile, SMS messaging, electronic notice boards, chat rooms, etc). Access to these channels could be from devices connected within the individual homes (e.g. personal computers, set-top boxes), from mobile handsets, or from shared facilities such as a local Internet café or computer learning centre.

11.3.4 Housing Associations

Housing associations are very interested in the benefits, value and opportunities that digital home technology can bring to their business and their tenants. There are over 2000 Registered Social Landlords in the UK with responsibility for more than one million homes. These are non-profit-making organisations that own and manage homes to provide quality housing to people in need. The associations work closely with local authorities to achieve this and are often involved in a range of urban regeneration and property improvement programmes. In order to meet the demand for housing, the associations also undertake the building of new homes and the transfer of existing housing stock from local authorities.

In addition to providing somewhere affordable for people to live, housing associations are also very much concerned with supporting and developing their tenant communities. This is being achieved by providing the tenants with access to local services, training and education opportunities so that they can learn new skills, develop as individuals, and get themselves back into the job market. The associations can provide a focal point for local regeneration within the community and help combat the social and digital exclusion of their residents.

The IT services and facilities potentially available within the home are seen by the associations as a key enabler to supporting urban renewal, self-development and inclusion. Access by tenants to information and communication services from their home can have benefits in a number of areas.

- Communications between residents and the housing association

  Some examples of how the facility can be used include reporting and tracking building faults and repairs, dealing with individual rental account and service queries, advising on the availability of vacant properties, and passing on news and advice on matters having an impact on individuals or groups of residents. These facilities can also be used to encourage and enable tenants to become more involved in the running of their communities, such as participating in regional forums and residents' associations.

- Communications between residents and third parties

  The facility provides access to local government information and a range of education and training material to aid individual re-skilling and job opportunities. This also relates to the on-line publication of community information such as what's on, transport information, local news and deals in local shops.

- Communications within resident communities

  The resource can support residents' associations, regional forums and special interest groups, but also communications of a more informal nature such as chat rooms and self-help groups. This type of communication can also help to identify and encourage the involvement of local champions who understand the problems of their community, have the support of the people, and act as a guide for the implementation of new services or initiatives.

Housing associations are constantly looking to provide improved services to their residents and this can take the form of property refurbishment (Figs 11.3 and 11.4) or new builds. As housing stock is transferred into an association, say from the local council or another housing organisation, the opportunity is being taken to identify and install digital-home-related services during the refurbishment work.

These activities are also being undertaken in the wider context of urban regeneration programmes and in collaboration with home builders, developers and local authorities on major building projects. Indeed it is now common for such developments to include a percentage of social housing which is integrated with, and to the same specification as, the private owner properties.

Two further areas of interest to housing associations are Telecare and the use of home technology to ensure individuals without bank accounts are not excluded from using IT-based services.

Fig 11.3 Property refurbishment — before.
[Courtesy Glasgow Housing Association]

Fig 11.4 Property refurbishment — after.
[Courtesy Glasgow Housing Association]

- Telecare

  The caring home or Telecare concept looks at how technology can be used to enable an increasing number of elderly and frail people to continue to live independently in their own homes. The Millennium Homes initiative [1] uses a variety of environmental sensors and devices, connected to a computer, to monitor the state and activity of the tenant. A number of functions can be monitored, such as movement within a room, if a bed or other furniture is occupied, the state of doors, locks and appliances, and whether gas and water taps are safe. If the system detects a potentially dangerous situation and cannot gain a suitable response from the individual, then it can call for assistance.

- Smartcard technology

  A common characteristic of the social housing community is the predominant use of cash and the lack of individuals holding bank accounts or credit cards. In order that individuals are not excluded from using IT-based services, particularly those that make use of eCommerce, the housing associations are interested in the potential use of smartcard technology to support the concept of an electronic purse and electronic payments. They are also attracted by the multiple-use capabilities of the smartcard to support tenant information, service access and loyalty bonus schemes.

In April 2001 a consultation was organised by the Building and Social Housing Foundation to consider how information and communication technologies could be harnessed for a more sustainable future [4]. The report included the following statement:

> '[The] emerging digital divide follows existing divides between poverty and wealth, sickness and health, knowledge and illiteracy. Whilst technology will not solve these basic problems, it does provide a tool that can be used to help reduce these divides and to provide opportunities for more sustainable development.'

It is therefore important to remember that technology in itself will not solve community problems and there must be an inherent need and benefit for it to make a difference. In the context of urban regeneration we cannot expect to solve the issues of digital exclusion by simply parachuting in the technology and hoping for the best. There must be a reason for it to be there, an underlying need, an overall benefit and — most importantly — it must be sustainable.

11.3.5 Local Authorities

The interest by local authorities in the digital home, and the underlying technologies and services, is very closely aligned with that previously described for the housing associations and will not be repeated here. Both are involved in providing housing for those in need, property refurbishment, urban regeneration programmes and bridging the digital divide. They also have the additional driver, set by the government, of delivering services electronically by 2005. The implementation of information access systems within the digital home, whether they are within private homes or social housing accommodation, will therefore provide a key route into their on-line initiatives.

One of the leading authorities to implement the electronic delivery of services is Bracknell Forest Borough Council. BT has been working with Bracknell Forest on the use of smartcard technology to support a wide range of services, including council tax payment, car parking, library loans, school meals and leisure facility bookings. Money can be stored on the card, together with card holder information for personal identification, loyalty schemes and access to the Council's intranet site [5].

In a separate smartcard initiative, parts of England will soon be able to vote electronically in local elections using mobile telephone text messaging (SMS), touch telephones, local digital television and via the Internet using home computers, and from local libraries and council-run information kiosks [6].

11.3.6 Home Owners

The ultimate users of the digital home are the private home owners and tenants. We have already seen some of the benefits and uses of the digital home technology and how it can support home working, home entertainment and home control. Detailed findings of how families make use of the technology in the home are presented in Chapter 10.

Different people will want the benefits of the digital home for different reasons. Whether it provides an essential aid to saving time, money or hassle, or whether they just want to have the latest technology around them. Whatever the reason, there must be a benefit and an associated value.

Once the key use or uses have been identified for implementing the services of a digital home, there are other factors that have to be addressed before it is taken up by the home owner. Such issues include complexity and disruption.

- Will I need a degree in computer networking or information technology before I can install it?
- Will I have to take up all my floor boards and remove the plaster from my walls in order to install the cabling?
- How do I get my old Windows 95 PC talking to my new Windows XP machine?
- Where do I go for support when it goes wrong?
- Are there any hidden costs?
- Is it future proof?

These fears can be addressed and broken down with the right support, advice and products. As in all areas, there will be enthusiasts and early adopters who are already installing technology into their homes and who have the time, knowledge and inclination to create bespoke solutions to meet their needs. However, for the mass market, the digital home will need to be cost effective, easy to install, easy to use and easy to maintain.

Three of the biggest benefits that the digital home can bring are flexibility, peace of mind and saving money:

- flexibility — services that are where you want them, when you want them and that can be upgraded or added to over time;
- peace of mind — added security when you are at home or away, knowledge that your family and property are safe, and independence for the old and the frail;
- saving money — time-shifting activities to benefit from cheapest energy supply, monitoring energy use, reducing waste and controlling your home climate.

11.3.7 Social Housing Tenants

We have already identified how housing associations and local authorities are looking to use elements of the digital home to help bridge the digital divide in their tenant communities. To recap, on-line and mobile communications from the home can provide tenants with access to local services, information, training and education material, and the real opportunity to become actively involved in decision making and running their community. Inclusion in every sense — but the biggest question here is not who will use the services or what they will be used for, but who will pay.

Within a community of people in need, on low incomes, receiving benefits and with little or no disposable income, it is very unlikely that having a digital home would figure highly, if at all, in their list of priorities. However, the housing associations and local authorities can act as a catalyst in the introduction of digital home services within their tenant properties as a means of improving social inclusion and improving their business processes. This would require careful consideration of a sustainable business model where service costs are subsidised by the association or authority, either directly or in conjunction with third party partners or suppliers. Some tenants may also be prepared to pay a small premium over the standard rent for certain services, e.g. if it provides them with additional security or personal care. Costs may also be offset against savings made elsewhere, such as in energy costs or the reduced need for in-home carers.

11.4 From Discovery to Delivery

Having identified the different customers, and some of the features and benefits of the digital home, how can we integrate their requirements to deliver an effective, integrated solution? An approach successfully used by BTexact Technologies is presented below.

- Business strategy and technology route map

  Most organisations have a vision and a business strategy, but are often unclear on how to align these with current and future technology. This stage helps to develop the strategy and sets out a technology route map for its implementation. A business, customer and user focused approach is taken to address the task from two viewpoints:

  — when the technology will be available to deliver our business strategy;

  — what the business opportunities are that arise from new technology.

- Demonstrators and prototypes

  Human factors and development expertise is called on to provide concept demonstrators and prototypes of emerging ideas and solutions. These offer rapid ways to visualise, implement and test physical solutions to business and technical issues.

- Development and delivery

  This stage takes the business requirements or concept demonstrators through to detailed solutions design, development, delivery, integration and support. BTexact Technologies has implemented some of the most complex and leading-edge solutions in the world, and it can call on this practical knowledge and experience for its clients.

- Technical and commercial research

  BTexact has a long history of leading technical research across a wide range of disciplines, backed up by an extensive knowledge of IT networks, systems and solutions. This is complemented by in-depth commercial and human factors research to help understand customer issues and identify new business opportunities (see Chapter 8).

- Partnerships

  The size, complexity and nature of the business means that no one company can address all the issues alone. BTexact uses its extensive network of contacts to form technical and commercial relationships with other BT business units and external partners to deliver and support integrated solutions.

This ‘discovery to delivery’ approach is currently being used on a major housing stock transfer and renewal programme with Glasgow Housing Association (GHA). After transfer, GHA will become the biggest Registered Social Landlord in Britain, with over 80 000 tenants at the point of transfer, and will spend over £4 billion over the next 30 years on tenants’ homes and the environment.

Key elements of the programme will be delivered by a Neighbourhood Renewal Action Plan which includes 'Glasgow@Home', an initiative to counter social exclusion through the provision of information-technology-based services and facilities to its tenants.

BTexact Technologies is helping GHA to set the strategy and direction for 'Glasgow@Home', identifying associated initiatives, services and solutions within Glasgow and across Scotland, and preparing a prospectus for submission to potential partners, suppliers and fund holders. An inclusive approach is being taken to define a business and technology route map for 'Glasgow@Home' that will have a significant impact on the lives of the individual residents and the local community.

11.5 Summary

Although homes have basically changed very little in recent years, technology is now becoming more readily available to support a range of home working, entertainment and control services that could transform how we live. Taken in isolation, any one of these services may not be enough to justify the installation of a home network. However, when combined, their benefits, value and attraction begin to add up, particularly if clear savings can be identified in terms of money, time and convenience.

Initial thoughts that the potential customers for digital homes are either 'executive' private home owners or technically proficient IT enthusiasts also have to be reconsidered. A whole range of interested customers have been identified, from home builders and developers through to housing associations and their tenants. The relationship between these different customers can be complex and this will have a major influence on the success, timing and take up of the digital home.

The key issues are not so much the technology itself, but the benefits it can bring and who will pay for it. In addition, technology for the digital home will need to be cost effective to buy, easy to install, easy to use and easy to maintain.

It is clear that the services offered by the digital home can provide tangible benefits to individuals, families, communities, businesses and organisations. Such services are not just for the rich, but for everyone in society, right down to the poorest and the socially excluded. Indeed, the major renewal and regeneration programmes being undertaken by housing associations and local authorities across the country may provide the 'kick start' needed to take the digital home into the mass market. They can provide the drive, focus, economies of scale and demand to stimulate significantly the emergence of the digital home.

In conclusion, the marriage between technology and the home can result in major benefits for everyone — 'from this day forward, for richer for poorer'.

References

1 Wolff, Prof. H. et al: '*Millennium homes: a technology supported environment for frail and elderly people*', Brunel Institute for Bioengineering (2000).

2 Hussey, S.: '*Building the wired home*', BT Technol J, **20**(2), pp 49-52 (April 2002).

3 CREC/KPMG White Paper: '*Smartcards — enabling smart commerce in the digital age*', — http://cism.bus.utexas.edu/works/articles/smartcardswp.html

4 Building and Social Housing Foundation Consultation: '*Harnessing information and communication technologies for a more sustainable future*', (April 2001) — http://www.mandamus. co.uk/bshf/

5 Bracknell Forest Borough Council: '*Bracknell Forest — the edge smartcard programme*', — http://www.bracknell-forest.gov.uk/edge/card/edgesmartcardprogramme.pdf

6 Department for Transport, Local Government and the Regions: '*May elections to trial online voting*', News Release 033 (February 2002) — http://www.press.dtlr.gov.uk/pns/newslist.cgi

Part Four

APPLICATIONS

S G E Garrett

This part outlines some of the issues relating to applications.

There have been great advances to the technologies used in homes. Digital computing, signal processing, communications, display technologies, etc, have all changed beyond recognition. Some of these changes have been described in preceding chapters.

However, technology alone will not change our lives; it is the *applications* that use the technology that will have most impact. Later on in this part, there are two chapters describing actual application technologies — 3-D virtual presence that can transform person-to-person communication, and new technologies for creating multimedia content.

But before that, there are two chapters describing some of the fundamental architectural issues about how to distribute application functionality in a distributed system. These chapters consider applications of *client/server* and *peer-to-peer* computing architectures. In this introduction, these terms are defined and explained.

In the era BC (before computer-networks!) issues of distributing application functionality did not arise — an application ran on a single computer. There was no other choice. For many of us, most of our computing experience lies in using PCs; very often PCs are not networked (especially in the home), or only limited use is made of the network. In most cases, an application runs entirely either on the PC (e.g. a word-processor application), or on a remote host (e.g. a Web server application). There are disadvantages to this 'either/or' approach. For example, applications running on a remote Web server can be rather slow. Writing a complex picture on the screen takes forever, as the remote host is sending complicated information over what is often a slow communications channel. Applications involving complex graphical images or complex interaction with the user are likely to be faster if run on the local computer. But applications that need centrally held databases (e.g. most Web servers) cannot be run locally, as the information is not held there. Hence, Web servers are inherently remote applications.

In recent years, many applications have begun to be run as *distributed processes.* Different parts of the application software run on different computers, possibly in different locations. This may be to locate the application parts in optimum locations (close to data or to the user, say) or simply to share a large processing task among multiple processors[1]. Architectures for distributed computer applications fall in two main categories: *client/server* and *peer-to-peer*. These terms are explained in outline here. It should be noted many practical systems are not pure implementations of one or the other category, but some sort of mix of the two architectural concepts.

- Client/Server Computing

  Most distributed processing systems are client/server. This is essentially asymmetric — the client controls the application, requesting services from a server. In general, many clients may be requesting services from one server (of course, the client may request services from multiple servers, too, but clients usually outnumber servers). The server(s) manage access to one or more resources, which can be accessed by one or more clients. The main advantage is that servers are specialist machines that centralise access to their services and resources. They can be optimised for their task, and are often built with high-speed hardware. They can control access (thus minimise security issues). Clients usually do not have direct access to the resource (e.g. data), but access it in a controlled manner. The disadvantage is that servers tend to be expensive. They are often built with high-performance hardware, and often need significant administration time. They can also be difficult and time consuming to manage.

- Peer-to-Peer Computing

  The principal difference between this and client/server computing is that peer-to-peer (P2P) is symmetrical — clients request services from other clients. If you like, clients behave as both clients and servers. Peer-to-peer is sometimes defined as the sharing of computer resources by direct exchange between client systems. The main advantage is low cost — client systems are invariably cheaper than hosts, and thus peer-to-peer is making use of cheaper resources. Also, because there is not the same concentration of resource, better use is made of

---

[1] An example of the latter is 'SETI@home' — a project that is analysing radio telescope signals in the hope of finding signs of extra-terrestrial intelligence. There is a vast quantity of radio telescope data, and the analysis would take a single computer many centuries. A project has therefore been set up at the University of Berkeley whereby people can download a small fraction of the data and a programme to analyse the data. This runs as a screen-saver, thus using spare time on users' PCs. This is a distributed computing project involving many thousands of volunteers' PCs (see http://setiathome.ssl.berkeley.edu/).

communications resources. They are also cheaper because, in general, there is no system manager or administrator. Also there can, in some cases, be better performance, as there is not the same concentration of communications traffic to one server. There are two main disadvantages of peer-to-peer — location of services and security. Because resources are more widely distributed and not centrally managed, how do you find out where they are? And it becomes much more difficult to work out whether a request is from a *bona fide* client. The location problem is sometimes tackled by providing directories — essentially, using a hybrid system with servers just to help you find out where resources are located. Such a directory, or index server, may also be used to help validate requesting clients to the clients hosting a resource.

Most of the communications in the digital home depend on distributed applications. The functionality and usability of the digital home depend on well-designed distributed application architectures. The following four chapters that make up this part give some insight into the work currently going on to address distributed applications architectures.

Chapter 12
Clients, Servers and Broadband

The first chapter in this part is about *client/server* architectures. It has been said that '... broadband is more bandwidth than you can use'. In a world where the unspoken assumption is that providing more bandwidth will solve Internet access problems, these are provocative words. But the historical evidence shows that adding bandwidth merely increases demand — analogous in many ways to the building of more motorways having the effect of increasing traffic problems instead of alleviating them. The result is that the world's appetite for bandwidth continues to outstrip the supply, and the only apparent solution is a spiral of ever-greater demands for bandwidth.

To find an equitable solution to this conundrum, the whole problem itself was re-assessed, to see if there was an alternative. It is now believed that a combination of technologies can offer a scalable solution to providing the services that people evidently want, but without requiring unfeasibly large bandwidths for network access. This chapter describes a solution, which uses a mixture of client/server, client-side, and peer-to-peer approaches to provide a novel answer. By making better use of client capabilities, we can also make better use of available bandwidth to the home. With mobile computing putting new constraints on the likely bandwidth, a technique that allows us to exploit bandwidth effectively could be a powerful weapon for the future. It also allows broadband to reveal its true potential.

Chapter 13
Bandwidth-on-Demand Networks — a Solution to Peer-to-Peer File Sharing?

An important driver for the take-up of broadband services into the home has been peer-to-peer file-sharing programs like the now-defunct Napster, which enabled users to share music and video files freely. This puts network operators and service providers in an awkward position as they are under pressure to shutdown or limit access to such applications. The majority have been operating illegally without the permission of the content owner.

Current activities to protect copyright focus on either keeping the contents secure (digital rights management) or tackling the distribution mechanisms via the courts. Neither of these are complete solutions. This chapter describes a third way which utilises bandwidth-on-demand network capabilities to make content sharing more controllable. Although this will not prevent video piracy completely, it should discourage it sufficiently to minimise the problem. This should also appeal to network operators, as bandwidth is still relatively expensive. The network solutions developed to prove the concept build on the capabilities in the current ATM/ADSL network but an option for an all-IP network is also described.

The techniques described may also have a general application in providing a pay-for-what-you-use mechanism to allow home users to buy higher bandwidth when they need it.

Chapter 14
A 3-D Visualisation and Telepresence Collaborative Working Environment

One of the most far-reaching changes brought about by new digital technology in the home — and especially by broadband connection to the network — is the increasing opportunities for home working. Most people work in teams, and the need for communication within the team meant in the past that teams had to work in the same place, usually in an office. Growing communications capabilities — telephone, fax and more recently Internet connection — have meant that more and more people can spend a greater proportion of their time working from home. Web conferencing can provide not only voice capabilities but also a white board on the PC screen and document sharing. Now new virtual-presence techniques are meeting still more of the needs for team working.

This chapter describes one such development — an integrated media system consisting of a computer-based environment that supports the creation, sharing, distribution and effective communication of multimodal information across the boundaries of space and time. The work has been carried out under an EU IST project — VIRTUE (virtual team user environment), which is working steadily towards the realisation of most aspects and properties of such a system, with particular emphasis on a three-way semi-immersive telepresence videoconferencing scenario. In contrast to the traditional videoconferencing system that we know now, the outcome of the project is expected to demonstrate distinctive presence features and experience for the conference participants. These include views of full body sized realistically rendered images, eye-to-eye contact, gaze awareness, normal hand gesturing and direct body language among others. The purpose of this chapter is to reveal the current efforts in its related technical field, and the main objectives and scope of this project. One optional software system framework is outlined, illustrating some component technologies in 3-D computer vision analysis that are being developed. The application of these component technologies, notably the dense disparity estimation and the novel view synthesis in 3-D interactive video manipulation and visualisation, is widely expected.

Chapter 15
A Virtual Studio Production Chain

New digital technology in the home, and broadband network connections to the home, offer unimagined opportunities for new applications. For the home, entertainment applications appear to offer the most profitable opportunities for product and service providers. Creating multimedia entertainment content and applications is vital, and developers are looking for new tools to create this content.

This chapter describes work to develop a virtual studio production chain within the collaborative 'Promethus' project. It discusses the technologies that are being investigated and, in particular, the research in progress in the content and coding laboratory within BTexact Technologies. Also presented are the historical changes in the production of real and animated performances and how Prometheus and current trends can have an impact upon this. Two areas are identified where further study is required to ensure believability and coherence when virtual performances are created from reusable digital components.

12

CLIENTS, SERVERS AND BROADBAND

M Russ and M A Fisher

12.1 Introduction

'Broadband is more bandwidth than you can use', Mark Bagley, Venation, 1999 [1]. In a world where the unspoken assumption is that providing more bandwidth will solve Internet access problems, then these are provocative words. But the historical evidence [2] shows that adding bandwidth merely increases demand.

Let us explore the metaphor of a castaway on a desert island (Daniel Defoe's fictional 'Robinson Crusoe' is a good example of the genre [3], or the true story of Alexander Selkirk [4]). Today's browser is the island, and it is one where time passes very slowly for the castaway. The castaway spends most of his time waiting, with very infrequent snippets of new information arriving on the tide as flotsam and jetsam. Communication with the rest of the world [5] does not happen much, and when it does, there is every chance that the castaway will not be noticed — maybe they cannot even get the fire started, or the leaves are not wet enough to generate thick smoke (Fig 12.1). Not quite the hi-tech image that the computer world would like you to have in your mind, is it? Is BTexact Technologies [6] providing the rescue ship?

So why is the combination of a browser and broadband access to the Internet not the compelling answer that many people have assumed that it would be [7]? There are a number of reasons that have been proposed for this:

- bandwidth — 'more is obviously better ...'
- transactions — 'forms are user friendly ...'
- content — 'content is king ...'

Viewed in context, each of these is only part of the answer. Getting bits into a browser faster might allow the faster downloading of files, and it may even make it possible to stream video on to the screen, but are these the real bottle-necks? If the actual problem has more to do with the fundamental divisions of functionality between a browser and the Web page server to which it is talking, then forms may not actually be the correct answer. One of the major reasons why 'surfing the Internet' can feel slow, prompting misappropriations of the 'World Wide Web' into the 'World Wide Wait', is that it takes a long time for all of the 'toing-and-froing' of transactions between the browser and the server. In the course of these transactions, images, scripts, text and links are gathered together, often from other Web sites. This can be rather like trying to tell a person how to prepare a meal over the telephone: 'Have you got any onions?' 'Hang on, I will go and look.' 'Have you got any garlic?' 'I think there is a clove somewhere around. I will poke around and see if I can find it. There is only one onion, by the way'. 'Ah, you really need two onions — can you pop out and get another one?' While generating long telephone calls, this method is not very efficient, and it ties two people into what can become an increasingly frustrating experience.

Fig 12.1 The island.

What you really need is some way to avoid all the multiplicity of transactions, and to get away from thinking that the process has to be a series of interactions. A recipe might be the way to achieve this in the case of a meal. For the browser, we need to be able to get the browser to do more of the work, rather than just doing the presentation of the questions to the user on behalf of the server. Giving the browser some autonomy is not the whole answer, of course. Not only do you need to change the way that you present the information in the browser so that the browser/server interactions can be reduced and simplified, or maybe even removed entirely, but you also need to make the information itself much more self-contained and self-reliant. Making two trips to the pantry to find out if there is any garlic and onion in the house can be avoided if the recipe knows that there is a list of items that are needed, and that they all need to be brought together at the right times if the recipe is to be completed successfully.

Rather than a recipe, the more familiar instance of this on the Internet is the form (Fig 12.2) — where filling in the fields correctly can often involve guessing the format required, and where the response to an incorrect answer is a message pointing you back to the form, asking you to fill in the field correctly.

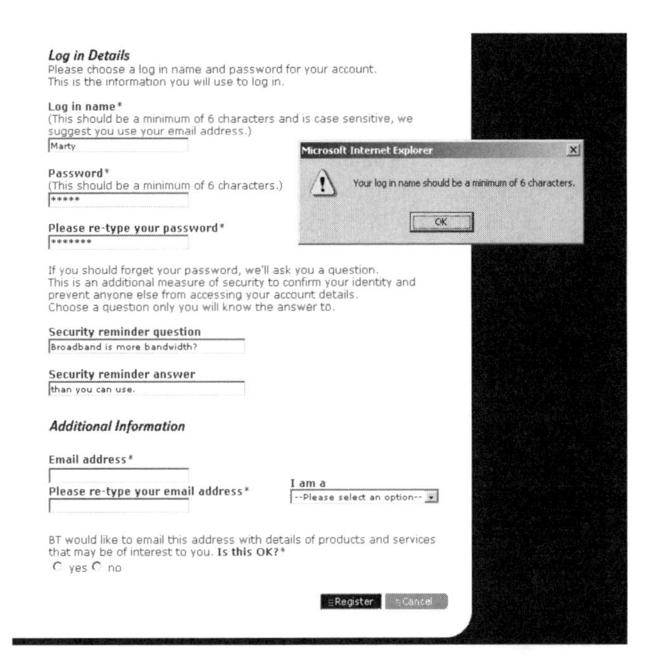

Log in Details
Please choose a log in name and password for your account.
This is the information you will use to log in.

Log in name*
(This should be a minimum of 6 characters and is case sensitive, we suggest you use your email address.)
Marty

Password*
(This should be a minimum of 6 characters.)
*****

Please re-type your password*
****

Microsoft Internet Explorer
Your log in name should be a minimum of 6 characters.
OK

If you should forget your password, we'll ask you a question.
This is an additional measure of security to confirm your identity and prevent anyone else from accessing your account details.
Choose a question only you will know the answer to.

Security reminder question
Broadband is more bandwidth?

Security reminder answer
than you can use.

Additional Information

Email address*

Please re-type your email address*

I am a
--Please select an option--

BT would like to email this address with details of products and services that may be of interest to you. **Is this OK?***
yes no

Register Cancel

Fig 12.2 A form from bt.com — while forms are gradually employing techniques that enable effective use of the client's processing power, the usage of 'client-side' processing in a broader context is less widespread.

Another aspect to this lies in the subtlety of the interchange. Painstakingly going through a list of required ingredients could be replaced by something much more akin to the complex exchanges in a conversation. So the recipe might be transferred from the Web page server to the browser, which then presents the required ingredients to the person. In the course of the gathering of the ingredients, the person might be unsure how to prepare the garlic, which requires a second interchange with the Web page server to get the summary of garlic pressing techniques. The summary, by the way, includes a short video, which is sent in the background as soon as the summary is requested. The server is also preparing the 'How to stop your eyes watering when peeling onions' tutorial for subsequent transmission, because there is a good chance that it may be needed. What you have now is something much more closely resembling a conversation than a series of robotic transactions. This is also very different to just filling in an interactive on-line form. There is a lot of implication, a lot of inference, a lot of preparation in anticipation, and more. Also this conversation is related to the content, as well as the content to the conversation. So communications and content are not separate entities — they are entwined together into a threaded dialogue that results in a meal being prepared, rather than a series of questions without a context.

All of which points towards the true 'browser of the future' [8]. In appearance it might well be very similar to today's busy screenful of information, but, in terms of the inherent functionality, it will be very different. It will be used to present contextualised fragments of content that are self-aware (I'm an onion, and there are several other ingredients) and self-organising (the recipe needs garlic, onions, mustard, pepper, and stock, for instance — and it seems that you've never used a garlic press before ...). Because the detailed interchanges between the person and the browser (or the server and its resources) are dealt with locally, there is no need for the time-consuming individual gathering of specifics in a fixed sequence. Instead we can have a rapid and flexible assembly of ingredients, technique guides and cooking instructions in a way that minimises the dependence on the transaction timings, and maximises the usability.

Also note that this interchange is not just simple form-filling, where you are asked about onions, then garlic, then mustard ..., but is now much more oriented towards higher level concepts like how to work with the ingredients, how to prepare them efficiently, how to present the results, etc. In browser terms, this is not just a presentation of a single type of media, or one type of interaction, but several appropriate interchanges in context, or maybe even something which you could encapsulate as 'Inept cook learns how to use a garlic press...' and distribute to others. The broadband connection is used to send appropriate bits of content, complete with information about its context and its relations to other bits of guidance, and waiting time is no longer something that you are aware of. This is what the browser of the future will really look like. Indeed, something called the 'X Internet' is already being discussed by technology analysts like Forrester [9], and emerging next generation Web-aware programming environments, like Curl [10], encompass the same viewpoint.

What about our castaway? Well, it turns out that a very shallow (and pleasantly warm) sea surrounds the island, with lots of other islands that are easily reachable by wading for a few hundred metres or so (Fig 12.3). The eco-friendly fire is built from furry leaves which attract the morning dew and so are self-moistening, while the fire itself is triggered by a hi-tech doppler radar-activated ignition system. A Web cam has turned the castaway into a media star with fame ready to envelop him when he chooses to leave the island. It does seem that Andy Warhol's '15 minutes of fame' has now become commercialised [11], with appearing on TV seeming to be an obsession with large numbers of people. 2000 and 2001 saw a large number of TV programmes designed to exploit this desire to become famous by appearing on TV — Big Brother, Survivor, The Mole, Castaway, Pop Stars, and more. There is even a Star-Trek [12] influenced transporter-fridge that can be stocked with onions and garlic whenever the castaway thinks about cooking something from that recipe

Fig 12.3 The islands.

Web page that Friday sent the bookmark for ...; Friday is, of course, busy assembling a guide to 'Castaway cooking' to add to the Web site, which is read avidly by a large number of people [13].

12.2 Implementation

With the concept defined, what about the reality? It turns out that most of the required functionality is already available, but it is just not in the right place. Moving functionality around sounds hard, but often is not. Each of the required elements — content, client/server and network — are discussed in turn.

12.2.1 Content

The content is probably both the best and least understood part of broadband delivery. Having the ability to download or stream large data files quickly, and immediately, opens up thoughts of movies or high-quality music, but this perpetuates one of the intrinsic assumptions: that content is just the raw media. We will miss a huge opportunity if we continue with the premise that content has to be delivered as nice tidy large-scale blocks of media information — movies, CDs, TV programmes, etc. On the large scale, broadcasting content is well understood, although there are some challenges for broadcasting over an IP network instead of a dedicated SDH network. This large scale is all about big packages: think of it as War and Peace, the movie and the book, plus the TV 'Making of ...' programme — and these are things that are sent from a few publishers to a lot of subscribers or viewers. On the small scale, however, content is rather more like telephone calls:

there are lots of them, they do not last for very long, and they involve one-to-one conversations, or interchanges between small groups of people. This is much more oriented around scene-based interaction and description — you do not talk about War and Peace, instead you say 'you know that bit where the...'

Making the transition from broadcasting big lumps of content to interchanging small fragments of content is the first challenge. It also disintermediates from a single source of content to potentially large numbers of sources, rather like the way that some people use second-generation Napster [14] network-based file-sharing utilities like Gnutella [15] and Kazaa [16] to provide the source material for personal 'best of' collections of media rather than pre-prepared collections. Most of this activity is currently based around music tracks, but video is increasingly becoming a similar unit of trade in this area.

The nature of these fragments is also becoming better understood. They will contain the data traditionally thought of as the content — the sound, pictures and text, for example. As delivery of content becomes less monolithic two more features become important. Firstly, the fragments will become capable of presenting themselves and combining with other fragments in a variety of ways. This introduces some intelligence and makes them begin to resemble software components. Secondly, the fragments will have associated descriptions (or metadata) which will allow users, using appropriate software tools, to select and compose a set of content elements into a coherent whole. Metadata is often seen as a tool for describing complete items of media, but, when used in this context, it becomes the glue for assembling the fragments.

12.2.2 Client

Once content evolves beyond simple blocks of data, the key questions to be considered are where it should be manipulated and transformed. At the moment, nearly all the presentation-related processing of content takes place on the client. All the blocks of data are collected on the user's computer and combined.

A Web browser is a good example of a piece of software that has been designed to integrate lots of disparate fragments of content into a composite whole. Ten years ago, specialist software was used to help magazine designers do page layouts [17], and very sophisticated software was required to produce the sort of animated multimedia experience that was obtained from CD-ROMs. The browser has put astonishingly complex page-rendering capability on to just about every computer screen, and with the micro-browsers found in mobile devices, on to a large number of small, portable, and personal screens too. Web browsing has become such an integral part of the way that computers are perceived by their users that some operating systems now have browser functionality built into them [18, 19]. Browsers are frequently referred to as being thin-client technology, implying that the client is a simple, minimalist device that serves just as a display device for content delivered by a much more powerful server device. In fact, browsers might be imagined as being thin, but they are large, complex and increasingly sophisticated pieces of software. Some devices will always be limited in processing power and storage, making them incapable of supporting such sophisticated software. This indicates a need to share the responsibility for composing and delivering content with other networked devices. The limitations of the final display device will influence the choice and format of content elements but manipulation may be better performed elsewhere, where the resource constraints are different. This begins to blur the distinction between client and server. In addition, a broader view needs to be taken of what a server is.

12.2.3 Server

Traditional Web servers generally provide a uniform service to users — they 'serve you with a Web page on demand.' Simple HTML applications behave in the same way, returning the same Web pages regardless of the type of browser requesting content. Dynamic Web page generation is now beginning to find a role by allowing the separation of the information in the Web pages from its presentation. A presentation-neutral representation (such as XML) is combined with a mechanism for generating custom HTML pages on the server. The information can then be matched to the characteristics of the client — whether it is a PC or a WAP phone, for example.

The way a user interacts with traditional servers is still quite inflexible. A server typically provides access to a specific set of information and this is delivered under the control of the provider, who has no knowledge of what the user wants to do with it. What we really need is a richer infrastructure for content and services so that users can define their own requirements and arrange for the necessary content fragments to be identified, manipulated and delivered.

A highly flexible approach to providing such an infrastructure is a network of general-purpose computing platforms. This is the essential vision of the 'Grid' — a coming-together of several approaches to distributed computing which is currently attracting a large amount of interest and public funding in both the USA and Europe [20]. While much current Grid work is concentrating on 'big science' projects linking supercomputers to address problems in particle physics and genetics, the implications of the technologies being developed are much broader. The target is to link together computing resources so that from a user's perspective the network appears to be a single computing platform with practically unlimited resources. There are few really new ideas needed to make the Grid vision feasible. Rather it is progress in high-speed networking and in computer processing speeds that are reaching the point where existing ideas can be applied on a global scale.

Various different kinds of computer are expected to form part of the shared network infrastructure. Specialised combinations of hardware and software would be hosted at a small number of locations and offered as a distinct service. There are also cases where it is not the specialised nature of the service that is important but rather the location. When the network is heterogeneous, then the locations where different parts of a distributed application actually 'run' become important.

General-purpose computing resources, such as CPU, memory and networked storage, would also be made widely available to provide computing power on demand. The major problem with deploying such an infrastructure today is management — how to make sure that users can be allowed to run useful programs on shared computers safely and without interfering with each other. This is an area in which BTexact is developing some promising solutions.

12.2.4 Network

The near-universal acceptance of IP as the basis for communication of data between devices makes it possible to view networks across the world as part of a single system [21]. When browsing the Web, users perceive that all sites are basically the same — they can all be accessed in the same way. When they browse the Web from a mobile terminal, the expectation is of a similar experience. One of the appealing features of the Web is the illusion of uniformity and simplicity, which stems from the use of common protocols such as TCP/IP and HTTP.

In fact, there is great diversity in the underlying network technologies which is generally hidden from users, and even developers, of networked services (for example, ATM is widely used for transporting data which appears to be carried via TCP/IP as far as the end connections is concerned). Traditionally, the Internet has been regarded as providing a mechanism for transporting packets between hosts. There is considerable advantage in hiding the full complexity of the network and working with this simplified abstraction. This is directly analogous to the use of high-level programming languages in place of machine code for programming computers. However, the characteristics of the underlying networks are becoming much more diverse. A typical application will use a set of network connections where the maximum bit rate ranges over several orders of magnitude. In addition, variations in other important parameters such as latency, jitter, congestion and error rate can have a serious impact on the perceived performance of a networked service [22].

Hiding this variation from developers will seriously limit their ability to produce attractive, reliable applications. However, just making this information visible is not enough in itself. The ability to make use of this information to optimise performance must be made available. Here the word 'performance' is being used in the sense of a combination of latency and transmission measures, plus the client and server processing times, and not the often-quoted capacity measures. The ability to program shared-network nodes allows this problem of bit-rate variation to be solved. In this case, it is the location of the processing node that gives an advantage — allowing, for example, different protocols to be used to traverse different parts of the network.

12.3 Putting it all Together — the Future

This chapter has presented a view of what the network infrastructure might look like in the near future. Client terminals with a wide range of capabilities are supported by a rich set of shared computing and content resources. Each time a service is used, a custom combination of content elements, processing and network connections is used. These parts all need to work together to provide a good quality of service to the user. Trade-offs will need to be made between computation and communication depending on which resources are available. The ability to find the best resources is essential to take advantage of a flexible network (Fig 12.4). Server-side caching and content distribution networks are playing an increasingly important role in the Web. This means that the same fragments of content will typically be stored in many different locations. The metadata describing each of the fragments will allow us to enhance Web search techniques so that the right set of information resources can be located, wherever they may be stored. Similarly, to find suitable computing and communication resources they must be described by metadata in a standard form.

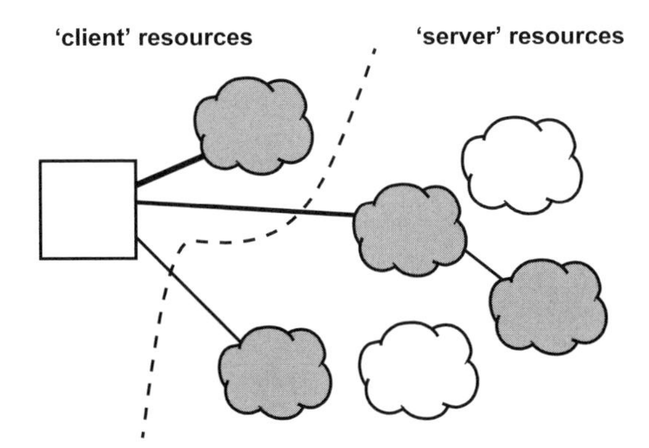

Fig 12.4 Resource assembly — local ('client') and remote ('server') resources being brought together for an instance of a service (the white clouds represent as yet unused resources).

The flexible infrastructure described here allows service providers and users to find out what resources are available and to dynamically control their selection and configuration to make best use of them.

This, then, is an answer to the problem posed at the beginning of this chapter. While bandwidth by itself will not satisfy users' expectations, a programmable network and a new way of building applications can. The best possible performance can be extracted from slow networks while real benefits can be obtained from deployment of broadband — and here 'performance' really means latency and speed rather than the more traditional capacity measures. This combination of communications and content is a key element in making broadband a truly compelling experience. It can be summed up as 'The better the network and the content, the better the service'. Perhaps the closing sentence of this chapter should be the sentiment that in the foreseeable future '... broadband is more network and content than you can use.'

References

1 Venation — http://www.venation.com

2 Historial sources —
http://www.web-caching.com/why-research.html
http://www.techbuddha.com/upcoming/0900preview.html
http://www.worldvisions.ca/~apenwarr/useless/oac/bandwidth.html
http://www.seas.upenn.edu:8080/~gaj1/bandgg.html

3 Defoe, D.: '*Robinson Crusoe — The Life and Strange Surprising Adventures of Robinson Crusoe*', originally published in 1719 — http://www.online-literature.com/defoe/crusoe/

4 Alexander Selkirk — http://wy.essortment.com/alexanderselkir_rehj.htm

5 UCL — http://www.geog.ucl.ac.uk/casa/martin/atlas/atlas.html

6 BTexact Technologies — http://www.btexact.com

7 Discovery — http://www.discovery.org/articleFiles/PDFs/BroadbandOrBust.pdf

8 Russ, M., Kegel, I. C. and Stentiford, F. W. M.: '*Smart Realisation: Delivering Content Smartly*', Journal of the IBTE, **2**(4) (October-December 2001).

9 Forrester — http://www.forrester.com/ER/Marketing/0,1503,214,00. html

10 Curl — http://www.curl.com

11 This is Your 15 Minutes of Fame — http://www.your15mins.com/geton.htm

12 Star Trek — http://www.startrek.com/

13 NUA Internet Surveys — http://www.nua.ie/surveys/how_many_ online/index.html

14 Napster — http://www.napster.com

15 Gnutella — http://www.gnutella.co.uk/about/

16 Kazaa — http://www.kazaa.com

17 Quark — http://www.quark.com/products/xpress/

18 Microsoft — http://www.microsoft.com/presspass/trial/feb99/020199. asp

19 Zdnet — http://www.zdnet.com/pcmag/features/memphis/memphis1. htm

20 Nature — http://www.nature.com/nature/webmatters/grid/grid.html

21 World Wide Web Consortium — http://www.w3.org/

22 Network performance statistics —
http://rescomp.stanford.edu/~cheshire/rants/Networkdynamics.html
http://rescomp.stanford.edu/~cheshire/rants/Latency.html
http://www.stuartcheshire.org/papers/LatencyQuest.html

13

BANDWIDTH-ON-DEMAND NETWORKS — A SOLUTION TO PEER-TO-PEER FILE SHARING

J A Clark and A Tsiaparas

13.1 Introduction

One of the most exciting aspects of the Internet is that it contains information about almost every imaginable subject. High-bandwidth availability with broadband networks is providing a powerful and popular means for the same situation to become true for video content. We describe here a vision where your broadband connection gives you access to an unimaginably vast library of almost every programme and film ever made. For example, if you are doing research on a particular subject, a quick search will turn up every documentary and programme made about that subject. Or you could watch obscure programmes from your childhood, with no need to wait until it appears on one of the many gold TV channels being beamed to us — they could be only a mouse click away. This chapter therefore suggests a way to make '... broadband more network and content than you can use', the final vision presented in Chapter 12.

So how is this vision going to come about? The entertainment industry does not have the resources or the prospect of profits to make it worthwhile for them to digitise all their back catalogue of contents and store them on huge video-on-demand servers. It is very likely to occur in the same way that MP3 and peer-to-peer applications like the currently unavailable Napster encouraged people to digitise their existing CD and record collections and share them with other users. They benefited by downloading music they did not have or had not got around to digitising. When Napster was at its peak it was possible to download virtually any music you wanted. Now that it has been restricted by legal action, its contents are far smaller but the music is still out there and relatively easy to find via alternative systems. In fact, the legal action encouraged more users to try out the service and evidence shows they are now utilising a number of alternative systems, sharing even more content than before. This demonstrates that despite the force of law and best technical endeavours of software developers, content is virtually impossible to protect.

So, is the same thing going to happen with films and TV programmes? Almost certainly it will, once users get access to high-bandwidth networks. What can be done to encourage consumers to use paid for and legal peer-to-peer applications instead of anarchic free alternatives? Perhaps the solution is right under our noses because at the moment this scenario is not too much of a problem with video as bandwidth is still limited. Most users on narrowband modems do not want to wait two days to download a movie from the Internet which they can get for a small fee from a nearby video shop. Perhaps limiting bandwidth may be the best way to reliably restrict end users to legal peer-to-peer (P2P) applications.

It is unlikely that people will pay extra for a broadband Internet connection that is only a little faster than their existing dial-up service. Therefore perhaps the next best thing to 'bandwidth all the time' will be 'bandwidth on demand'. We will explain the bandwidth-on-demand network options that are possible using enhancements to today's broadband technology and how a typical application such as a file-sharing program would use these capabilities.

13.1.1 Peer-to-Peer Versus Client/Server Services

It is said [1] that peer-to-peer computing created a revolution in computing systems, much like the World Wide Web in the 1980s. The first popular peer-to-peer system on the Internet was Napster. It encouraged the sharing of user content among the multitude of people that used the application worldwide.

Peer-to-peer computing is the sharing of computer resources and services by direct exchange between systems. Resources range from information and processing cycles to remote disk storage for files. Peer-to-peer enables the clients to communicate directly between themselves, thus acting both as clients and servers, assuming the most efficient role in the network. This direct exchange reduces the load on dedicated servers, allowing them to concentrate on specialised services.

Peer-to-peer computing is not new. Companies in the 1970s were researching architectures that would now be labelled peer-to-peer. The success of peer-to-peer systems nowadays is due to three reasons:

- computing power is inexpensive — modern personal computers (PCs) at clock speeds beyond 1 GHz allow the development of sophisticated applications;
- bandwidth is abundant in some networks — high-speed connections using fibre-optic technologies and broadband xDSL technologies provide bandwidths of at least 2 Mbit/s and technologies that will offer at least 10 Mbit/s, and potentially 1 Gbit/s, at cost-effective rates, are now becoming available;
- storage is also abundant — hard disk drives (HDDs) offer 10-60 Gbytes of information storage while 100 Gbyte and even 1 Tbyte HDDs are available.

For these reasons, peer-to-peer technology can flourish in the modern networking world. Two models prevail in the design of peer-to-peer systems. Pure peer-to-peer computing uses the direct communication between peers (see Fig 13.1).

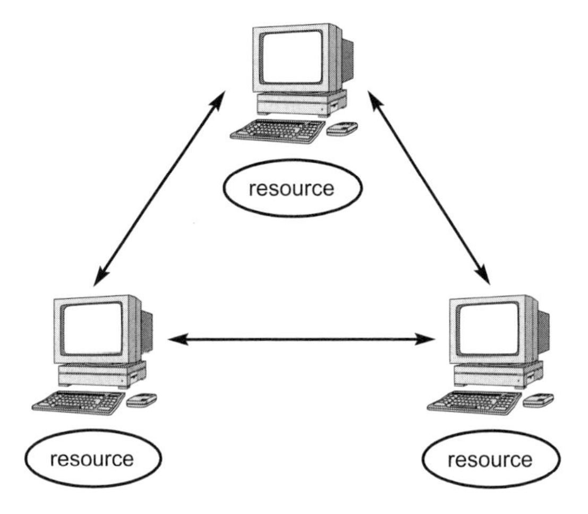

Fig 13.1 Pure peer-to-peer computing.

Indexed peer-to-peer computing relies on an index server which stores the addresses of peers to provide the connection (see Fig 13.2). Because this category of peer-to-peer system is more controllable, as all users need to access the index server, this was the model chosen for our legal bandwidth-on-demand P2P system.

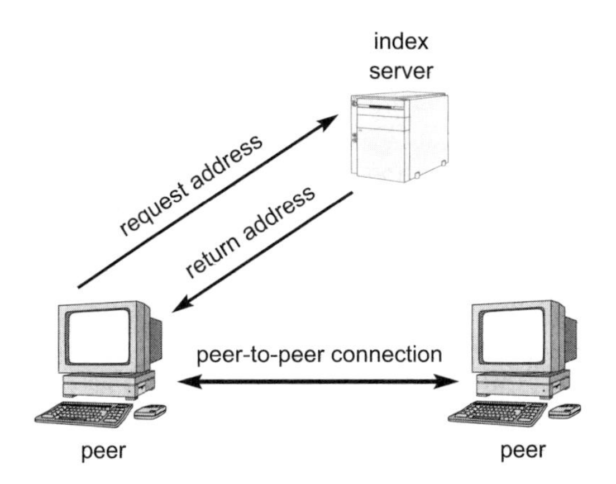

Fig 13.2 Indexed peer-to-peer computing.

The notion of client/server systems is centralisation of management. Peer-to-peer computing is the exact opposite. The peers must be sufficiently autonomous. There is less centralisation. In a pure peer-to-peer system, even an index server is not required.

Peer-to-peer systems enable the operation of servers with no systems administration. Users with little computing background use the systems to share resources. Client/server systems require administration to operate and secure the servers.

On the other hand, client/server systems are more efficient than peer-to-peer systems. This is due to the fact that client/server systems use specialised hardware and connections to operate (Fig 13.3), whereas peer-to-peer systems operate on normal personal computers. This also makes client/server systems expensive, as they need to be engineered for 100% reliability and peak loads.

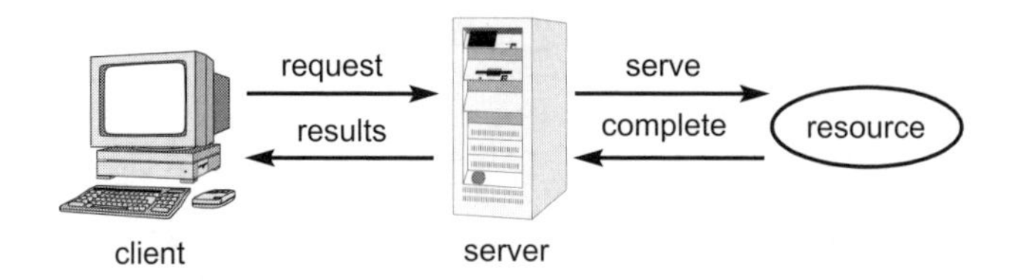

Fig 13.3 Client/server access to resource.

Security issues prevail in peer-to-peer systems. Securing a single server has proved to be a tremendous task, usually undertaken by highly trained personnel. Peer-to-peer systems, without dedicated system administrators, do not provide the same levels of security. Ensuring the privacy of the users' contents and withstanding attacks is a large problem in guaranteeing success for peer-to-peer.

13.1.2 Creating Digital Content

One of the most exciting aspects of the Internet is that it contains information about almost every imaginable subject. High-bandwidth availability with broadband networks is providing a powerful and popular means for the same situation to become true for video content. The vision here is that your broadband connection gives you access to an unimaginably vast library of almost every programme and film ever made. For example, if you are doing research on a particular subject, a quick search will turn up every documentary and programme made about that subject. Or you could watch obscure programmes from your childhood, with no need to wait until it appears on one of the many gold TV channels being beamed to us — they could be only a mouse click away.

Could file swapping of movies and TV programmes be the fabled broadband killer application? When Napster was at its peak, many people would have convinced themselves that they needed the extra bandwidth to download that track in under a minute as opposed to waiting thirty minutes, so it is thought that Napster was responsible for a large number of users moving to ADSL.

From a telecommunications operator's perspective, peer-to-peer services should definitely be encouraged as the users can see the benefits of extra bandwidth. However, the problem facing operators, as they attempt to move towards video services, is the lack of unique content to make the network more attractive than competing cable, satellite and terrestrial TV systems.

It is unlikely that a lot of content will be digitised by service providers as the revenue is not guaranteed for more obscure content and they would not want to fill their expensive video-on-demand servers with content that is only accessed infrequently.

Most computers sold today are able to digitise content to a degree of quality that is really only dependent on how big you want the end file to be. All content can be recorded in analogue form by connecting the video output into a video card with a video-in connector. A lot of people are digitising their old VHS videos this way (some of which may even have originally started as home cine recordings). Even Microsoft Windows now includes easy-to-use video editing software.

It is also relatively easy to store DVDs on your hard drive, and then compress them to a range of different qualities and sizes. Although DVDs are encrypted, the encryption was broken back in 1999 and any enforced encryption change now will render every current DVD player obsolete. The story goes that the encryption was broken by Linux developers trying to find a way to play DVDs on PCs equipped with their favourite operating system. It was done by reverse engineering a commercial PC DVD viewer. With hindsight it may have been better to release Linux DVD players, but the chances are that the challenge to break the encryption would have proved hard to resist and it would have been broken eventually.

One of the most common sizes to which DVDs can be compressed is 650 Mb. Movies of this size can fit on to an ordinary CDR for archiving or sharing. One of the most popular compression systems is DivX and this format looks set to become the equivalent to MP3 in the video world [2]. The process of compression does take a fair time but the end results are pretty impressive. Mastering the DVD decryption software can be tricky, but the download installs the DivX viewer 'Playa' (Fig 13.4) — so any user could be downloading and watching movies in under an hour.

Now that we have a way to digitise video and software to easily share it around, all that is then required is a way to keep the sharing legal to ensure the entertainment industry gets rewarded for what they produced and own.

13.1.3 Potential Solutions to Copyright Theft

Most solutions to restrict copyright theft focus on encrypting the contents so that only authorised users can view them. This area is normally referred to as digital rights management (DRM). However, because of the ease with which people can create their own digital versions, this is only a partial solution.

At the moment the industry is targeting the worst offenders, monitoring the most popular P2P systems and identifying who is most active and has the most valuable content. From their IP address they can identify the Internet service provider (ISP) with whom they are logged in, and therefore it is just a matter of getting the ISP to look up who was on-line at the time with that IP address and their name, address and

Fig 13.4 DivX Playa viewer.

telephone number are now known to the authorities. A stern legal letter is then usually all that is required.

The main winners in this scenario are the lawyers, but, in the process, network operators, ISPs and the movie industry are being distracted by activities outside their main business.

13.2 Limiting Bandwidth to Trusted Applications

The radical solution proposed here is that, as it is not possible to restrict P2P applications using existing methods, end-user bandwidth should be limited to make the downloads unacceptably slow so that only the most dedicated user would be tempted to use pirated movies. This by itself will not be a solution, as people would not be prepared to pay more for an always-on ADSL connection if it was hardly any faster than their old analogue modem. Therefore the service will need to include a bandwidth-on-demand capability to provide end users with bandwidth when they need it.

The technology used to develop the demonstration service and test out the theories was based on an ATM-based ADSL network. This is because there is a bandwidth-on-demand capability available (but not offered as a service) on one type of ADSL multiplexer (DSLAM) installed in BT's network, and standard Windows software can be written to use this capability. A full description of the network is given in section 13.4, but first consider the service from the end-user perspective.

13.3 Service Description

Let us assume that BT changes its current ADSL network so that the basic service is 256 kbit/s downstream (from the network to the user) and 128 kbit/s upstream. The current service is 512/256 kbit/s.

Therefore to download a 650 MB movie from another user (who also has 128 kbit/s upstream) over the always-on IP connection would take: 650 MB x 1024 (to kB) x 8 (to kbit/s) /128 kbit/s = 41 600 secs = 11.5 hours (Fig 13.5).

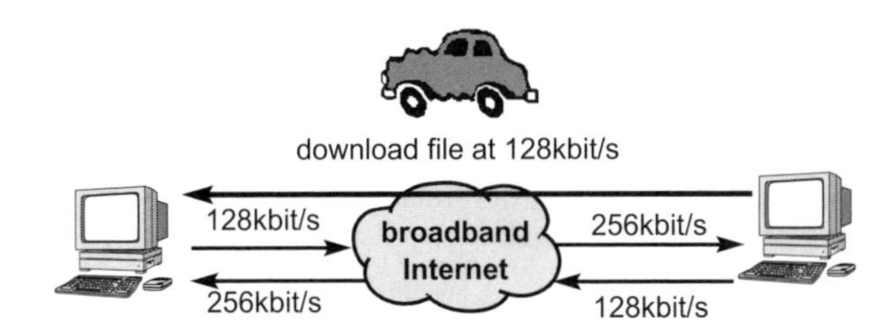

Fig 13.5 Download over always-on connection.

Now assume the bandwidth-on-demand connection can take full advantage of bandwidth available for users near the exchange, that could be 5 Mbit/s downstream/512 kbit/s upstream. To download from this user if there was no congestion, users could receive the file at 512 kbit/s. This will result in the download taking a more feasible 2.5 hours (Fig 13.6).

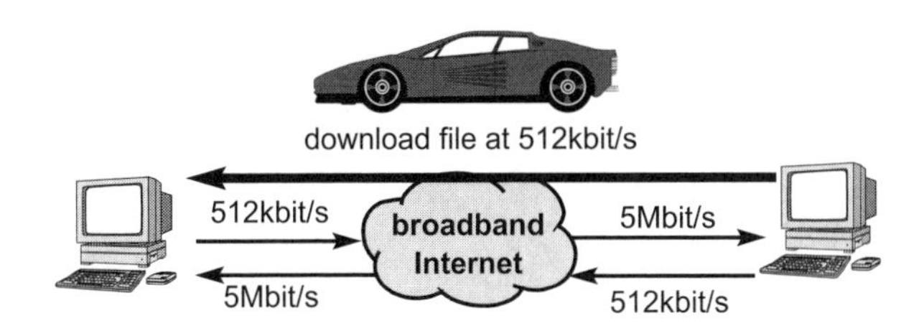

Fig 13.6 Download via higher bandwidth connection.

If it was a 2-hour movie, you could probably start watching it after perhaps 10 minutes buffering, and so it would be almost as quick as the video-on-demand service. A proper VoD server, however, would stream the data encoded at rates of at least 2 Mbit/s instead of 600 kbit/s, as in this example. The time is further reduced if higher upstream connections are available, for example SDSL will provide 1.8 Mbit/s bi-directionally.

Therefore, to summarise, using a bandwidth-on-demand peer-to-peer application, the end-user choices given in Table 13.1 would be available to download a movie from another user.

Table 13.1 End-user choices.

	Description	Time (to download 650 MB movie)	Cost for movie
1	Download over the always-on IP connection, using P2P software	**11.5 hours** (assume other user has 128 kbit/s upstream connection)	**Low/Free**
2	Download over a high-bandwidth virtual connection set up via authorised P2P software.	**2.5 hours** (assume other user has 512 kbit/s upstream connection)	**Medium**, split between Telco and content owner.
3	Download over a high-bandwidth guaranteed virtual connection set up via authorised P2P software.	**20 minutes** (assume 2 Mbit/s connection from SDSL user)	**High**, split between telco and content owner.

Obviously the movie industry would not like to see the free option being made available but users would just download an alternative system that does not offer any revenue opportunities at all. It was therefore concluded that the legal offering should match all the features of its competitors including free but slow downloads. These three options are shown in Fig 13.7, the search result screen of the test software that was developed to prove the concept.

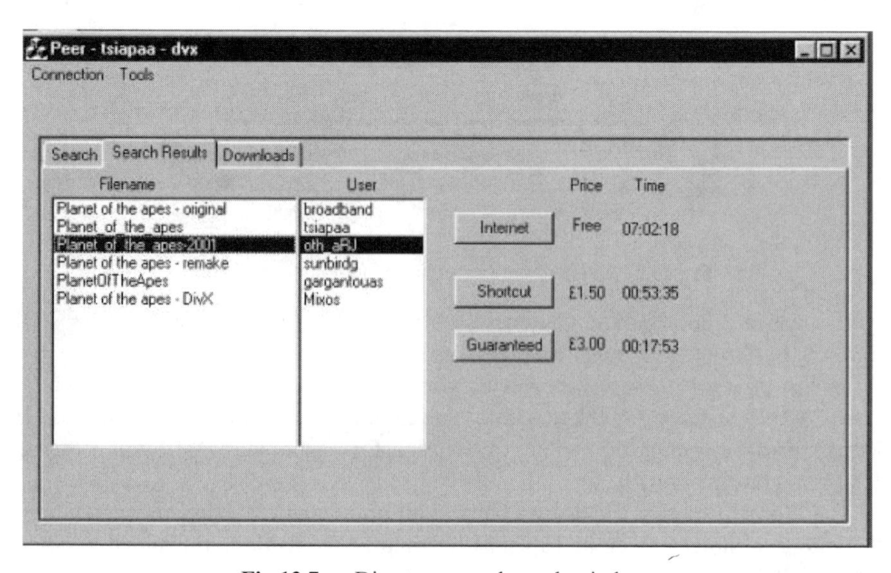

Fig 13.7 Directory search result window.

Figure 13.7 shows the results of a search for the film 'Planet of the Apes'. When you select a film the three download option buttons will be highlighted depending upon which network capabilities are available and the upstream speed of the peer holding that file.

The 'Internet' option will always be available as it is based on the ubiquitous IP protocol. The 'Shortcut' button indicates that this film can be downloaded from a peer on the same ATM access network; it will utilise an unspecified bit rate (best-effort) connection which will be able to use spare capacity not available to the always-on Internet PVC. The 'Guaranteed' button also downloads the film from a peer on the same ATM access network, but this time it will set up a constant bit rate (guaranteed) connection which takes priority over best-effort connections and in this case the other peer has a 1.8 Mbit/s upstream SDSL connection.

13.4 Network Solutions

13.4.1 ATM — Bandwidth on Demand

The ideas and software described in this chapter were developed on an ATM-based ADSL network that replicates BT's currently deployed broadband network, with one major enhancement — the use of ATM switched virtual circuits (SVCs). These provide a mechanism for supplying the bandwidth-on-demand capability that is key to the new peer-to-peer ideas described here.

They are a capability that has been added by some vendors to the ADSL access multiplexer (DSLAM) in current use, although SVCs are not a feature currently available to BT end customers due to their complexity, as explained here.

13.4.1.1 Current Broadband ATM Network

The current broadband ADSL network relies on ADSL line rates and ATM to provide different QoS to different end users. In effect, it guarantees that video users' traffic gets a higher priority than Internet users; it also means that businesses and home users that pay for a 2-Mbit/s Internet connection can download four times more quickly than those who only pay for a 500 kbit/s service. Data from each user is assigned to a permanent virtual circuit (PVC) that is dedicated to the end user and provides a connection across the ATM network to a broadband access server.

Figure 13.8 shows a high-level view of the current broadband network. The management system sets up PVCs when a new service order comes in and the QoS is decided at this time — based on what broadband product was ordered. How the data is treated by the ATM switch and DSLAM depends on what traffic type was assigned to it. At the moment constant bit rate (CBR) is used for video, and unspecified bit rate (UBR) for Internet (IP) connections. Queues in the switch and DSLAM will prioritise data on the different PVCs according to traffic type and

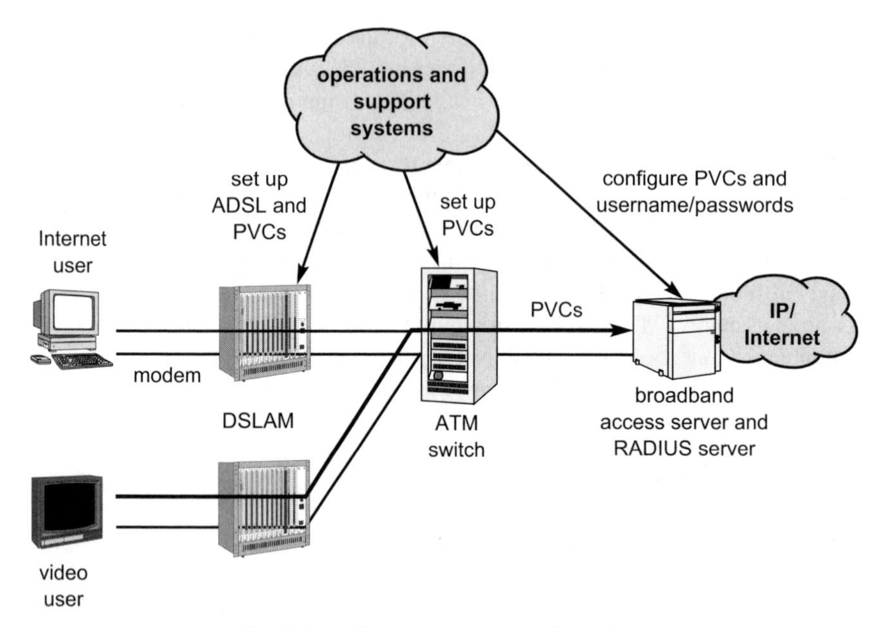

Fig 13.8 Current network configuration.

parameters, such as requested bandwidth. They will also discard data if it exceeds the agreed rate at times of network congestion. This means that if the video connection is for 2 Mbit/s, cells on this PVC will only be thrown away if the rate exceeds 2 Mbit/s. If the rate is less than or equal to 2 Mbit/s, at times of high network loading, cells from other UBR VCs will normally be discarded.

The broadband access server is the gateway between the circuit-switched ATM network and the packet-routed IP world. This server terminates the ATM cells whose payload includes IP/PPP data. It should be noted that the ATM network will be acting as a throttle on Internet traffic as it is usually provisioned in such a way that not all users can download data at their full rate at once. This usually happens in the connections to the access server, which act as a concentrator for PVCs from a number of exchanges.

13.4.1.2 End User Set-up VCs

With an ATM access network that supports SVCs [3], the end user sets up their connections when they need them and clears them down when they have finished with them. The process involves sending into the network a set-up message that contains the traffic type and bandwidth required. Each network component, such as the DSLAM and switch, checks whether it can support the traffic required over its links and switching fabric, and, if acceptable, forwards the set-up to the next network component and so on until the end destination is reached. If the end resource is able to accept the new call, it responds with a connect message, which is sent back across the network and the data connection is now established.

This considerably reduces the load on the network management system as a lot of processing is now done in the network itself.

Putting this together gives the three options shown earlier in the broadband Napster software (Fig 13.7) screenshot routed over the network as illustrated by Fig 13.9:

- option 1 is via a 128 kbit/s always-on PVC to the Internet — this is the maximum rate assuming the network is not fully loaded;
- option 2 is via a best-effort SVC across the local ATM switch — the data throughput rate is not guaranteed but the user will get a much faster transfer as it is not bound by the 128 kbit/s PVC, but can peak at the other user's maximum upload rate;
- option 3 is via a guaranteed SVC across the local ATM switch — the data rate is guaranteed and if the other peer has no other uploads, it can be set at their maximum upstream rate.

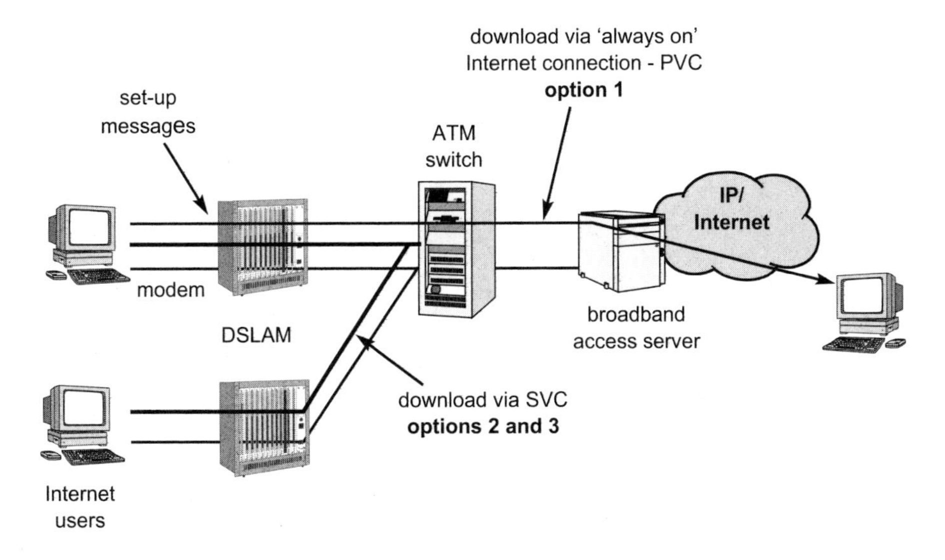

Fig 13.9 Data route for three QoS options.

13.4.1.3 Software Solution — ATM

The software that enables peer-to-peer communications over the bandwidth-on-demand network exploits the Windows Winsock 2 API to communicate with the directory service and other peers. Winsock 2 is a programming interface, specified by Microsoft™, which allows networking over IP, ATM, and other protocols.

Winsock 2 is an open API, allowing users to develop custom applications for the network.

IP, a connectionless, best-effort protocol, provides the basis for the Internet. A PVC connection is set up to provide access to the broadband Internet. The software uses IP in order to communicate with the directory server, exchanging addresses and lists of content.

It can also upload and download content from peers in the Internet — the IP-only download solution. The directory server is placed in the Internet in order to allow the greatest coverage of users to the software. Both peers subscribed to the broadband Internet and users with normal Internet accounts can share content with the application.

The power of SVCs is exploited when the software opts to download content with a high-speed connection from a peer. Both peers must be subscribers to the broadband Internet on the same ATM access network in order to take advantage of the SVC capabilities. In this case, the network uses ATM, a connection-oriented protocol, as well as UNI (user-network interface) signalling, to set a route between the two users.

Figure 13.10 depicts the relationships between the software and Winsock 2, as well as between Winsock 2 and the network. The solid lines symbolise the IP/PVC connections, while the dotted lines symbolise the ATM SVC connections. The IP protocol packets are first carried over a PVC on the ATM network to the Internet, and then, using the IP destination address in this packet, routes it to its IP-based destination.

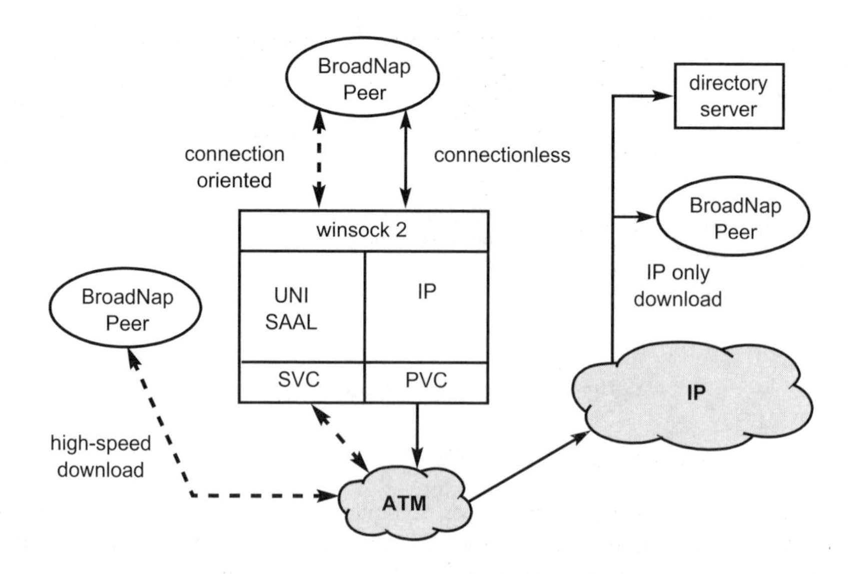

Fig 13.10 Network-to-software architecture.

13.4.2 IP-Bandwidth on Demand

With over 3 million ATM-based ADSL lines deployed it looks likely ATM will now be around for the foreseeable future, but there will be a number of non-ATM access technologies over which it would be useful to support bandwidth on demand. Gigabit Ethernet is emerging as the most popular interface for direct fibre connections into businesses and some homes/blocks of flats. So what are the options for bandwidth on demand in this scenario?

At first sight the 'RSVP' QoS [4] mechanism looks promising as it works in a similar way to SVCs, as the application sends a message to the network to reserve bandwidth. This has been proven in small implementations and probably could be deployed on a large scale over core IP/MPLS networks, but there is no way to restrict the bandwidth of always-on connections. Therefore when the network is lightly loaded, P2P users will be able to download for free at high speed and therefore the incentive to pay for content is significantly reduced.

An alternative IP QoS mechanism is DiffServ [5]. Each packet is marked with a priority and the router handles higher priority packets faster and discards the lowest priority packets when the network is congested. If this was offered as a commercial service to normal Internet users, it would be essential that a differential charging mechanism was in place to stop everyone sending everything at the highest priority. The usual process is to count packets of different priorities for each user and calculate additional usage charges based on these packet counts.

Figure 13.11 shows a router supporting three priority classes of IP traffic. The lowest priority is used for the always-on data for which there will not be a charge; the other two classes will be charged on a packet count/size basis. Two QoS classes are suggested to enable low latency data like VoIP to be prioritised over less latency-dependent traffic like video streaming, which can be buffered. The way to limit always-on data would be to configure the router with class-based queues that

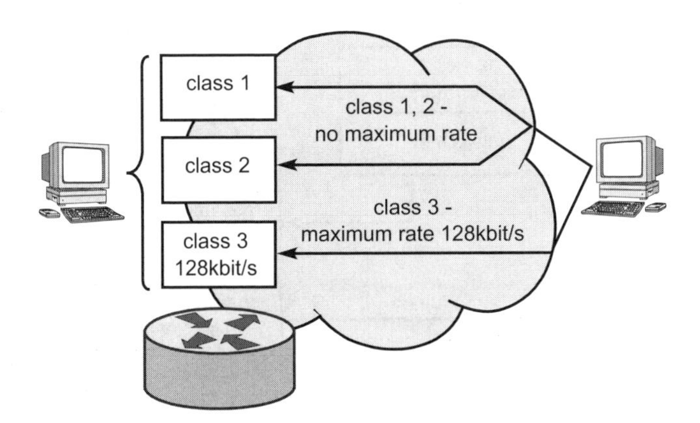

Fig 13.11 Differential QoS.

police the data. Using committed access rate (CAR), excess class 3 packets will be marked as out of contract and discarded; hence, in this example, the free packets would be limited to 128 kbit/s. If the peers wanted to exchange data more quickly, they would mark the packets with a higher QoS class. Billing could then be done as described in a later section for either of the bandwidth-on-demand network solutions.

13.5 Software Solution — IP

To utilise the proposed IP-based bandwidth-on-demand solution, the application needs control over the type of service (ToS) bits in the IP header. This is also a feature of the generic QoS components in the Winsock 2 API.

The application, when communicating with the directory server or transferring data using a best-effort (class 3) connection, will tag the packets as class 3 when sending them out. When the peer is transferring data at high speed to another peer it will tag the IP packets it sends out at a higher QoS (class 2). If appropriate, traffic shaping will also be applied to send the packets at the agreed rate, ensuring no packets are dropped later in the network.

Figure 13.12 depicts the relationships between the application software, Winsock 2 and the network.

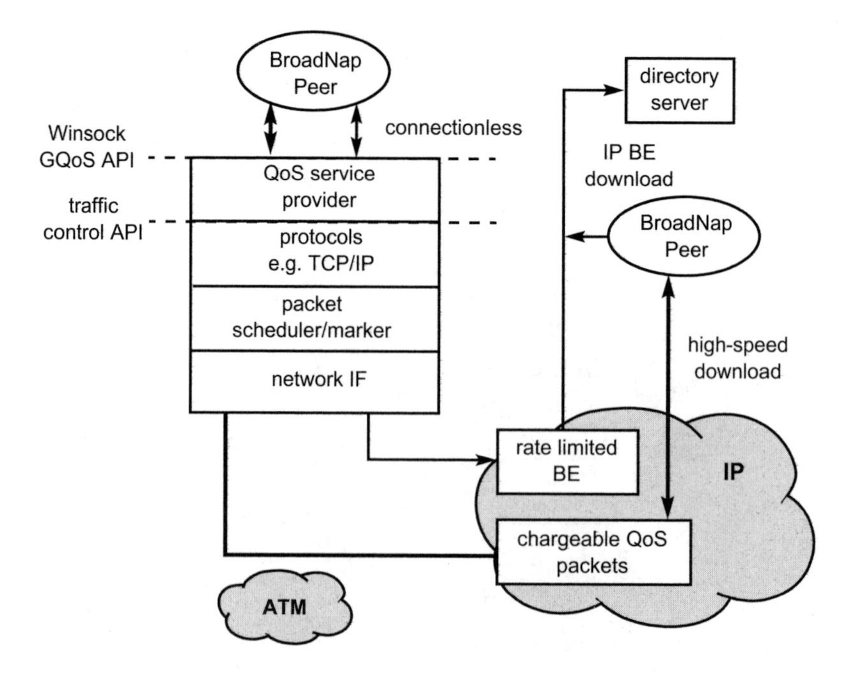

Fig 13.12 Network-to-software architecture.

13.6 Billing Solution

The billing for usage is done by one of two options depending upon what type of bandwidth-on-demand network is used. With the ATM SVC network, every time a high-bandwidth connection is established the signalling message to set up the connection is logged, and when the connection is released a call record is generated that contains the duration of the connection, how much bandwidth was provided, and its traffic type.

This call record can then be sent to a billing engine which calculates what the actual charge should be and adds it to the user's bill. For example, a 1 hour call at 2 Mbit/s could be charged as 2p a minute, so they would get a bill for £1.20. While a 30 minute call at 4 Mbit/s could be charged at 4p a minute, so they would also get a bill for £1.20, which would reflect the use they have made of the network.

For an IP bandwidth-on-demand network, the number of packets of each priority is counted in the router. The packet counts would then be converted to the equivalent of a usage record and sent to the billing engine for calculation of the end-user bill.

The obvious flaw in this is, of course, that if the telco charges £1.50 for the high bandwidth connection, then surely people are just going to go and write their own P2P software that uses the bandwidth-on-demand capability. They will then be able to download the movies for only £1.50, with the network operator still getting their money, but the content owner not — this is not therefore a viable solution.

The potential solution to this issue is to make the cost of high-bandwidth connections very expensive by default so that it is not worthwhile writing your own software. If you are using software approved by the network operator as secure, it becomes feasible to charge a different rate and perhaps collect money on behalf of third parties. The P2P application described here would use the directory server to generate a set of download records (service calls) which would be compared in the billing engine with network call records to produce a realistic charge for network use. How this could work is shown in Fig 13.13.

The three-stage process is detailed below.

- Network records

  The network generates call or usage records that detail every on-demand high-bandwidth connection. These are sent to the billing engine.

- P2P server records

  The P2P server generates service records for every valid download made via its software. These are also sent to the billing engine.

- Billing algorithm

  The billing engine runs an algorithm that looks for matched call records. If it finds a match, the end-user bill will be reduced to the correct amount, and this

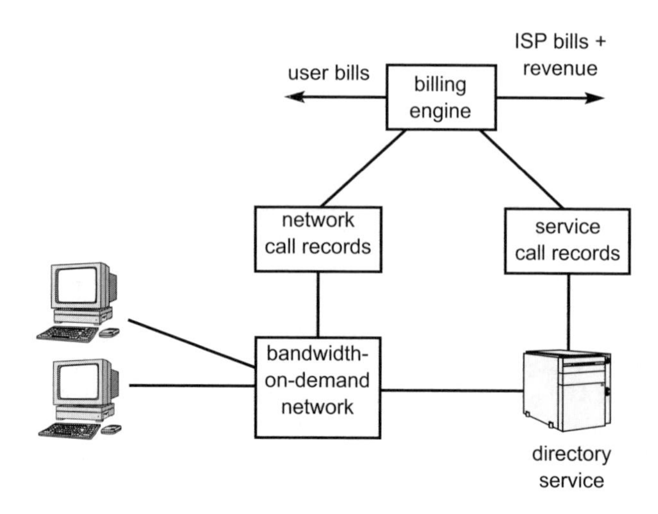

Fig 13.13 Billing on bandwidth-on-demand networks.

amount will include some money for the network service provider and for the content owner. If no match is found the end user is charged the default high amount as it is assumed they are using the connection for something high value.

This is, in effect, a premium rate number network but in reverse — every call is expensive unless to an approved location or made via approved software.

This system would be applicable to other services as well, not just P2P applications. For example, a videoconferencing server would act as a central point for control of videoconference calls and generate records appropriately. The server could even act as a gateway connecting small bandwidth-on-demand networks together over the wide area.

Other applications where QoS is beneficial include conventional VoD servers, games servers, TV streaming and even interactive shopping sites which wish to differentiate themselves by providing a more graphical interactive site that requires a bandwidth boost.

13.7 Controlling Your Content

Many grand Internet-based services have found out that content is the most important way to attract people to your site. Similarly if the quality of what you download is poor, people will go elsewhere or stop using the service. It is especially important, when you are charging for a service, that it is of the right quality. The problem with the peer-to-peer model is that it is particularly hard to ensure that the content is what you really want and the price is what you are prepared to pay. The problems foreseen include people passing one film off as another, inserting pornographic material in the middle of a children's film or even trying to spread viruses around.

It will not be cost effective to have a team of people checking all content before it is put on the system. That may cost almost as much as digitising the content ourselves. What would be needed is some form of end-user quality control. So when an end user has watched a film they rate it according to various factors such as video and audio quality, lack of breaks and most importantly being what it says it is. Search results would contain the highest rated versions first, and if any concerns were raised perhaps highlight them to enable the cost to be reduced although there may need to be some management control here to avoid people giving everything a low rating to reduce the cost of the service. Users whose ratings are proved to be wrong could also be barred from making further ratings.

In summary therefore, if you devolve the quality checking to end users and provide a secure refund mechanism in the event of dissatisfaction, this P2P high-content network could virtually run itself and provide a reasonable revenue stream. This system could be the best way of getting the content that is gathering dust on people's shelves in the form of video cassettes into a more usable digital format. For example, the BBC has been able to retrieve long-lost radio and TV programmes after appeals to the general public, proving that the content is out there.

13.8 Satellite, Cable and Digital Terrestrial TV Opportunities

Would the ideas here be applicable outside the area of telecommunications broadband networks? Almost certainly — set-top boxes (STBs) are now being supplied with hard disks to store content broadcast to them. This increases the content available to users by storing programs they may like to watch or the latest blockbusters to free up near-VoD channels. It would not be a significant change to add a video feed input so that end users can digitise content and send it to their STB, perhaps for a reduction in next month's rental if the content is not currently available on the network.

Getting this content off the user's STB and into another user's STB or the broadcast network will require higher bandwidth upstream connections. Cable STBs already have this — so satellite and terrestrial digital TV STBs may need to add broadband ADSL connections and new software. Then they too will be able to build a very high content video library with their end users' help.

13.9 Summary

It is hoped that this chapter will provoke a debate on whether too much bandwidth could be a problem for the development of the broadband network. The telecommunications world is now at a junction and has to decide which route to take.

- Scenario 1 — continue the existing drive to higher bandwidth networks

  We could continue down the current path to give everyone ADSL/SDSL/VDSL/ Direct Fibre always-on IP access at ever higher bandwidths, thus providing a network that will be a dream for video pirates as we will make it easier and easier to share content for free [6, 7].

  The advantage of this is that broadband telecommunications networks will end up with far more content than competing satellite networks could hope to provide at a far cheaper cost. Unfortunately, the likely end result will be that the network operators and ISPs become embroiled in legal action from the content owners trying to get the offending users removed from the network, or high churn as ISPs cut off offenders who then just join another ISP. Also, the very expensive content systems we are now putting in place will become worthless as they could be bypassed for free content from other end users.

  Companies providing digital rights management may be able to secure content they digitise but that still leaves a lot of unsecured content that is already sitting on people's shelves on video cassettes and DVDs just waiting to be digitised.

- Scenario 2 — telcos recognise the revenue opportunities for peer-to-peer and bandwidth on demand.

  What would be far better from a telecommunications operator's perspective is to build a network that encourages legal peer-to-peer trading where money goes to the appropriate content owner, while at the same time making illegal video trading so slow or expensive as to discourage it. This is the proposition here and that process can now be demonstrated on today's broadband network equipment with existing applications.

For years people have been trying to find a way to charge for bandwidth dependent upon the value of the content rather than the actual number of bits transferred. The network proposed here appears to offer a solution. Charging highly for bandwidth, unless you work with the telecommunications operator to ensure that your application is secure and your content is legal, is a new model that should perhaps be evaluated further.

References

1 Peer-to-peer Working Group — http://www.peer-to-peerwg.org/

2 Movie Encryption Information and Software — http://www.divx.com/

3 ATM Forum: '*UNI 3.1 Specification*', — http://www.atmforum.com/

4 IETF: '*RSVP Specification*', RFC 2205 (September 1997) — http://www.ietf.org/rfc.html

5 IETF: '*DiffServ Specification*', RFC 2474 (December 1998) — http://www.ietf.org/rfc.html

6 '*Pirated Movies Abound on the Web*', (August 2001) — http://newsbytes.com/

7 '*Coming Soon to a Monitor near you — Broadband should be outlawed*', Daily Telegraph — Connected (September 2001).

14

A 3-D TELEPRESENCE COLLABORATIVE WORKING ENVIRONMENT

L-Q Xu, B Lei and E Hendriks

14.1 Introduction

The concept and the realisation of a sense of telepresence via computer-mediated audiovisual media and other sensors has been a focus of interest that attracts considerable efforts from both computer science and human factors researchers and practitioners over the last decade [1, 2]. The idea that a system is capable of bridging time-space gaps, and bringing a real sense of touch and interactive action to people physically located apart but in a shared collaborative working environment [3], is powerful and commercially desirable. This development trend has been largely fused by the rapid advance of key enabling technologies including digital video processing and coding techniques, 3-D computer vision [4, 5], powerful micro-processors, high-speed broadband networking, and large flat high-quality displays [6]. These technologies provide more opportunities for an integrated system than ever before. Specifically, driven by the belief that 'seeing is believing', there have emerged increasing demands for a more realistic experience of visual communications/presence in contrast to the traditional audio/video-conferencing. Various telepresence scenarios [7] have since been proposed, such as telemedicine, virtual tours [8], and teleconferencing [9], potentially offering some revolutionary new means of working, improving considerably the quality and effectiveness of services in many different sectors.

One of the key technical developments that is at the heart of such a system is the integration of computer vision and computer graphics techniques [10-12], especially in dynamic 3-D scene modelling and novel view generation [13]. A brief account of the current efforts and various technical advancements is given below.

- 'Through the window'

  The development and continuous improvement of the 'through-the-window' visual communication paradigm has provided, for example, the desk-top videoconferencing system, and the audiographic conferencing system with shared white boards, applications, and simplified synthetic 3-D avatars [14]. With such a system, not only can people talk to each other, but they can also see their remote counterpart(s) or avatar(s) and how they behave in certain predefined ways. This kind of system is normally limited to a small display window or low resolution images (hence, small volume of information) and it often inhibits much of the interaction that would otherwise take place if the collaborators were collocated in the same room.

- Virtual reality

  On the other hand, virtual reality (VR) with a head-mounted display (HMD) brings people into a shared immersive virtual environment in which people can see 2-D video-based avatars of each other and interact with various synthetic objects. The main problem with this paradigm is the disconnection between the real world and the virtual world 'inside the display', and the disconcerting factor of wearing headgear and behaving differently from normal activities. Other representative efforts in using avatar technologies for virtual space conferencing can be found in Ohya [15].

- Viewing 3-D video objects

  The emergence of multi-view image coding and auto-stereoscopic 3-D display technologies makes it possible to perceive 3-D video objects without wearing any glasses. It is achieved through image-based rendering with novel view interpolation [16-18]. The system facilitates certain eye-to-eye contact and motion parallax effects. It is limited, however, to a 'head and shoulders' scene with a small-sized display and two-way point-to-point single-person communications [19]. Body gestures, postures and some socially important phenomena are not taken into account.

- Collaborative working

  The availability of a large room-based spatially semi-immersive display environment, like CAVE [20] and CABIN [21], brings a new experience of closeness to the collaborative working. It is appropriate for groups of people located in two sites to inspect the same data and learn to do the same task. But a system like this provides little direct person-to-person interaction between the parties involved in the tasks; neither do they see the real 3-D scene in correct perspective.

- Image-based rendering

  The appearance of high-quality 3-D image-based rendering, e.g. more photo-realistic 3-D 'talking heads' with emotion-expressing ability, is visually more compelling, but for conferencing purposes it suffers from the same drawbacks as those described in the 'through the window' development above.

In view of the above analysis, the concerted efforts of the EU IST VIRTUE (Virtual Team User Environment) project [22] have been focusing on the following three main objectives, addressing the technology, system integration and human factors tasks in a coherent manner, which extended broadly the activities in the third point above, dealing with the advances in viewing 3-D video objects:

- to develop the innovative technology necessary to produce a convincing impression of presence in a semi-immersive teleconferencing system;
- to integrate and demonstrate a telepresence system incorporating such innovative techniques;
- to investigate the human factors involved in maximising the effectiveness and realism of telepresence, and to use them to drive the design of the system.

The scenario envisaged by VIRTUE at the present time is a three-way single-person isotropic videoconferencing system with a seamless mixture of the virtual and real-world scene. The virtual scene and remote participants' rendered images are presented in correct perspective and change dynamically relative to the view point (head position) of the local participant — the effect of so-called motion parallax. A mock-up scene of such a system is shown in Fig 14.1.

Fig 14.1 The mock-up of one VIRTUE station where the real table is extended seamlessly into a virtual table in the display. The full-size remote participants are rendered as arbitrary 2-D video objects and their synthesised looks will change in line with the local participant's head position. The two cameras mounted on the left and top-left of the screen provide two video streams for 3-D analysis and view synthesis for the left viewer in the display; likewise for the right two cameras. The eye-to-eye contact, normal habitual hand gesturing, and gaze awareness are expected to be maintained.

A more illustrative networked system of three stations is shown in Fig 14.2.

The chapter is organised as follows. In the next section, an outline of the software system architecture for the VIRTUE project is given, presenting functionally each block module description. In section 14.3, a core component of the system responsible for 3-D scene analysis is described. The hierarchical adaptive curve matching (HACM) based disparity estimator offers accurate and robust dense-disparity maps that are crucial for the quality of rendered virtual views. In section

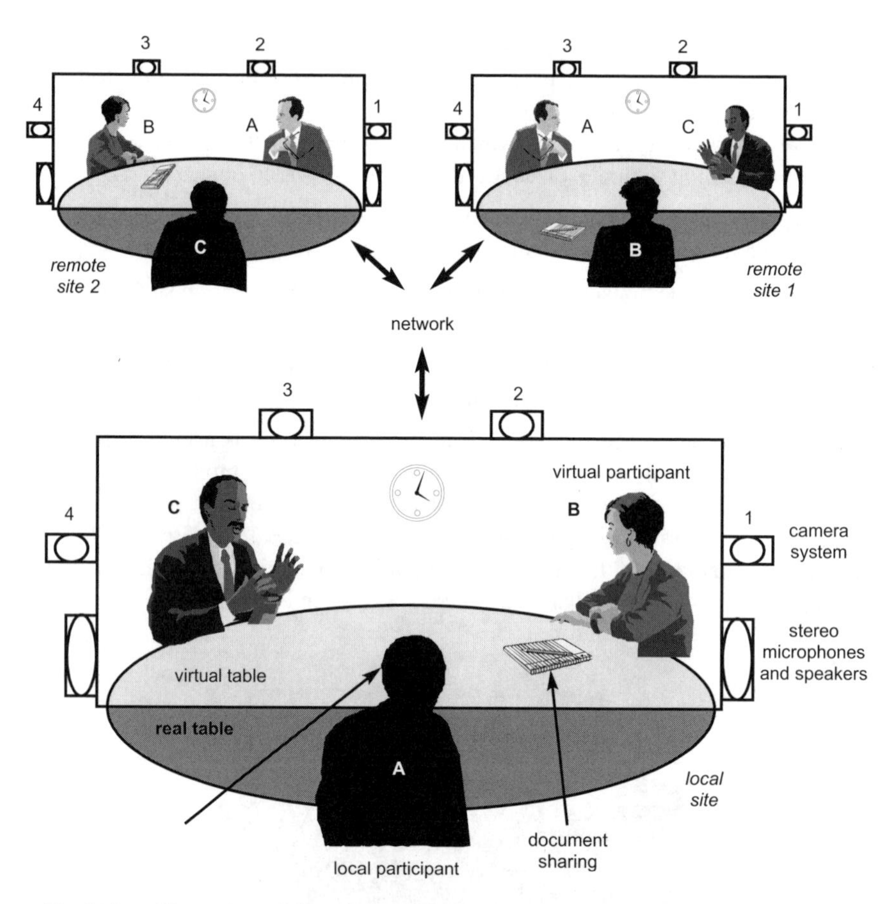

Fig 14.2 Illustration of the networked VIRTUE stations in a three-way telepresence videoconference session. The techniques behind the creation of motion parallax cues in 3-D vision as viewed by each participant are the focus of the current investigations.

14.4, sampled experimental results are given to evaluate the performance of the algorithm, when applied to stereo images of both static and dynamic scenes, using a newly proposed image-based rendering framework for novel view synthesis. The chapter concludes in section 14.5 with discussions of the outstanding issues and future research direction.

14.2 System Overview

Given the illustration in the mock-up diagram shown in Fig 14.1, in each of the VIRTUE stations we treat the video information captured by the left pair of video cameras independently of the right pair of cameras. Each stereo provides the video source for 3-D analysis and rendering of the local participant for the remotely located left and right counterpart in the display, respectively. This is appropriate in the current isotropic three-way single-person conferencing set-up, in which the eye-to-eye contact can be realised based on head tracking. Therefore, in the following discussions, we will consider working on the right pair of stereos acquired only. Figure 14.3 shows the typical pair of frames of the scene captured by camera 2 (left view) and camera 1 (right view) shown in Fig 14.2 in a test videoconferencing sequence. Figure 14.4 shows schematically the block diagram of the software system architecture for the VIRTUE station.

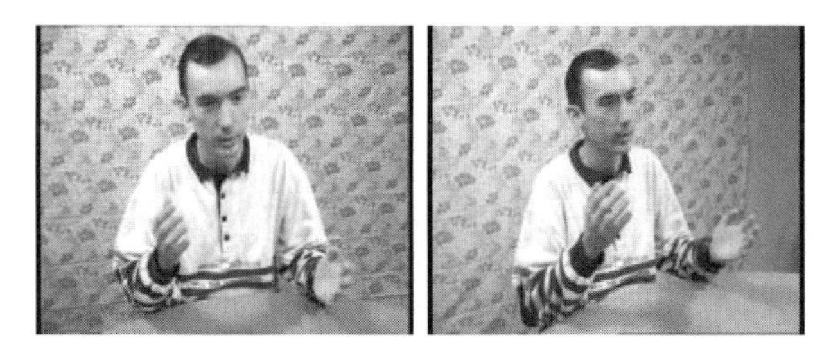

Fig 14.3 The typical frames of the conferencing scene from the left camera (2) and right camera (1), respectively, in which normal body and hand gestures are allowed, and the background of the scene could be arbitrary.

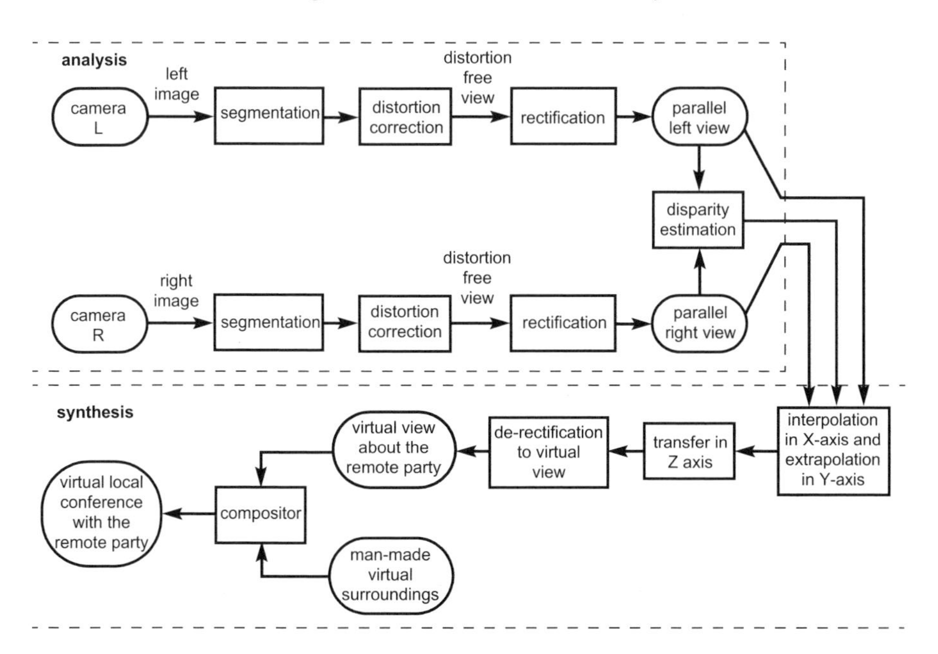

Fig 14.4 The schematic diagram of the software system architecture for the VIRTUE station.

Functionally the system architecture consists mainly of two processing blocks. The analysis block is to extract and represent necessary 3-D information of the foreground figure in a very efficient manner from the pair of calibrated stereo cameras. The synthesis block aims to reconstruct the dynamic scene information from a virtual viewpoint in line with the head position. It is noted that in the current discussion we do not specifically prescribe a point in the processing chain at which the output of a module in the analysis block should be transmitted over the network. There are various choices that are behind such a decision. The network bandwidth available, the encoding/decoding mechanism employed, the final visual quality desired, and the computational complexity afforded by the real-time requirement in the final system implementation, all have an impact on this assessment.

The chain of processing in Fig 14.4 is explained as follows. The stereo video streams captured are firstly segmented through a foreground/background segmentation module. The segmented foreground, the human subject, is compensated for radial distortion and rectified so that the resulting two stereo views form a parallel set-up and only horizontal disparity thus exists. An efficient disparity estimation module is then applied to output temporally coherent dense-disparity maps in which the holes due to occlusions are properly filled in. Given a new viewpoint found using the head tracker and the 3-D calibrated camera information, in the synthesis block, an image-based rendering approach — a multi-step novel view synthesis framework — is adopted to synthesise the desirable new virtual view image of the scene in correct perspective.

The two core techniques involved in the above process are the disparity estimation and the novel view synthesis. The former provides us with the 3-D information embedded in the stereo views that have been acquired, while the latter manipulates the 3-D scene in the way we desire, based on the 3-D information. More detailed information about the image-based rendering view synthesis can be found in Lei and Hendriks [23]. A candidate for the disparity estimation is described in the next section.

14.3 The HACM Disparity Estimator

In stereo vision, the disparity estimation, or correspondence problem, that finds a pair of matching image points, which are the perspective projection of the same 3-D point, has been a long-studied issue. The feature-based approaches normally give rise to disparity measurements for a set of sparse feature points that are suitable for explicit 3-D model reconstruction, while the area-based approaches output dense-disparity maps that lend themselves to image-based rendering applications. For the latter the correspondence problem is compounded by a number of difficulties, typically:

- inherent ambiguity — the search is done in 2-D image space, and is thus sensitive to the uniform and repetitive patterns appearing in the images;

- occlusions — parts of the 3-D scene may only be seen by one of the two cameras, hence incomplete information is available;
- photometric distortion — due to lightening and surface reflectance properties, the same 3-D point may appear in the two cameras with different image properties;
- projective distortion — the appearance of the same 3-D object changes between the two stereo view images as a result of the perspective projection, e.g. a circle in one image may become an ellipse in another image.

There have been various solutions proposed in the past to deal with these problems, and they are useful in one way or another in certain applications. We propose a novel approach — the HACM — for dense-disparity estimation, which is suitable for stereo imaging of both static and dynamic scenes, and ideal for image-based rendering. The schematic diagram of the algorithm is given in Fig 14.5.

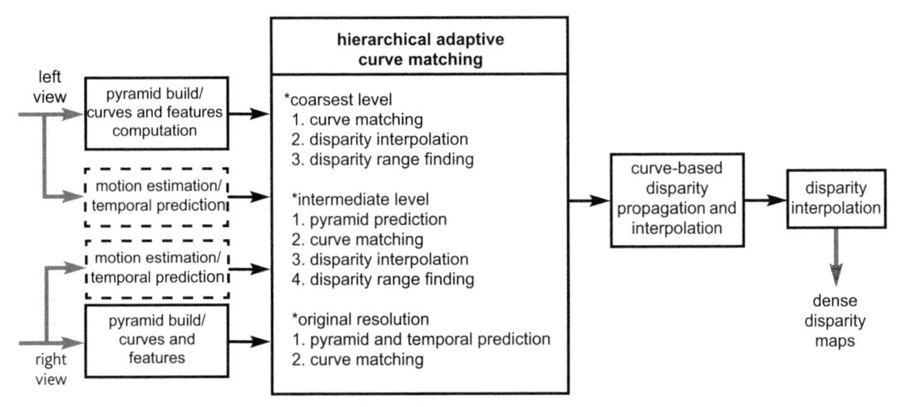

Fig 14.5 The schematic diagram of the hierarchical adaptive curve-matching-based disparity estimation.

The traditional well-studied block-matching-based disparity estimation algorithms normally encounter difficulties at choosing appropriate matching window size. A solution to this is to use an adaptive-sized matching window, but it is invariably computationally intensive [24]. The proposed approach manages to circumvent both problems successfully. The core to this approach is the derivation of, for each pixel, a unique adaptive curve representation that characterises the local surface curvature changes in shape, direction, horizontal mean and variance, all at a slight overhead cost. The matching is carried out in a hierarchical manner that incorporates three ideas to constrain the search range and improve its efficiency and robustness. Firstly, for each pixel, use is made of the available spatial disparities from its immediately previous neighbouring pixel on the same scan line, and the neighbouring pixels on the adaptive curves intersected at this pixel. Secondly, the disparity predictions from the corresponding coarser level pixels already computed are considered. Thirdly, in the case of the dynamic video sequence, the motion-compensated temporal-disparity prediction from the previous pair of stereo frames is incorporated to satisfy the temporal coherence constraint. The constraints of ordering, continuity, and consistency have also been properly incorporated in the matching process, detailed discussions of which are beyond the scope of this chapter. The following sections outline the main ideas and illustrate, with examples, the outcome associated with each processing step.

14.3.1 Adaptive Curve and Features Representation

The derivation of an adaptive curve representation for each pixel is the basis of the proposed approach. This is obtained by comparing the intensities of the pixels successively in the 8-connected neighbourhood from the scan line immediately above and below the current pixel and linking together the most similar pixels, respectively. The curve goes up to $h = 3$ pixels vertically on each end from the current pixel, giving a maximum length of $2h + 1$ pixels. The curve effectively characterises the intensity profile and the physical shape contour around the pixel. It will serve as the aggregation support segment for later cost-function computation in the matching stage. Figure 14.6 shows a test image and a marked area around pixel (308,141) for which an adaptive curve is to be found, as shown in Fig 14.7.

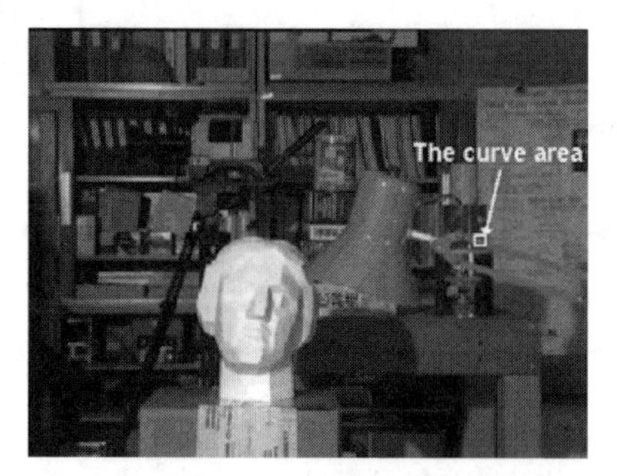

Fig 14.6 An example image with a marked area for a pixel's curve determination.

54	39	35	31	30	26	22
101	90	77	57	44	37	31
110	106	106	100	94	77	58
109	108	109	107	106	103	99
101	105	108	107	108	106	108
59	69	91	101	106	107	109
41	37	47	56	72	88	99

Fig 14.7 The locally adaptive curve found for the pixel (308, 141).

In addition to the curve derived and associated intensity information, two discriminative features are also computed for each pixel. One is the mean value along a short horizontal segment, $2h + 1$-pixel long say, centred on the pixel; the other is a modified variance along that same segment. Figure 14.8 shows the two computed results, respectively, for the image shown in Fig 14.6.

Fig 14.8 The horizontal segment-based mean intensity map and modified variance map for the test image shown in Fig 14.6.

14.3.2 Hierarchical Matching

Given this rich set of representations for each pixel that is located at each level along the image pyramid, we can now perform an efficient coarse-to-fine matching process to find the desired dense-disparity map. The central block of Fig 14.5 describes the necessary steps involved at each level of the image pyramid. They are briefly discussed here.

14.3.2.1 Curve-Based Pixel Matching

The local matching cost function employed for measuring the (de-)similarity between a pixel (x, y) of a curve representation $A(x, y)$ in the left view image I and its matching candidate pixel $(x + d, y)$, that has a horizontal disparity d and *a* curve representation $A'(x + d, y)$ in the right view image I' is shown as equation (1).

$$C(x, y, d) = \frac{1}{S_{A(x,y)}} \cdot \sum_{(m,n) \in A(x,y)} \frac{(I_A(m,n) - I'_{A'}(m',n))^2 + (M_A(m,n) - M'_{A'}(m',n))^2 + (V_A(m,n) - V'_{A'}(m',n))^2}{\left| I_A(m,n) \cdot I'_{A'}(m',n) \right|} \quad \text{...... (1)}$$

In equation (1) $A(x, y)$ serves as the aggregation (support) segment for the pixel (x, y); $S_{A(x, y)}$ is the length of the segment. Note that for clarity we have used A and A' to denote, respectively, the two corresponding curve representations $A(x, y)$ and $A'(x + d, y)$. In the right view image I', a pixel's horizontal axis m' assumes the index along the curve A' while its vertical axis position is n.

In this cost function we have incorporated the impact of all the three feature representations of corresponding pixels — the original intensity (I, I'), the mean (M, M') and the modified variance (V, V') based on short horizontal segments as discussed in section 14.3.1. The denominator in the cost function is used to remove the possible scale difference existing between the left and right view images.

In order to improve the efficiency and robustness of the matching process, for a pixel (x, y), the range of its initial disparity d can be largely constrained. This can be achieved using both spatial and temporal (in the case of dynamic scene) disparity predictors.

The spatial predictors are coming, respectively, from the pixel immediate before the current pixel, the pixel one line above and on the same curve $A(x, y)$, and the pixels whose curves intersect the pixel. The value d is chosen as the disparity value whose matching cost is the minimum among the set of predictors. In carrying out the above analysis we have also taken into account the visibility problem and the continuity and ordering constraints, so that the disparity maps computed at this stage are consistent, or suffering from less erroneous matching.

14.3.2.2 Hole Filling by Disparity Interpolation

Following the above analysis, the disparity maps obtained will inevitably contain holes for which there are no corresponding disparity values as a result of mismatching or occlusions. These holes can now be filled in, based on the existing consistent disparity values. The interpolation scheme includes two steps — firstly, thanks to the adaptive curve representation we derived, we can naturally carry out disparity propagation and interpolation along these curves, and then we simply fill in holes along scan lines by linear interpolation.

14.3.2.3 Disparity Range Determination

In the hierarchical matching strategy, for the disparity search at the coarsest level of the image pyramid, the user can define a reasonably large search range $[d_{min}, d_{max}]$. While for the finer resolution levels, we use the disparities obtained at the lower resolution to specify the search range in which a fine disparity value can be found. This process is influenced by the way in which the original image pyramid was built, specifically the Gaussian filter and sampling scheme used.

14.3.2.4 Pyramid Prediction

Similarly, the estimated disparity map at the lower resolution level can offer an approximation about the intended disparity map at the immediate higher resolution level.

14.3.2.5 Temporal Prediction in the Case of Video Sequence

For a pair of synchronised stereo video sequences acquired from a dynamic scene, apart from the stereo spatial correspondence between the left and right view images within each frame pair, there also exists temporal correspondence between two consecutive frames within each sequence. This provides another source of information that can be exploited to improve the quality of disparity estimation. The temporal coherence is very important to the quality of synthesised virtual scenes as will be demonstrated in section 14.4.3. One way to maintain the temporal continuity of the estimated disparity field is to impose the temporal constraint equation [25], which we have adopted in this study, albeit in a simplified form.

14.3.3 Towards Final Complete Dense-Disparity Maps

14.3.3.1 Curve-based Disparity Propagation and Interpolation

Due to the specific way that a curve is derived, as discussed in section 14.3.1, we can argue that, at the original resolution level, all pixels within one curve are most likely to reside on the same physical object without discontinuity. This enables us ideally to propagate and interpolate the disparities along curves. Note that if a pixel has already been assigned a disparity value, then we call it valued, otherwise we say it is unvalued:

- propagation — if a pixel is valued, then we assign its disparity value to all unvalued pixels on the curve centred on it (an update is necessary if the new assigned disparity value has a lower matching cost);
- interpolation — if the current pixel is unvalued, then we may be able to locate two valued pixels on either side of its curve, in which case, we can interpolate a disparity value for the current pixel based on these disparity values.

14.3.3.2 Left-Right Inter-View Matching Co-operation

By employing spatial disparity predictors discussed previously, we have in fact realised inter-view co-operation in the matching process (Lawrence and Kanade [26] provide an example of accounting for the co-operation in the post-processing stage). In addition to considering neighbouring pixels in the same view, we have also incorporated left-right inter-view co-operation in the matching process. It is briefly described here.

For a current pixel α in the left view image the matching process gives rise to a disparity value d_α and the associated cost c_α. Assume its corresponding pixel in the right view image is α' and it has a currently assigned disparity $d_{\alpha'}$ and associated cost $c_{\alpha'}$. If $c_{\alpha'}$ is larger than c_α, then $c_{\alpha'}$ and $d_{\alpha'}$, for the pixel α', will be replaced by

c_α and d_α respectively. In doing this, the left-right view consistency can be enhanced, thus improving the quality of the final disparity maps.

14.3.3.3 Final Disparity Interpolation

Following the above consistency checking (and disparity propagation and interpolation along curves at the highest level), there remain holes with no disparity values assigned in the disparity map. These holes are mainly caused by occlusions [27] in the 3-D scene. They are filled simply by linear interpolation along scan lines.

14.4 Experimental Studies

In order to demonstrate the effectiveness of the techniques described and the applications that it can afford in 3-D visualisation and especially the videoconferencing scenario of the VIRTUE system, a variety of stereo images and video streams of synthetic and real 3-D scenes have been analysed. Specifically, the quality of estimated disparity maps and the synthesised virtual views based on the view synthesis tool-kit [23] are evaluated. Some examples are discussed in this section.

14.4.1 Synthetic Stereo Images

A pair of synthetic stereo images has been created using a ray-tracing technique from real images. The synthetic 3-D scene consists of one planar and three curved surfaces placed at different distances from cameras, and four real images are mapped on to those four surfaces, respectively. The ray-tracing technique was then employed to produce the left and right view images (top row in Fig 14.9), together with the ground-truth left-to-right disparity map and the real middle-view image (middle row in Fig 14. 9). (Note that the disparity value is shown as the intensity of the image for visualisation purposes, the brighter the pixel, the larger is its corresponding disparity value.) This pair of stereos has several interesting properties that challenge a disparity estimator — the large occluded areas (the baboon face and the ball behind it), containing both texture and uniform areas, and large changes of surface curvatures.

The left-to-right disparity map, computed using the HACM, together with the synthesised middle view, is shown in the bottom row of Fig 14.9. Comparing these results with the ground-truth, we can see that, apart from small trivial details and problems that appeared on the left part of the baboon image due to very large occlusions, the curved surfaces of 3-D objects have been recovered very well. Furthermore, the synthesised middle view bears no appreciable difference or artefacts from the correct middle view.

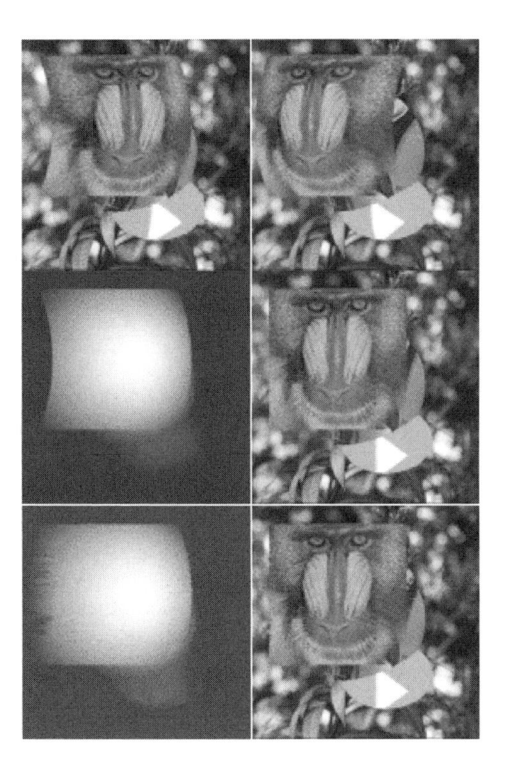

Fig 14.9 Top — the left and right view of a synthetic stereo pair produced by mapping four real images on to three curved surfaces and one planar background placed at different distance from cameras, respectively. Middle — the round-truth left-to-right disparity map and the correct middle view generated. Bottom — the actual left-to-right disparity map computed using the HACM and the synthesised virtual middle view.

14.4.2 Real Stereo Test Images

We next illustrate the results obtained for a benchmark stereo test image — the 'Head' images — we used before. Fig 14.10 (top row) shows the original stereo pair. This is a complex and carefully composed indoor scene for deliberate challenges. The various objects in the scene are placed at different distances from the cameras, hence exhibiting many different object layers; the scene also contains thin but visually important objects like lamp stems as well as uniform and curved surfaces.

The computed left-to-right disparity map along with the synthesised middle view is given in the bottom row of Fig 14.10. It can be seen that with the new algorithm, due to imposing both the locally adaptive curve features and the spatial disparity predictions, not only are the fine details (e.g. lamp stems) recovered, but also the inter-scanline consistency is maintained. The main objects and their layers, as

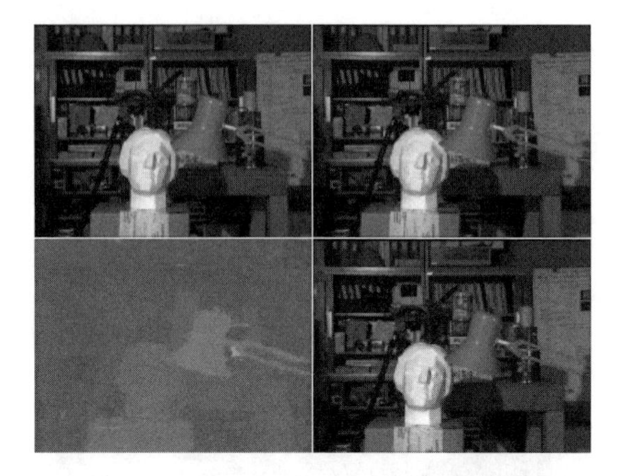

Fig 14.10 The 'head' stereo: (top row) original left and right view image; (bottom row) the left-to-right disparity map computed using the HACM, and the synthesis middle view image.

indicated by the different range of disparity values, are clearly seen. The synthesised middle view preserves perfectly the scene information.

14.4.3 VIRTUE Test Sequence

For the static scene stereo images discussed so far, no temporal information is involved, and the temporal disparity predictions in the HACM are not used. For the dynamic scene, such as the videoconferencing situation sought by VIRTUE, however, the temporal coherence in the synthesised views is very important in removing some annoying visual effects and discontinuities. The camera set-up, in which the video sequences, typified by the two images shown in Fig 14.3, are captured, is shown in Fig 14.11.

The specific difficulties associated with this set-up are as follows:

- large variation of the viewing angle due to the converging camera set-up, thus the appearance (sizes, reflections) of the hands in the two views differ considerably, giving difficulty to the uniqueness assumption [27, 28];
- large occlusion due to the fact that the right hand is waving in front of the body;
- fine details need to be recovered accurately such as all fingers.

Selected computational results for the first frame pair and the fourth frame pair of the video stream are given in Fig 14.12. The original left and right views shown in row 1 and row 3, respectively, are segmented, rectified and cropped images, which are the input to the disparity estimation module. It is observed that most parts of the 3-D scene have been correctly detected and recovered. The disparity maps and the synthesised middle views are very good and visually acceptable. The black areas in

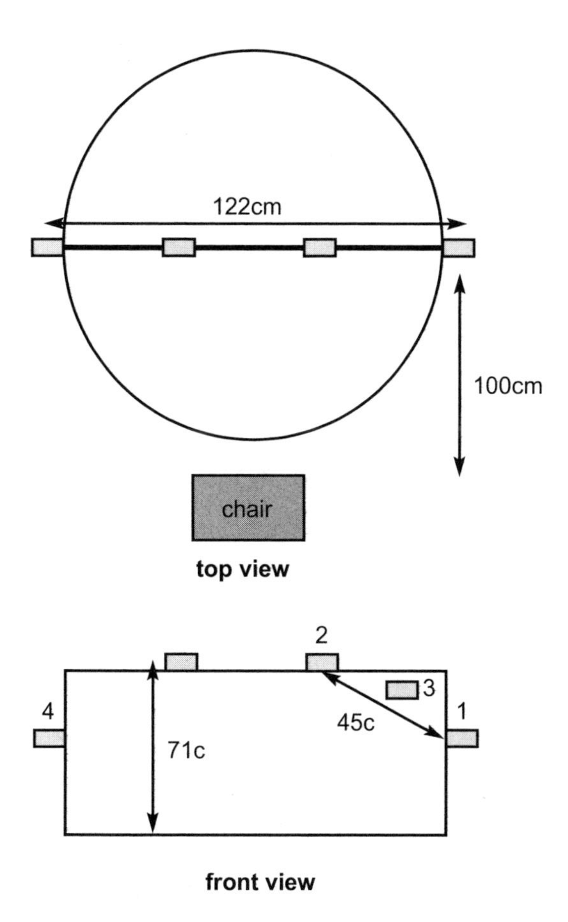

Fig 14.11 The set-up used for capturing a pair of stereo videoconference sequences. A local participant sits on the chair by the round table, and is viewed by four cameras from different locations and angles. The pair of stereo sequences reported in the experiments come from camera 2 (left camera) and camera 1 (right camera).

the disparity maps are mainly due to occlusions. Fingers in the disparity map of the first frame are almost distinguishable. In the 4th frame, because the sizes of all fingers change too much, the algorithm encountered difficulties in matching single fingers correctly. However, the hand as a whole was marked clearly from the body.

To show the functioning of the motion information, we have included in Fig 14.13 the results obtained for the 4th frame pair without employing any motion information and temporal disparity constraints in the HACM. It is apparent that, with the motion information, the estimated disparity maps are smoother and most small outliers have been removed.

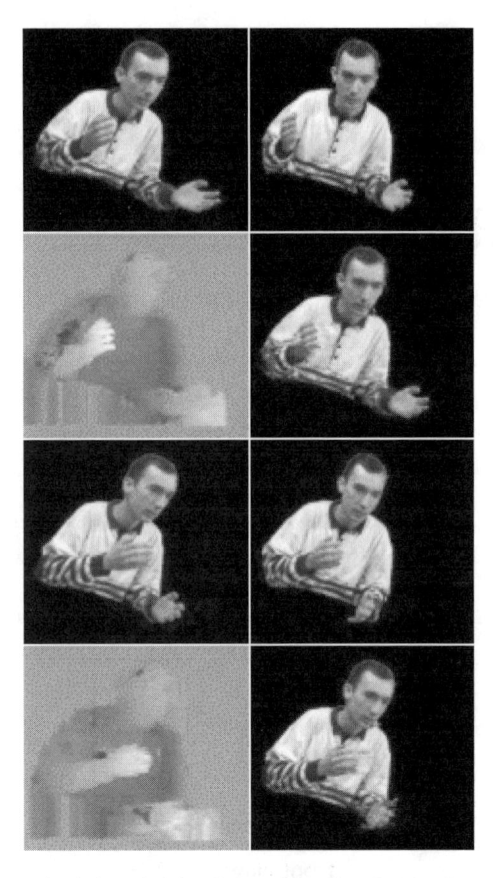

Fig 14.12 Row 1 — the left and right views of the first frame of a stereo video sequence. Row 2 — the disparity computed and the synthesised virtual middle view. Row 3 — the left and right views of the fourth frame of the stereo video sequence. Row 4 — the disparity computed and the synthesised virtual middle view.

Fig 14.13 The estimated left-to-right disparity map and the synthesised middle view image for the original fourth frame shown in Fig 14.12 (row 3) without using motion information and temporal disparity prediction. Comparing this result with that of Fig 14.12 (row 4), the problem of inconsistent disparity values along the moving right arm is obvious in this case.

14.5 Summary

We have discussed in this chapter the recent trend in developing an advanced videoconferencing system that moves towards being a seamless integration of a real and virtual world environment with high-quality 3-D telepresence, while focusing on outlining the scope of research work and software system architecture for the VIRTUE project. The entire VIRTUE system design is flexible and conceived largely on the basis of exploiting state-of-the-art technologies in the computer vision, and image and video processing fields guided by the results of human factor studies, rather than physically placing multiple displays, cameras and human subjects. Specifically, we have chosen to demonstrate a core component for 3-D scene analysis, an efficient technique for robust disparity estimation from stereos of still and dynamic scenes, and its application in 3-D visualisation and videoconferencing applications when employing an image-based rendering approach for reconstructing new (and temporarily coherent) virtual views.

In the course of working towards realising VIRTUE's main objective of a three-way realistic conferencing prototype demonstrator with a high quality of presence, continued and parallel effort has been invested by the consortium in developing novel technical solutions behind all the system modules shown in Fig 14.4. Special and foremost emphasis has been put on real-time performance. These include:

- investigating an alternative hardware-ready disparity estimator [29];
- the multi-step novel view synthesis framework with compact 3-D information representation and an ability to account for scene occlusions [23].

Promising performance has also been achieved on:

- foreground and background segmentation based on adaptive reference templates and invariant colour information;
- a vision-based hybrid head tracker for view point determination;
- simultaneous two-hands tracking to combat hand occlusions and enhance the quality of 3-D information acquired.

These are being combined with the intensive work in PC-centred real-time platform development and implementation, and overall system integration and network transmission.

References

1 Sheppard, P. J. and Walker, G. R. (Eds): '*Telepresence*', Kluwer Academic Publishers, Boston (1999).

2 Lanier, J.: '*Virtually there — three-dimensional tele-immersion may eventually bring the world to your desk*', Scientific American (April 2001).

3 Raskar, R. et al: '*The office of the future: A unified approach to image-based modelling and spatially immersive displays*', ACM SIGGRAPH'98, Orlando, Florida (July 1998).

4 Ichikawa, T. (Ed): '*Immersive Interactive Technologies*', IEEE Signal Processing Magazine, **18**(3) (May 2001).

5 DePiero, F. W. and Trivedi, M. M.: '*3D computer vision using structured light: design, calibration, and implementation issues*', Advances in Computers, **43**, Academic Press, pp 243-278 (1996).

6 Benton, S. A. et al (Eds): '*3-D Video Technology*', Signal Processing: Image Communication (Special Issue), **14** (1998).

7 Mulligan, J. and Daniilidis, K.: '*View-independent scene acquisition for telepresence*', Proceedings International Symposium on Augmented Reality, Munich, Germany, pp 105-110 (October 2000).

8 Onoe, Y., Yamazawa, K., Takemura, H. and Yokoya, N.: '*Telepresence by real-time view-dependent image generation from omnidirectional video streams*', Computer Vision and Image Understanding, **71**(2), pp 154-65 (1998).

9 Fuchs, H., Bishop, G., Arthur, K., McMillan, L., Bajcsy, R., Lee, S., Farid, H. and Kanade, K.: '*Virtual space teleconferencing using a sea of cameras*', Proc of the 1st International Symposium on Medical Robotics and Computer Assisted Surgery, Pittsburgh, PA (1994).

10 Lengyel, J.: '*The convergence of graphics and vision*', Computer, **31**(7), pp 46-52 (1998).

11 Isgro, F., Trucco, E., and Xu, L-Q.: '*Towards teleconferencing by view synthesis and large baseline stereo*', Proc of 11th Int Conference on Image Analysis and Processing (ICIAP'2001), Palermo, Italy (September 2001).

12 Xu, L-Q., Loffler, A., Sheppard, P. J. and Machin, D.: '*True-view video conferencing system through 3D impression of telepresence*', BT Technol J, **17**(1), pp 59-68 (January 1999).

13 Strintzis, M. and Hendriks, E. A. (Eds): '*3-D and Stereoscopic Visual Communication*', IEEE Signal Processing Magazine (Special Issue), **16**(3) (May 1999).

14 Mortlock, A., Machin, D., McConnell, S. and Sheppard, P. J.: '*Virtual Conferencing*', in Sheppard, P. J. and Walker, G. R. (Eds): '*Telepresence*', Kluwer Academic Publishers, Boston, pp 208-226 (1999).

15 Ohya, J.: '*Virtual space teleconferencing: real-time reproduction of 3D human images*', Journal of Visual Communication and Image Representation, **6**(1), pp 1-25 (March 1995).

16 Panorama project summary — http://www.tnt.uni-hannover.de/project/eu/panorama/overview.html

17 Izquierdo, E. and Kruse, S.: '*Image analysis for 3-D modelling, rendering, and virtual view generation*', Computer Vision and Image Understanding, **71**(2), pp 231-251 (1998).

18 Jancene, P., Neyret, F. et al: '*RES: computing the interactions between real and virtual objects in video sequences*', INRIA Technical Report — www-rocq.inria.fr/syntim/ analyse /video-eng.html

19 Ohm, J-R, et al: '*A real-time hardware system for stereoscopic videoconferencing with viewpoint adaptation*', in Benton, S,. A,. et al (Eds): '*3-D Video Technology*', Signal Processing: Image Communication (Special Issue), **14** (1998).

20 Cruz-Neira, C. et al: '*Surround-screen projection-based virtual reality: The design and implementation of CAVE*', ACM SIGGRAPH '93, pp 132-142 (1993).

21 Hirose, M., Ogi, T., Ishiwata, S. and Yamada, T.: '*Development and evaluation of immersive multiscreen display CABIN systems and computers in Japan*', Scripta Technica, **30**(1), pp 13-22 (1999).

22 EU IST Framework V Programme, VIRTUE — http://www.virtue.eu.com (January 2000-January 2003).

23 Lei, B. J. and Hendriks, E. A.: '*Multi-step view synthesis with occlusion handling*', in Proc of Vision, Modelling and Visualisation (VMV01), Stuttgart, Germany (November 2001).

24 Kanade, T. and Okutomi, M.: '*A stereo matching algorithm with an adaptive window: theory and experiments*', IEEE Transactions on Pattern Analysis and Machine Intelligence, **16**, pp 920-932 (1994).

25 Hendriks, E. A. and Marosi, G.: '*Recursive disparity estimation algorithm for real time stereoscopic video applications*', in Proc of ICIP '96, pp II-891-II-894, Lausanne, Switzerland (1996).

26 Lawrence, Z. and Kanade, T.: '*A cooperative algorithm for stereo matching and occlusion detection*', IEEE Transactions on Pattern Analysis and Machine Intelligence, **22**(7), pp 675-684 (July 2000).

27 Bobick, A. F. and Intille, S. S.: '*Large occlusion stereo*', International Journal of Computer Vision, **33**(3), pp 181-200 (1999).

28 Ishikawa, H. and Geiger, D.: '*Occlusions, discontinuities, and epipolar lines in stereo*', Proceedings of 5th European Conference on Computer Vision (ECCV'98) (1998).

29 Kauff, P., Brandenburg, N., Karl, M. and Schreer, O.: '*Fast hybrid block- and pixel-recursive disparity analysis for real-time applications in immersive teleconference scenarios*', 9th Int Conference on Computer Graphics, Visualization and Computer Vision, Plzen, Czech Republic (February 2001).

15

A VIRTUAL STUDIO PRODUCTION CHAIN

J M Thorne and D J Chatting

15.1 Introduction

The Prometheus project is seeking to create a virtual studio production chain. Prometheus is a three-year collaborative LINK project under the Broadcast Technology Programme funded by the UK DTI and EPSRC, led by the BBC. The project includes markerless face and body tracking, actor and clothing model animation, scene construction and three-dimensional display technologies. It is seeking to build an entire production framework to encapsulate these technologies.

This will open up new possibilities in the creation, distribution and display of multimedia content, and promises a revolution in which the director and viewer of performances will have unprecedented powers. Actors, scenes and props become components to be arranged at will.

This chapter considers these technologies, discussing the challenges that actors and directors now face and the new powers of the viewer. It also reflects on the historic changes in 'entertainment experiences'.

15.2 The Prometheus Project

The Prometheus partners are the BBC, BTexact Technologies, AvatarMe, Snell and Willcox, UCL, Queen Mary University of London (QMUL), De Montfort University and the University of Surrey. The BBC as lead partner is aware of the limitations of existing 2-D film media and is here trying to create an equivalent 3-D production chain. The project addresses everything from creation and animation of models through to 3-D displays. BTexact Technologies is responsible for the markerless real-time motion capture of the actors' faces and the creation and animation of photo-realistic 3-D head models. A more detailed overview of the project is given by Price and Thomas [1].

Photos or short sequences of frames from multiple cameras are used to automatically create realistic 3-D avatars of the actors complete with bone structure and facial features that can be animated. Markerless real-time motion tracking is then employed to capture the movements, voice and facial expressions of the actors in the studio. This tracking data is used to animate the avatars. Real-time cloth simulations drape the avatars in the latest fashions and virtual studio techniques place them in the scene. MPEG-4 [2] compression and streaming transmits the complete performance to the end user who can explore it in full 3-D.

MPEG-4 is the newest standard to be released by the Moving Picture Experts Group, providing further leaps in audio and video compression technology and a platform for true multimedia experiences. An MPEG-4 stream contains information about a 3-D stage on which are placed images, video, audio sources and 2- or 3-D mesh objects. All of these can be animated. In addition, MPEG-4 defines specific ways of encoding the human form and its movements, paying particular attention to the intricacies of the face. It also includes the Internet concepts of hyper-linking and scripting, thus enabling sophisticated interaction with the viewer. MPEG-4 shares many things in common with VRML [3] and is of a similar file format to QuickTime [4].

While there are still some exciting challenges left to address, many of the initial technological hurdles have been overcome.

A novel editing studio is being developed whereby the director can change the lighting, the set, the clothes worn by the actors, or even replace one actor model with another. For non real-time performances the director may ultimately edit the motions and gestures of the actors — removing a limp here and adding some more anger there. Moreover, because of the use of MPEG-4, the viewer gains the ability to make choices as they watch and choose between threads as the performance unravels.

Full 3-D body models can be assembled from still images and tracked without markers against a blue screen. A cross-platform avatar tool-kit has been developed to interpret the animation streams and manipulate the avatars.

Simulated cloth exhibiting the properties of its real-life counterpart can be draped over these avatars, though for complex cloth the intense computation required limits the speed to less than real time. MPEG-4 compression codecs are being evaluated and integral imaging techniques and displays have been developed to render 3-D scenes so that the viewer can inspect the scene from an infinite number of viewpoints.

Photo-realistic head models can be created from two photographs. Head movements and changes in facial expressions can be tracked in real-time in a normal office environment without placing markers on the face, and MPEG-4 compatible animation streams can be derived to represent this.

15.2.1 Creating Photo-Realistic 3-D Head Models

Sufficient information to create a realistic head model can be obtained from just two photographs of the face — one from the front and the other from the side (Fig 15.1).

Image processing techniques can be applied to these photos to automatically locate important feature positions and the shape of the head. A generic head model can then be deformed to look like the person by comparing the locations of these feature points in both the photos and in the generic head. The photos of the person can then be projected on to this conformed head and stitched together to form a seamless texture. The resulting heads are easy to animate because we already know the locations of the important features. The Prometheus head is MPEG-4 compatible.

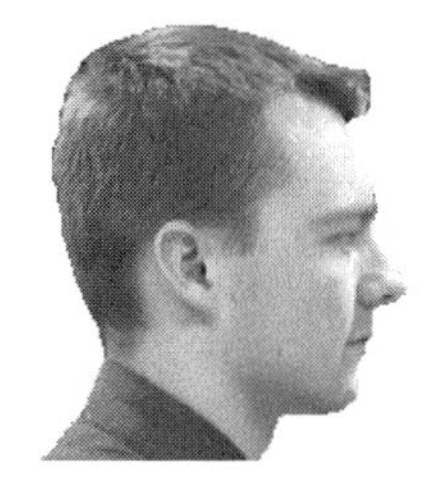

Fig 15.1 A generic head model constructed from two photographs.

15.2.2 Markerless Face Tracking

The challenge of face tracking is to capture the expressive qualities of the face, which manifest themselves visibly through the movement of features and wrinkling of skin, then to store and replicate them on a computer-generated puppet. Traditionally these changes have been captured using markers placed on the face; however, through using computer vision techniques, markerless systems are becoming viable.

The Prometheus face tracker (see Fig 15.2) uses a single camera to capture the actor's performance in real time, from which it derives the head's orientation and the facial expression. This can be encoded as an MPEG-4 stream and be used to animate the head previously created. The tracking algorithm is based on earlier work in BTexact by Machin [5] and Mortlock et al [6]. With a real-time constraint, the tracking data will be inherently probabilistic and unreliable; in addition, with a single camera depth, information is not readily available. For these reasons the interpretation of the raw tracking data is crucial. Where frames are missing or features adjudged unreliable, estimates must be made of reality. However, we must be aware that we are manipulating a very sensitive communications mode, human facial expression.

The human face can convey the extremes of emotion, with a small variation of muscle combinations. If we fail to interpret the tracking data correctly, the result can have a huge impact on the meaning of a performance. Interpretation is the focus of our current research.

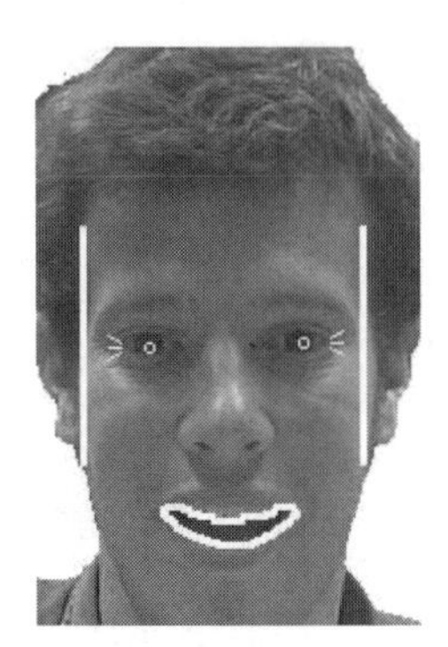 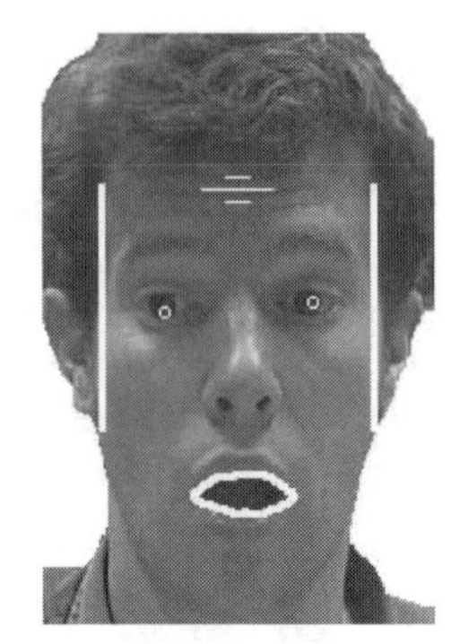 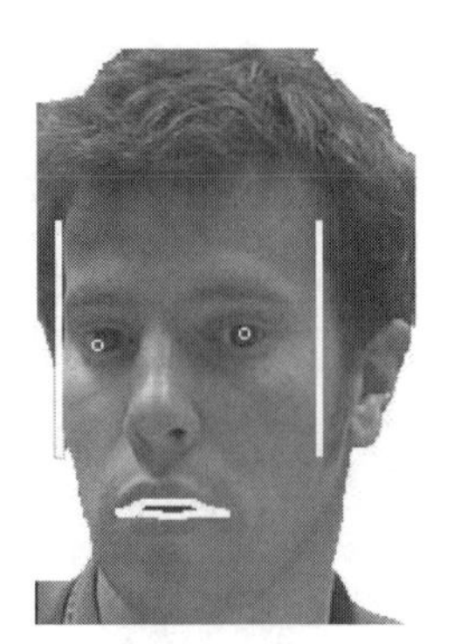

Fig 15.2 Screen shots from the face tracker.

15.3 Considering the Future of Entertainment Experiences

How do technologies like MPEG-4 and projects like Prometheus develop and extend current possibilities for programme makers and consumers and magnify challenges that have existed in the industry for many years? This section first considers the history of entertainment experience and then examines these future challenges more thoroughly.

15.3.1 A Brief History of Theatre, Film and TV

In the beginning there was theatre. Here actors and audience interacted in a live 3-D arena. The story unfolded before the viewers, adapting and changing in a continual feedback loop with the audience. If the director had allowed for it, the story could diverge down any one of a number of routes (and even if they had not, a few well-aimed tomatoes could dramatically alter the course of a performance). Simple tricks interwoven with the story could provide the impression of the impossible happening. By changing clothes and their style of acting (voice, mannerisms) actors could pretend to be different people. Under cover of a curtain, a quick change of set and lights could conjure up a different place or time. Hidden trapdoors, ropes, props, magic tricks and simple effects (thunder, smoke) could further allow the suspension of reality. However, everything had to be planned carefully in advance with much rehearsing since in the live performance everything has to be right first time.

Along came film — in essence nothing more than a discrete sequence of images. These are replayed fast enough to fool the human eye into accepting the impression of continuous movement. The recorded nature of film however, gives the director great gains in flexibility. They can split up and rearrange the images — to remove and replace bits that went wrong the first time, or to alter the order and pace of story. The soundtrack of the film can also be independently rerecorded, so that effects can be added or language changed.

The ability to make the camera lie created the world of visual effects. Carefully scaled models can create impossible places or events — giant star cruisers or long lost cities. Films themselves can be blended together — for example, the teleportation effects in Star Trek were accomplished through the blending of the actor, the empty stage and some illuminated aluminium foil strips.

Then came TV and video which not only brought the viewer new freedom to watch what they wanted when they wanted it but also imposed tighter production schedules and smaller budgets. This prompted new ideas in the reuse of media.

In both TV and film, however, the viewers are restricted to seeing just what the director intended, they cannot change their point of view and cannot alter the story. Theatre still has not been replaced — the atmosphere and 'liveness' can still hold an audience enthralled.

In parallel with all this was the development of animation — from the first animators, such as Winsor McCay and his hand-drawn animation *Gertie the Dinosaur* in 1914 [7] (see Fig 15.3), through Walt Disney's lavish productions and Hanna-Barbera's low-budget TV shows, to the computer-generated wonders of *Toy Story* and *Final Fantasy*.

Fig 15.3 Gertie the Dinosaur.

Throughout this history there has always been a close connection with live action. Winsor McCay conversed on stage with his creation Gertie on screen behind him. Whenever a close-up shot was used of the puppets in *Captain Scarlet* the hands seen were those of real people. Films like *Who Framed Roger Rabbit* made cartoon characters the stars of live action films while *The Labyrinth* brought together actors and the animatronic creations of Jim Henson. The latest blend is the entirely virtual character Jar Jar Binks in the new *Star Wars* films. However annoying we find him, it cannot be doubted that the gap between reality and animation is becoming increasingly thin.

15.3.2 The Future

It would also seem that people are shifting their attentions away from TV and film and into gaming; in its opening weekend the game Resident Evil 2 took more money than any film showing except *Titanic*. People are demanding more from their media, they want interactivity — games, out-takes and access to extra information. With advances in avatar creation technologies, for instance the AvatarBT installation at the Millennium Dome, we are able to create low-cost photo-realistic avatars. A popular application of these avatars is as characters in computer games, allowing people to play as themselves or their heroes. Personal avatars are explored further in Chapter 16. Consumers are able to put together components to customise their own game experience.

Technologies like MPEG-4 and projects like Prometheus promise the same sort of customisation for other entertainment experiences, closer to today's cinema. As actors, scenes and props become components to be arranged at will, the traditional role of the actors, directors and viewers changes and a new broadcast medium is created. This will have a profound effect on the type of experiences that can be delivered to the consumer. There are many ways in which this can be used creatively by 'experience' producers.

Imagine a soap opera in which the consumer can choose to follow any single character from a group scene. In a football experience you can choose to take the perspective of a player or even the ball. With avatar technology the consumer is able to swap different actors to play each part, including models of themselves.

Directors' powers are also enhanced. They are now able to manipulate the performance of their actors as never before. Every muscle movement is within their control, every glance, shrug, to be edited without any intervention from the original performer. Every element of the set is interchangeable, the dress of the actors, the lighting, the camera position, etc. Scenes, props and performances can be integrated from archives into a new consumer experience, at minimal cost. This is already possible in a limited way — for instance, the synthesised meeting of Tom Hanks and John F Kennedy in the film *Forrest Gump* (1994) — but at considerable expense. Furthermore, the director is able to use and edit an old performance with a new actor. The director may choose to produce an interactive consumer experience or a conventional film, from this process. Objectised media facilitate a broad range of new opportunities for the director and consumer, but there are a number of challenges which remain unresolved.

15.3.3 Disembodied Acting

Considering a scenario where the director or user is able to select any actors (including themselves) as the lead in an action experience, a number of important questions are raised. For this application it is not enough to simply change the actor model. To give a convincing performance as an identifiable actor, the physical model must be a high-quality facsimile, the voice must be a good mimic, the appropriate vocabulary and turns of phrase should be used, and the characteristic expressions and mannerisms need to be replayed too. This is the skill of the impersonator. In fact, it is possible for some performers to adopt the persona without changing the way they look, the effect is achieved with voice and action alone. The television mimic Rory Bremner can 'look' like Tony Blair, without looking like Tony Blair.

In order to facilitate this level of customisation, mannerism and presence need to be describable too. From the recording of an original performance with its associated characterisation, a new sequence needs to be synthesised with an alternative characterisation. There are two general methods of computational impersonation we could apply from the speech synthesis field:

- to capture and store a suitable set of atomic elements of the target person's performance and to recombine these into a new performance — this requires a complete performance database for any given actor and sophisticated interpolation between elements;
- using a generic model with a very rich set of parameters to alter the performance — this requires that a generic model be created and a sufficient set of parameters chosen, while avoiding the need for a complete library of performance (assuming mimic parameters can be generated from a set of sample performances).

In both methods a representation which allows the classification of instances of performance is required — in the first, to find corresponding elements to replace, and, in the second, to select the appropriate model for that behaviour. In speech this could be a phonemic representation, while in facial expression this could be a FACS (facial action coding system) [8] or an extension. However, this representation needs to be abstract enough to allow a wide interpretation — for instance 'character A intends to greet character B'. This intention is then expressed depending on the individual characterisations, using different language, facial expression and body posture, dependent on the individual character. It is debatable whether these intentions are derivable from the original performance by capturing raw motion and sound samples.

Assuming that it is possible to derive a high-level description of the pure performance or disembodied performance, we can consider how this can be combined with a characterisation model. A comical illustration of this is the 'Enchefferizer' [9]. The Enchefferizer converts ASCII text into a sentence uttered in the style of the Muppet's Swedish Chef, using a set of transformation rules. These rules can be considered to encapsulate a characterisation of the Swedish Chef (see Table 15.1). We can imagine an extended set of rules to encode the vocabulary of Tony Blair. The source dialogue needs to be sufficiently abstract to drive any arbitrary character. Again this raises the question as to whether these descriptions can be generated automatically from a sample performance. They would need to describe both the verbal and non-verbal performance.

Table 15.1 'Enchefferized' text sample.

Source text	Enchefferized text
Colourless green ideas sleep furiously	*Culuoorless greee idees sleep fooreeuoosly*

There is therefore a need for a language to describe a performance, in such a way that it can be impersonated by any actor. A number of current schemes partially address this need.

HumanML [10] is an XML and RDF schema specification, which embeds human characteristics including emotions, intentions, motivations and allusions within conveyed information. Another example is SABLE [11], which is an XML/SGML-based mark-up scheme for text-to-speech synthesis allowing attributes of the speaker and their delivery to be described.

Laban Movement Analysis (LMA) [12] allows human movement, specifically dance, to be transcribed in Labanotation. LMA analyses five components of motion — body, space, shape, effort and relationship. Body considers which parts of the body are moved during the sequence. Space describes how the body moves through space — the locale, directions and paths of motion. Effort looks at how rather than what the person performs, the dynamic quality of the movement. Shape refers to the way in which someone moves in space, which Laban believed reveals a person's inner thoughts, feelings, drives and emotions. Relationship is the interaction between the performer and their environment.

The EMOTE (expressive motion engine) work, at the University of Pennsylvania [13], is an example of a computation model of LMA which allows the animator to alter the effort and shape of a model.

A description of facial expression, independent of the morphology of the target face, has been addressed by Noh and Neumann [14] among others. MPEG-4 also seeks to encode abstract facial animation parameters (FAPs) which can be applied to any arbitrary MPEG-4 face model. The FAPs encode the required displacement of control points on the face which are then expressed in units derived from the dimensions of the target face. The problem has also been considered for body motion by Gleicher [15], where motion capture data is 'retargeted' from one character to another, using their physical features.

If it were possible to capture and represent all the modes of performance, automatically or in an assisted way from a sample performance, it would be plausible to consider computer-based mimicry.

15.3.4 Dislocated Performances

In animation there has always existed a dislocation of the director and the performance, unlike film making. How do you direct a scene when it takes an animator a week to finish a suggestive wink?

Jane Horrocks, an actress who supplied the voice of a chicken in the animated film, *Chicken Run*, made an interesting observation regarding the divorce of her voice from her character. She commented that she would have spoken differently if she had been aware that the character had such prominent teeth.

With current tracking technologies, such as the Prometheus face tracker, a 'face-over' scenario is considered where the voice, body and face capture for a single performance may not be collocated and the director or viewer may wish to change any aspect of it at any time in the future, i.e. the performance is dislocated from the character's embodiment.

Live action with real and virtual characters creates a problem of dislocation too. How do actors cope with acting in empty studios imagining the virtual characters with whom they are interacting? 'It's really difficult', said Ewan McGregor, of Star Wars Episode II: *Attack of the Clones*, commenting that: 'You just have to remember to look up there,' referring to acting with invisible alien co-stars. Humans are very sensitive to inconsistencies in eye gaze.

Augmented reality systems may address some of these problems in the short-term, enabling more believable and sympathetic performances. There will exist the potential for a viewer or director to change individual features of an actor, giving the possibility to mix-and-match characteristics. Here there emerges the problem of characteristic interaction. In this case, the embodiment of the actor and their mannerisms become dislocated, so for example the voice can be changed irrespective of the sex of the character. There are inter-dependencies between characteristics of humans which, if they are violated, create unbelievable characters.

There is a need to develop a language in which dependence between features can be described and a set of mechanisms built to resolve conflicts. So, in the *Chicken Run* example, the character's voice could be changed if the mouth was changed.

15.4 Summary

Technologies like those being developed within Prometheus will bring great flexibility to the director and viewer of future entertainment experiences. Features of the identity of a performance (animated or real) are becoming increasingly parameterised and represented in digital electronic form. These features can be decomposed, duplicated, changed, recalled and reassembled.

The challenges we have identified are those of disembodied acting and dislocated performances. In the first, we must develop novel methods of representing and extracting pure or disembodied performances and applying new characterisation to them, while in the second, we must ensure coherence between the interdependent features of a performance even when that action is assembled from multiple discrete components.

References

1 Price, M. and Thomas, G., A.: '*3D virtual production and delivery using MPEG-4*', International Broadcasting Convention (IBC 2000), Amsterdam, IEE Conference Publication (2000) — http://www.bbc.co.uk/rd/pubs/papers/pdffiles/ibc00mp.pdf

2 MPEG-4 —http://www.cselt.it/mpeg/standards/mpeg-4/mpeg-4.htm

3 VRML: '*International Standard ISO/IEC 14772-1:1997*' — http://www.vrml.org/Specifications/VRML97/index.html

4 QuickTime — http://www.apple.com/quicktime/

5 Machin, D. J.: '*Real-time facial motion analysis for virtual teleconferencing*', Proc of the Second Int Conf on Automatic Face and Gesture Recognition, IEEE Comput Soc Press, pp 340-344 (October 1996).

6 Mortlock, A. N., Machin, D. J., McConnell, S. and Sheppard, P. J.: '*Virtual Conferencing*', in Sheppard, P. J. and Walker, G. R. (Eds): '*Telepresence*', Kluwer Academic Publishers, Boston, pp 208-226 (1999).

7 Gertie the Dinosaur — http://www.dinosaur.org/Gertie.htm

8 Ekman, P. and Friesen, W. V.: '*Facial action coding system: A technique for the measurement of facial movement*', Consulting Psychologists Press, Palo Alto, California (1978).

9 Hagerman, J. posted the original chef.x program to Usenet in 1993 — an on-line version of the current Enchefferizer can be found at — http://www.cs.utexas.edu/users/jbc/home/chef.html

10 HumanMarkup — http://www.humanmarkup.org/

11 SABLE — http://www1.bell-labs.com/project/tts/sabpap/sabpap.html

12 Laban/Bartenieff Institute of Movement Studies (LIMS) — http://www.limsonline.org/

13 Chi, D., Costa, M., Zhao, L. and Badler, N.: '*The EMOTE model for effort and shape*', Proc of SIGGRAPH 2000, ACM Computer Graphics Annual Conference (University of Pennsylvania), pp 173-182 (2000).

14 Noh, J. Y. and Neumann, U.: '*Expression Cloning*', Proc of SIGGRAPH 2001, ACM Computer Graphics Annual Conference, pp 277-288 (2001).

15 Gleicher, M.: '*Retargetting motion to new characters*', Computer Graphics Proc, Annual Conference Series, pp 33-42 (1998).

Part Five

THE FUTURE DIGITAL HOME

S G E Garrett

This last part gives two views of the future. Leading visionaries at BT's Adastral Park give their insights into aspects of the future for the home environment.

Chapter 16
Personal Virtual Humans — from Inhabiting the TalkZone to Populating the Digital Home

One of the fundamental goals of graphics has always been to visually create a three-dimensional person that is indistinguishable from a real person. This target is only slightly short of being reached, as was demonstrated when Columbia Pictures released *Final Fantasy* in 2001, an animated science fiction film with high-definition emotive characters. Through what has been a cultural and technological convergence, we are now starting to see software tools and techniques that can generate lifelike characters while not sacrificing the human judgement and artistic skills that are needed in character animation. Computer games, films, and the Internet are now starting to use virtual humans, which will become ever more realistic. However, a new challenge has been set — people now do not want to just play a computer game with a visually realistic character or see a television programme with a vivid computer-generated character, they want to see themselves in that computer game or film.

The future is towards individuals having their own virtual clones, which they can utilise in computer-generated worlds and applications. This chapter discusses the advantages of having a personal computer-generated character, and also describes several systems that BTexact Technologies has successfully developed and deployed to generate them, as well as some of the applications for which they can be used. The chapter finishes by providing a glimpse into the future of what we can expect over the next few years, with the advent of this new exciting technology.

Chapter 17
Hype and Reality in the Future Home

What will stay the same and what will change in homes of the future? This chapter tracks and analyses various streams of thought, past and present, looking ahead to emerging technologies which extend and develop our thinking on what might be core to the future home.

16

PERSONAL VIRTUAL HUMANS — FROM INHABITING THE TALKZONE TO POPULATING THE DIGITAL HOME

D Ballin, M Lawson, M A Lumkin and J Osborne

16.1 Introduction

A virtual human is a graphical representation of a human being; BTexact Technologies is actively working in the field of virtual humans, particularly avatars. The word avatar comes from the ancient language of the Vedas and of Hinduism, known as Sanskrit. Avatar means a manifestation of a spirit in a visible form typically as an animal or human.

The idea of an avatar became synonymous with a graphical representation of oneself in a computer-generated environment, after the science-fictional novel *Snow Crash* was launched, written by Neil Stephenson [1], an author whose ideas were ahead of their time.

The first graphical computer system that used avatars was called Habitat, a networked virtual world developed for the Commodore 64 by Lucasfilm Games in 1985 [2]. Here players were represented by 2-D cartoon animated figures, crude by today's standard. Habitat avatars (see Fig 16.1) were controlled by a joystick, and could pick up objects, players could talk to one another by typing, and speech would appear in a word balloon.

In general, avatars can take on any form, from a simple 2-D picture as used in Microsoft's Chat (formerly Comic Chat) or V-Chat [3, 4], to a 3-D rendered model that symbolically represents the user as in Blaxxun's Interactive or Active Worlds' AlphaWorld [5, 6].

Fig 16.1 Scene from Habitat.

However, BTexact Technologies has taken the concept of an avatar a step forward, by enabling people to project their own body image on to their avatar, and create personal avatars.

These personal avatars not only represent the user symbolically but also physically resemble them. In the year 2000 over 271 000 people had their body images scanned in the AvatarBT scanning booths at the TalkZone in the Millennium Dome. In a short two-minute process a visitor was able to obtain their full body 3-D photo-realistic representation. These avatars were then used in a range of entertainment applications in the TalkZone, and other applications that were available to download at home.

16.2 Overview

There are several techniques of creating three-dimensional personal avatars. The two methods described here both require access to either images or the actual person — one is through photogrammetry techniques, the other is through laser scanning.

Photogrammetry reconstructs virtual humans from a series of images of the subjects. The process usually starts by indicating where the edge of the subject exists. Photogrammetry software reconstructs a simple version of the geometry by analysing and comparing perspective lines and shading information in several related photographs of the same person.

Laser scanning is another effective technique in modelling three-dimensional humans. This technique typically uses a linear laser scanner to collect three-dimensional point data that is then converted into a polygonal mesh. Although it is faster and more accurate than photogrammetry, it is expensive and needs a skilled procedural modeller during post-production on the data.

This chapter covers examples of both photogrammetry and laser techniques of virtual humans created by BTexact Technologies. Section 16.4 discusses the photo-realistic personal avatars that were developed for the Millennium Dome, while section 16.5 discusses a stylised personal virtual human that was created as a virtual newsreader. Section 16.6 describes the avatar applications at the TalkZone, and section 16.7 shows some of the highlights of an evaluation of the exhibit.

16.3 The AvatarBT Exhibit

16.3.1 Introduction

The Millennium Dome was designed as a showcase to celebrate the achievements of the last thousand years and, more importantly, to take a glimpse of the future to come. British Telecommunications sponsored the communications zone known as the 'TalkZone' [7] within the Dome, and was given the task of telling the story of communications and showing new and exciting technologies that will affect the future way we live and work. Several exhibits were conceived through a set of feasibility studies in 1999, one of these was known as AvatarBT. The AvatarBT exhibit was to be developed where visitors in the TalkZone could be 'scanned in' and in a few minutes see their own avatar in an exhibit. It carried a strong message to the public about the power of communications and how communications technology will have a positive impact on their lives in the future. It was exciting and new, showing BT as a world-leading innovator, and fitted within the overall message of the TalkZone. However, the project was using experimental technology that was unproven in terms of being used for entertainment, and, although the concept was simple, the technology was complex and it needed to be made simple for public consumption.

Thus began the AvatarBT project. What follows in this section is an explanation of the technology that evolved from AvatarBT and how it was received by the general public.

16.3.2 The AvatarBT Scanning Facility

The AvatarBT scanning process [8] is reminiscent of a passport photograph booth. The device, referred to as the avatar booth, shown in Figs 16.2 and 16.3, is rugged, self-contained and compact, with an internal area for a person to stand. The scanning process takes a little under two minutes from start to finish and can be either operated by the subject themselves, in the same way as a photo booth operates, or by a facilitator who operates the booth for the subject. In the case of the TalkZone, there were three 'avatar booths', all of which used facilitators to guarantee that the required throughput of visitors was met. Once the avatar had been created, the subject was able to immediately view it on a screen. The avatar was also made available via the Internet so that it could be downloaded at a later date from a Web page and used on their desktop PC [9].

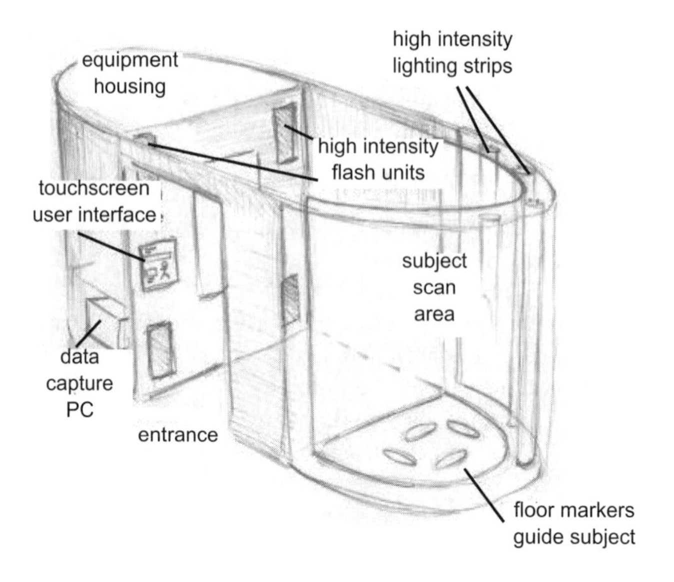

Fig 16.2 Design of the avatar booth.

Fig 16.3 Prototype of avatar booth.

16.3.3 AvatarBT Algorithm

An overview of the personal avatar creation algorithm is shown in Fig 16.4. It is based on taking four orthogonal whole body images of the user. It should be pointed out that this chapter documents the process for creating as it was done for the TalkZone — further developments have seen improvements in the technique and hardware, but the fundamental concept is the same.

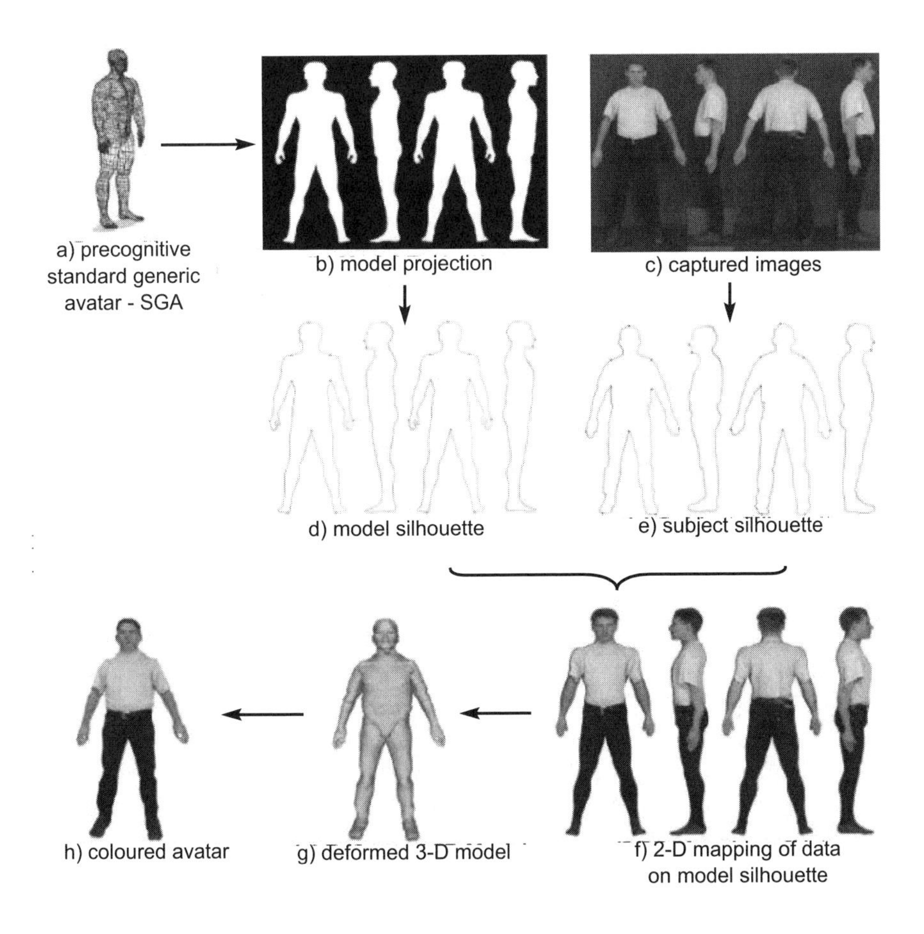

Fig 16.4 Overview of the algorithm for creating a personal avatar from orthogonal pictures in the booth.

16.3.3.1 The Standard Generic Avatar

The system is based on a 'pre-cognitive' 3-D humanoid [10, 11], a pre-cognitive being a starting template of a typical humanoid form; this is then deformed by data calculated from the scanning process to give a 3-D humanoid shape that represents the body of the subject that was scanned. As the computer is only required to process changes to the model and does not need to build a new model from scratch, the scanning system requires far less processing and is much faster than a system that does not use a pre-cognitive. The AvatarBT pre-cognitive is known as a 'standard generic avatar' (SGA), and conforms to the standard of the Humanoid Animation Working Group (H-Anim), which defines a virtual humanoid structure [12]. The SGA consists of around 2600 polygons arranged in the shape of an average height, averagely proportioned, non-sexed human being and has a wire mesh model, as shown in Fig 16.5. Four synthetic images have been projected in advance from the SGA from four orthogonal views (front, left, right, back), as shown in Fig 16.4(b).

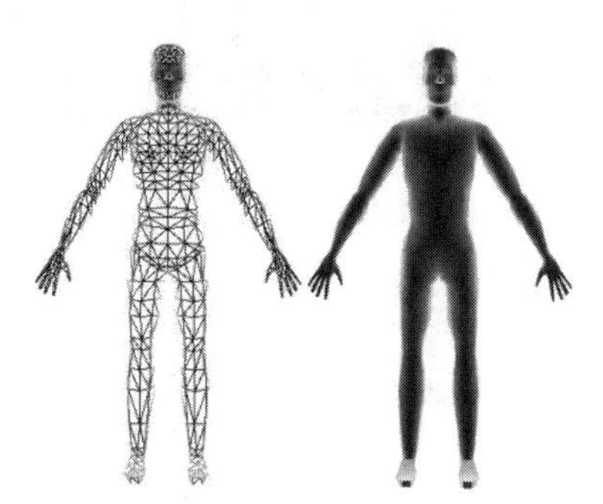

Fig 16.5 Pre-cognitive standard generic avatar (SGA).

16.3.3.2 Capturing the User

To create a virtual model of a person in the avatar booth it is necessary to capture four orthogonal images of the subject [13] shown at Fig 16.4(c). The subject needs to stand in four predetermined poses, (front, left, back and right) as shown in Fig 16.6; to help the subject with this, clear foot markers are shown. To get a good avatar the subject uses these poses and removes baggy clothing, that may cause both the armpits and crotch to be obscured, and any extraneous objects on their person, as well as tying back long flowing hair. The system will automatically recognise if certain key features are obscured or if the subject is incorrectly posed at run time, and ask for a rescan.

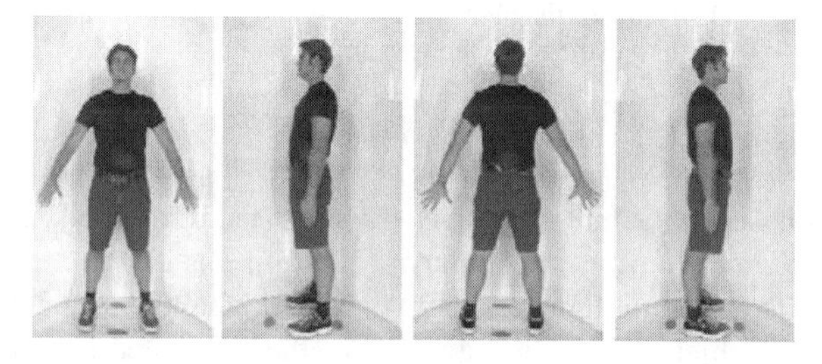

Fig 16.6 The four poses required for scanning.

During each pose the avatar booth takes two pictures of the subject using two digital cameras giving two sets of four images, making a total of eight images which are used to build the avatar. The first set of four images, referred to as 'silhouette images', are taken by a Kodak DCS 560, the second set of four images, referred to as 'texture images', are taken using a Fuji DS 330 (see Fig 16.7). For each pose two pictures are taken simultaneously so as to reduce the time required for the scan. Recent developments mean that we now only use one camera.

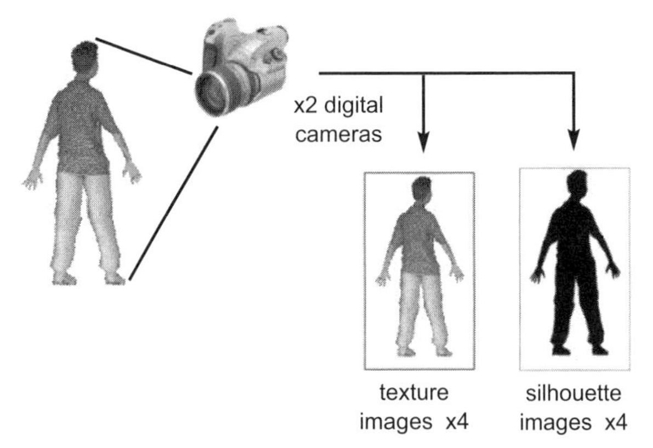

Fig 16.7 Image acquisitions.

Typical raw silhouette images (as shown in Fig 16.8(a)) are dark solid images of the subject, set against the bright white background of the luminescent back panel. The brightness of the back panel is important to gain the correct lighting conditions necessary for photogrammetry.

The raw texture images are taken using a high-intensity flash pulse that cancels the brightness of the luminescent back panel giving a very high-resolution digital colour image of the subject against a white background, as shown in Fig 16.8(b).

SGA silhouette data has already been accurately projected from the model projection (Fig 16.4(b)), to reach the step in Fig 16.4(d). Silhouette extraction is performed on the raw silhouette image, between the image of the person and the background, constructing a contour of pixels that form the outline of the person, so as to get to the step shown in Fig 16.4(e).

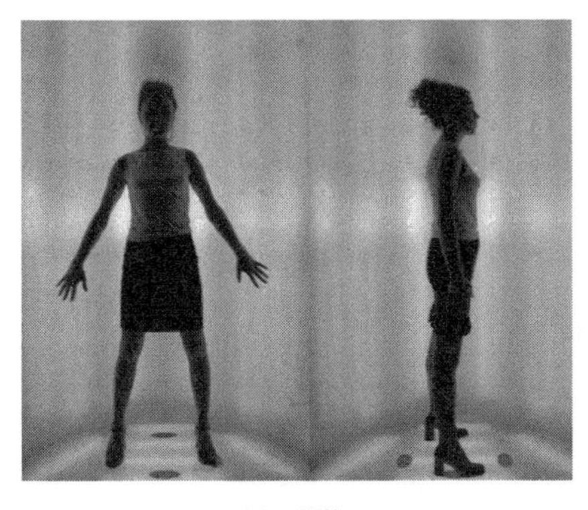

(a) Silhouette.

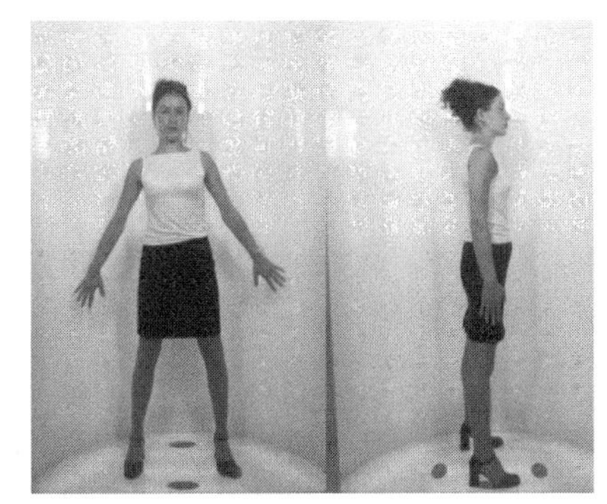

(b) Texture.

Fig 16.8 Raw image examples.

Alignment of the key points between the SGA model and data ensures that separate functional parts of the SGA model (arms, legs and head) are correctly mapped to corresponding parts of the captured image silhouettes. Figure 16.9 shows a silhouette with all the feature points highlighted. Five specific feature points are known as extrema points, and are the dots shown in Fig 16.9 on the tip of the skull, hands, and feet. However, the precise position of where the extrema points lie on the contour of the human shifts according to the pose, weight, and size of an individual. However, the other feature points (of the crotch, armpits, and shoulders) are extracted more accurately, and the algorithm uses these reference points to shift the extrema points giving a more accurate position of where they are on the contour of the body. This procedure is applied to all four profiles of the subject's silhouette that we have. Once we have this information we can judge where the arms, torso, head and legs are. The next step (shown as Fig 16.4(f)) is a fine one-to-one mapping of data points between the model silhouette and subject silhouette; this also helps to provide us with the information on where colour data should be mapped.

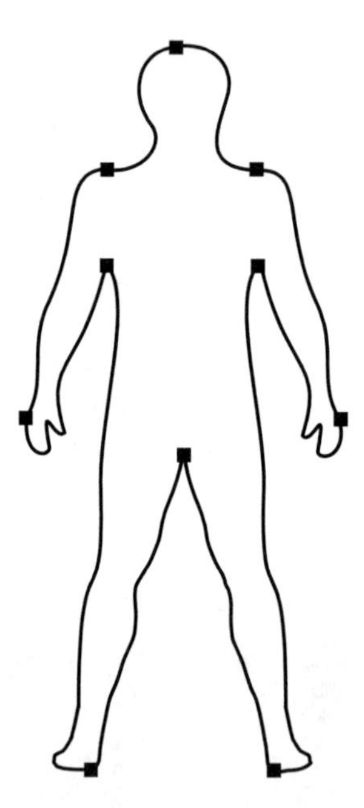

Fig 16.9 Feature and extrema points.

A 2-D to 3-D mapping is then performed as outlined in section 16.3.3 which combines the orthogonal images generated at Fig 16.4(f) with the standard model 3-D generic human shape to estimate the subject's profile to create a 3-D mesh model of the subject (Fig 16.4(g)). This mesh is then textured so that we have a fully reconstructed model and it is saved as a personal AvatarBT (Fig 16.4(h)).

16.4 Vandrea — The Virtual Newscaster

16.4.1 Stylised Virtual Humans

A virtual human does not have to be photo-realistic like the scanned AvatarBT ones to look visually convincing. An example of such a virtual human is the popular games character Lara Croft [14]. Although the textures on her face and body are synthetically generated she looks believable. Avatars whose geometric features are highly precise, but whose synthetic skin may not conform to what typically looks natural but is still of a high resolution, are referred to as stylised virtual humans. An example of such a virtual human is Vandrea, a virtual newscaster (see Fig 16.10).

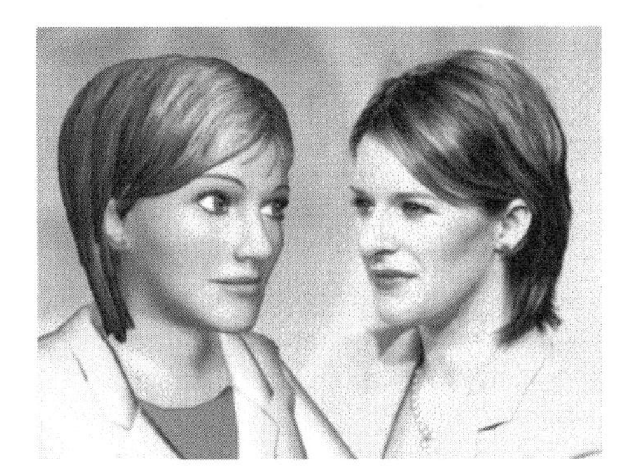

Fig 16.10 Andrea Catherwood sizes up her cyberclone.

16.4.2 Creating Vandrea

Vandrea was a project that BTexact Technologies led in collaboration with ITN, Channel 5, and Televirtual [15]. Vandrea is a stylised virtual human, a high-definition character that delivers the latest news from the desktop of your personal computer.

Unlike the avatars from the BT avatar booths, Vandrea is not a traditional avatar in that she is not a digital puppet, Vandrea works automatically, responding and acting to real-time data. To make sure that the virtual newscaster would be easily accepted, it was important that the general public could relate to the virtual character. Humans are neurologically wired to recognise faces [16], and this, in conjunction with our memory, helps us immediately to associate a face with a task or role. For this reason Vandrea is based on a real person, news anchorwoman Andrea Catherwood. Hence the character's name Vandrea — virtual Andrea. Andrea Catherwood, shown in Fig 16.11, joined the UK's Channel 5 newsroom in January 2000 as presenter of the flagship News and Sport. By modelling Vandrea on a popular real newscaster, it allowed people to psychologically associate the role and tone of the character as one that would soberly present the news.

Fig 16.11 Andrea Catherwood is laser scanned at a Channel 5 studio.

16.4.2.1 Laser Scanning Process

Vandrea was created using a mixed process of laser scanning and image capture. The top half of Andrea's body was digitised using a laser scanner that determined the precise distance to the surface of the subject. The surface of Andrea was scanned using a class 2 laser measuring the contours of the subject using time-domain reflectometry from the scanning laser beam. The sampling space defined the accuracy of the model, and the scan involved measuring many thousands of points on the surface of the body to build up a very detailed graphical model of Andrea. To get a full three-dimensional model, it was necessary that the scanner be rotated around the subject. This meant the subject had to stay still throughout the process. As well as laser scanning the face, several high-quality pictures were taken. These were used as textures, and applied to the model to give the correct colour and feel of her skin, hair, and clothes.

16.4.2.2 Lighting Conditions

Lighting conditions are critical when digitising someone with any scanning device, particularly if you want a high-definition model, as was required with Vandrea. This is particularly the case when scanning the face and head, where light is more easily reflected from the surface of the skin, and shiny hair. Any undesirable reflections from the face, hair, or skull will show-up as little spikes on the model. Scanning the eyes has exactly the opposite effect, as the eye's lens absorbs photons from the coherent beam of light. This has the equally undesirable effect of showing up as pits on the scanned model. It was for this reason that it was decided to laser scan Andrea

Catherwood in a Channel 5 television studio (shown in Fig 16.11), an environment well suited to the minimisation of reflections and shadows of presenters.

16.4.2.3 Vandrea Post-Production

Once Andrea had been scanned, the hard manual work of post-production took place. This contrasts with the automatic generation procedure of the AvatarBT booth — here the required level of accuracy needed to be higher. The various 3-D meshes of her head and shoulders had to be assembled into one model. This involved carefully smoothing joins together. Particular attention was focused on the mouth — due to people's natural critical misgiving if a mouth looks incorrect. It was also necessary to create something analogous to a muscle structure for the mouth. Therefore several nodes on the facial mesh were defined so that when they were moved concurrently, at different rates of torsion, a defined number of mouth articulations would occur.

Texture samples were then applied carefully to the wire mesh model. Conceptually it is rather like wrapping up a wicker model in Christmas wrapping paper, except that the wrapping paper happens to be the clothing, face, hair colour, etc, of Andrea. To make sure there were no creases in the final model, edge blending software algorithms were applied to the textures. The whole post-processing procedure was done in close association with Andrea Catherwood herself. Figure 16.12 shows a cut-away image of Vandrea showing the wire frame structure beneath.

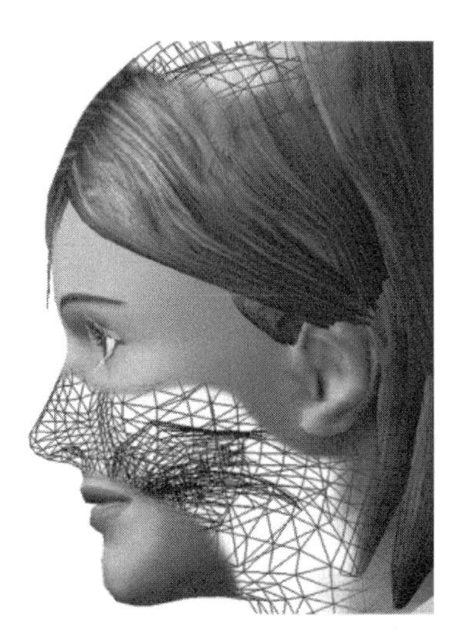

Fig 16.12 Composite image of Vandrea showing the wire mesh of the face.

16.4.3 Talking Vandrea

16.4.3.1 Vandrea's Synthetic Voice

If Vandrea was going to be credible as a newscaster she needed not to just look the part, but also sound it. Her computer-generated voice needed to sound warm, properly cadenced and smoothly synthesised. People would not tolerate listening to news with an electronic voice [17]. For that reason BTexact partnered with Lernout and Hauspie as the chosen text-to-speech (TTS) engine technology provider, using their RealSpeak^{TM} TTS engine. At the time of the work, this TTS engine generated synthetic speech that faithfully resembles human speech more accurately than any other [18]. Most TTS engines use a formant synthesiser that creates totally digitised or synthetic speech; no human recordings are used. Although the pitch and duration of words may be varied easily with these, it is perceived as sounding robotic with natural inclinations.

The RealSpeak^{TM} TTS is based on concatenation algorithms, where actual human voice units which are indistinguishable from one utterance to another — phonemes [19] — are stored and used to convert any text into speech. In-depth language-specific linguistic knowledge provides intelligent pronunciation of a wide range of variable input. The first phase of the TTS is to convert the text into a phonetic representation with markers for stress, accents and pauses. The second phase is a process of digital signal processing that maps the phonetic representation into the correct utterances that are to be spoken.

16.4.3.2 Vandrea's Lip Synchronisation

The phoneme markers that are extracted in the first phase of the TTS process are mapped to visemes [19]. A viseme is the mouth shape that is naturally required for a human to perform a particular utterance. These visemes drive the mouth articulation, and are in synchronisation with the voice, which uses exactly the same phonetic mapping.

16.4.3.3 Vandrea 2-D and Vandrea 3-D

Vandrea was available in two forms. Vandrea 3-D was a high-resolution three-dimensional desktop virtual reality solution, that ran on a high-end specification personal computer, with an Open-GL accelerated graphics card. It required a large install and was distributed for a trial on a CD-ROM. Vandrea 3-D takes up the full screen and is immersive, and engaging (see Fig 16.13). It is a second generation virtual human, suited not just for reading out the news on your PC, but for use in public locations on large screens such as train stations, airports, etc, for broadcasting news, advertising, and marketing. There was also Vandrea 2-D, a lightweight version that could run on low-end personal computers. This has a file size of about 4 to 5 MB and was made available for download from the Channel 5 Web site. Vandrea 2-D uses a less sophisticated text-to-speech engine, and was based on the Microsoft Agent Technology [20]. The Vandrea 3-D installation also automatically included the 2-D installation. When Vandrea 3-D started, a quick diagnostic test of your personal computer was performed. This checked that it was capable of running Vandrea 3-D; if this test failed, the user would be notified that their PC was not suitable and Vandrea 2-D would be activated instead.

Fig 16.13 Vandrea 3-D.

16.4.3.4 Creating the Illusion of Life

Vandrea needed to verge on being believable, believability [21] being a basic yardstick for virtual characters — an ill-defined concept used for evaluation purposes. It might be considered the degree to which a virtual character supports or undermines the user's overall sense of presence.

It is important to understand that this is not the same as naturalism, indeed naturalism and believability may conflict under some circumstances [22], for example, humans can believe in a virtual human flying, but it is not a natural thing for one to do.

For Vandrea to appear credible as a person she needed to have natural fluid movements of the body while reading the news out. Realistic but subtle shoulder, torso, and neck gestures were encoded as well as secondary facial behaviours such as blinking. Even when Vandrea was not talking she had idle gestures. All attempts were made to model these gesticulations faithfully on the mannerisms of Andrea Catherwood.

16.4.3.5 The Vandrea Service Model

Vandrea was a local application that resided on the desktop of a PC, and worked without the need for a Web browser. An ASCII file with up-to-the-hour news was delivered to the Channel 5 Web site, from ITN (the UK's Independent Television News service). When Vandrea was activated she would request the latest news script from the dedicated Channel 5 Web site to read out. This solution provided a narrowband delivery of news content, where the local virtual presenter does all the processing. The small text script was only one file that was a few kilobytes. Figure 16.14 shows the service model of Vandrea. This type of solution means that even with a slow modem connection you could have the news read out, without any delays, interruptions, or need for buffering.

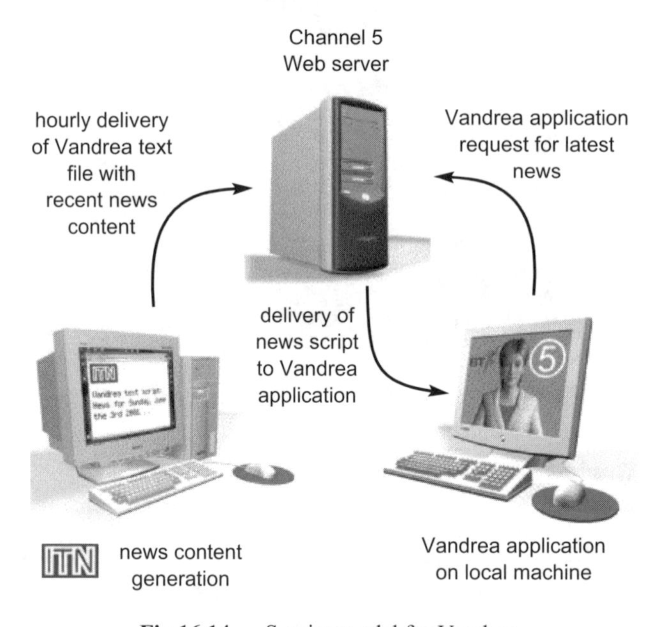

Fig 16.14 Service model for Vandrea.

16.4.4 Social Interfaces

Vandrea reflected the start of a growing trend to try to humanise the computer interface. Current graphical interfaces typically rely on the WIMP (windows, icon, mouse, pointer) model, which involves the user pointing and clicking on objects that have visual metaphors, such as files, folders, a desktop. In the case of the Web browser a simple back/forward metaphor is used. However, despite breaking down adoption barriers, WIMP interfaces have taken the power away from the user, and, as Sanders and Scholtz [23] point out, in some ways this is a backward step as 'the user is forced to do direct manipulation once per object'.

The future will be to a more intuitive natural interface using conversation characters; this will provide a more social interface that corresponds to our everyday face-to-face real-life interactions. When interfaces conform to social and natural rules, no instruction is necessary. It is now well understood that humans tend to anthropomorphise their computer [24], even with dull command line interfaces. It is thus not surprising that people build stronger relationships with life-like computer interfaces [25].

Social interfaces utilising conversational characters are already being field-tested and are proving to be very powerful [26] — BTexact Technologies has already received commercial interest in exploiting a conversational on-line bank interface that is in the development stage.

16.5 AvatarBT applications

Unlike Vandrea that was tied down to being a virtual newscaster, the personal avatars from the AvatarBT scanning process were specifically designed not to be application specific. These were individual files that could be loaded into any avatar application where the user wanted to be represented. To demonstrate this, several applications were developed for the AvatarBT, some of which are described below.

16.5.1 AvatarBT Entertainment Applications for the TalkZone

Once the individual avatar had been generated it was shown in entertaining linear animation shows within the TalkZone. The first of these linear animations allowed the visitors to see their avatar participating in activities that they would not have been capable of in the real world. In this case their avatar was seen taking part in a scene from the ET movie in which the alien causes the bike and its passenger to fly through the night sky (see Fig 16.15).

Fig 16.15 Visitors experience their avatar flying over the dome with ET.

The second animation showed all the avatars created in the TalkZone as part of a virtual audience entertainment show about the future applications of avatars (see Fig 16.16). The show ran every two minutes and the avatar audience was updated with new avatars as they were created.

Fig 16.16 People watch their avatar in a show.

16.5.2 Avatar Applications that could be Downloaded at Home

Unlike the other experiences in the Millennium Dome, the TalkZone provided the participant with an opportunity to continue their experience once they had left, by downloading their personal avatar and its related applications for their personal computer. The avatars were made up of around 2600 polygons and, with their associated texture files, had a total file size of between 150 kbytes and 450 kbytes. This means that they were easily distributed across the Internet and did not incur a tedious download time, even via a modem.

A 3-D avatar viewing application was provided (Fig 16.17), allowing users to examine their avatar. When the project was launched initially, two linear

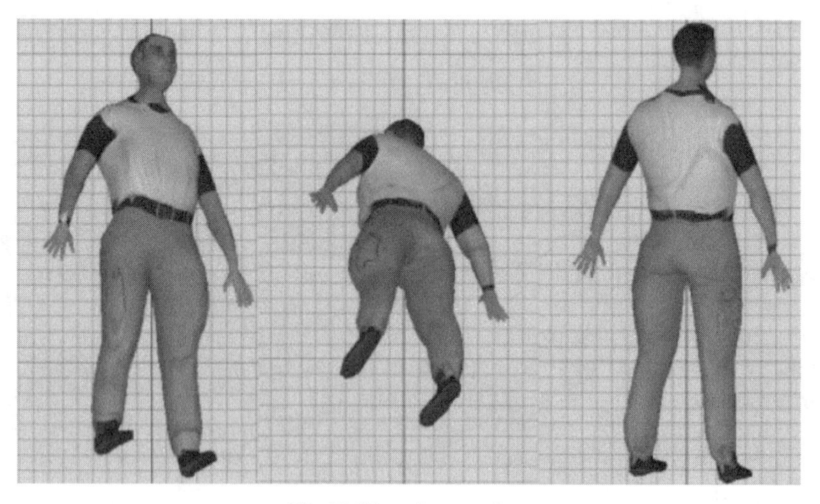

Fig 16.17 Avatar view.

entertainment animations for the avatars were available (since then more have been developed).

One involved their personal avatar performing amusing 1970s style disco dancing (Fig 16.18(a)), while the other showed their personal avatar skateboarding down a mountainside (Fig 16.18(b)).

(a) (b)

Fig 16.18 A 1970s dancing player (a), and a skateboard player (b).

16.5.3 Making the AvatarBT Personal Avatars Talk

The first iteration of the avatar scanning technology in the TalkZone produced avatars that did not have a functional mouth. Although a mouth was visible on the 3-D human models this was only a texture and the mouth was unable to be animated. A second generation of the scanning system in the TalkZone was introduced in September 2000 that was based on a new second generation of standard generic avatar (SGA2) (see Fig 16.19), which now included a simple but functional mouth. In order to achieve this the original SGA model used for creating all avatars required

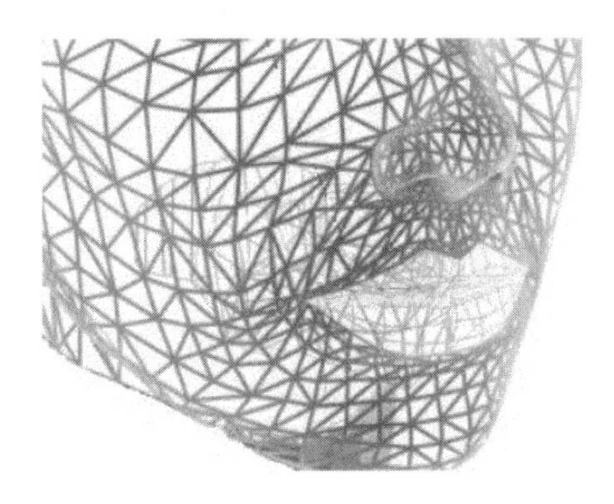

Fig 16.19 Generic avatar 2.

a re-distribution of polygons in the face to provide for the functional mouth and the addition of a new underlying node structure to which motion data could be applied to achieve mouth movement.

16.5.3.1 Avatar E-mail and Avatar Puppet

The first two AvatarBT speech-enabled applications were 'AvTalk' and 'AvPuppet' (see Fig 16.20). AvTalk is a free e-mail plug-in application for Microsoft Outlook that allows avatars to read their e-mail from the inbox using text-to-speech software, making use of technology similar to but less sophisticated than that of Vandrea. Default avatars are distributed with the application for those people who do not have their own avatar while those who do have an avatar are able to use it for reading e-mail. Further functionality allows friends, families and co-workers to collect one another's avatars and 'lock' each avatar to the associated e-mail address, so that e-mail arriving from a person is read out to you by their personal avatar.

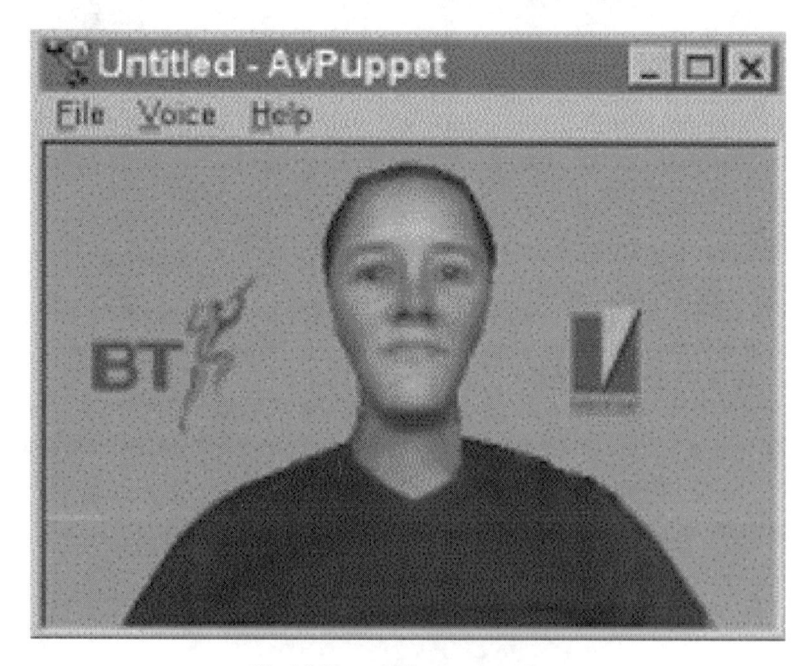

Fig 16.20 AvPuppet in action.

AvPuppet is bundled with the AvTalk download and allows people to use a microphone attached to their PC to speak through their avatar in real time. The avatar's lip movements are synchronised to the speech of the user so that the avatar becomes a digital puppet. Similarly sound files can be recorded and played back through the avatar.

Another feature of AvTalk, incorporated in order to alter the avatar's facial features and perceived attitude, is the inclusion of emoticons, which are now universally popular in both SMS and instant messenger applications. These are triggered in the avatar by means of the user typing markers in the text or changing the colour of the text. Currently the emotional range is limited to happy typed in yellow or a happy symbol ☺ or :-), sad can be typed in blue, or a sad symbol ☹ or :-(, and angry typed in red or by an angry symbol :-0 .

16.5.3.2 Personal Avatars for Computer Games

As pointed out in Chapter 15, new technologies are generating new entertainment experiences. The Playstation 2 is supposedly capable of drawing up to 66 million polygons per second — games certainly have come a long way since the days of Pong or Asteroids. 'Sony's Playstation 2 and Microsoft's X-box both move the entire graphics pipeline off the increasingly powerful central processor ...' [27], freeing computational resources for non-graphic processing activities. The computer games industry is highly competitive, but, now that graphics realism has almost reached the pinnacle, the industry is looking for new ways to improve games [28]. Including personal avatars in computer games adds an extra dimension that the games industry, with an ever-widening audience, will be quick to pick up on, as Datamonitor [29] points out ' ... if a new theme is found to be popular, it is turned into a brand ...'. With the games industry being valued at over US$12 bn [30], it is a key market for this new avatar technology. As section 16.6.2 points out, people's perception of personal avatars is closely related to games technology. Personal avatars open up a whole new arena of opportunity — you could put yourself in the hot seat as a demon driver in Gran Turismo, or shred courses on your mountain bike in ESPN Extreme Games. 'By putting your own physical self into a game, to fight aliens and save the planet, you'll experience a greater sense of involvement — and it's much safer than doing the real thing! It's a very exciting time for the gaming industry ...' [31].

Since the launch of the AvatarBT project, the most popular application, and most discussed by the press, has been the use of the personal avatars in computer games — the concept being that instead of playing a pre-supplied character such as Sonic the Hedgehog, a player's personal avatar allows gamers to play as themselves. Two games were initially chosen as they had an open development platform, and allowed the development team to easily and automatically convert the personal avatars from their proprietary format, created in the TalkZone, into the correct format to become a computer 'game skin'.

The first game chosen was Quake III (see Fig 16.21) — this was one of the most popular first person games in the UK at the time. Quake III allows multiple players to fight one another as individuals or teams across a network using a range of weaponry in a range of 'arenas'. By adding your own personal avatar into the game

Fig 16.21 Your personal avatar in Quake III.

it allowed team dynamics to improve, as players could easily recognise opponents and team members.

The second game developed to incorporate BT's personal avatars was 'The Sims' (see Fig 16.22).

This game, unlike Quake, depends on the players co-operating with each other; their characters, called 'Sims', act out their daily lives including participating in a polite conversation. It is a classic example from the God Game genera [27], in that these characters have significant autonomy to satisfy their goals, but God (the player) comes in and stirs things up by managing the individual and changing its environment, thus creating an interactive 'soap opera' style game. Now armed with their own personal avatars, families can play out this game just like living in reality.

It is worth pointing out at this juncture that 'The Sims' is a good example of designing non-verbal vocalisations into virtual humans [32]. Although the

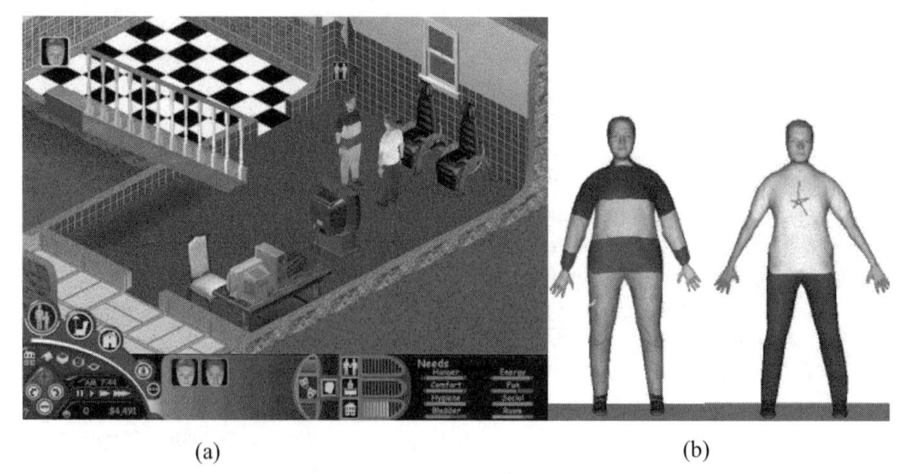

(a) (b)

Fig 16.22 Avatar Sims (a), Avatars converted into Sims (b).

characters talk only through thought-bubbles and gestures, it is quite easy to distinguish between polite conversation and angry exchanges by the frequency of non-verbal communication.

16.5.4 Creation of the AvatarBT Applications — Motion Capture

The disco dancer, and skate boarder were created using motion capture. Motion capture implies that a real actor generates motion data by recording the positions or angles of the joints; these can then be applied to virtual actors. Two main formats of motion capture are available — magnetic motion capture where receivers detect changes in the magnetic field, or optical motion capture which was used for the AvatarBT applications, where cameras detect reflective markers, that have been placed on an actor, to determine the three-dimensional position of the joints. As pointed out earlier, the SGA conform to a standard called H-Anim, which evolved from a need to define interchangeable humanoids in the Web-based Virtual Reality Modelling Language — VRML2 [33]. The personal avatars also conform to this standard. Figure 16.23 shows the joint skeletal structure. It consists of fifteen fundamental joints — the left and right ankle, knee, hip, wrist, elbow, shoulder, as well as the neck-base, skull-base, and pelvis.

When a personal avatar is loaded into one of the avatar entertainment routines, such as the disco dancer, the motion capture data extracted from the actor drives the joint structure of the avatar. This movement also drives the avatars in the other applications that BTexact Technologies has developed, such as Avatar Conferencing, which is similar to the BT Forum meeting place [34], where a 3-D environment offers a much more powerful visual reference to what is going on in a conference call, except in this case the meeting is not with default avatars but personal avatars.

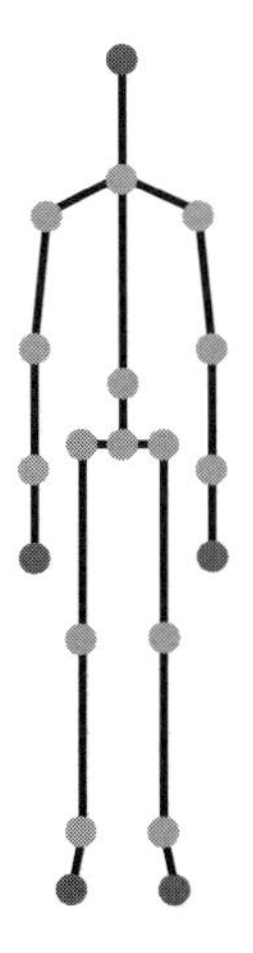

Fig 16.23 H-Anim skeletar structure.

16.6 User Evaluation of AvatarBT

16.6.1 Evaluation Overview

Two thorough evaluations of AvatarBT were performed; the first, prior to the AvatarBT system being launched in the TalkZone, was at SIGGRAPH '99 in Los Angeles, where an avatar booth was set up on a dedicated stand. The other was undertaken at the AvatarBT exhibit in the TalkZone at the Millennium Dome, where many thousands of visitors had their personal avatar created. What follows are summarised evaluations from SIGGRAPH '99and the TalkZone — comprehensive results can be found in Lawson [35].

16.6.1.1 SIGGRAPH '99 Evaluation

The annual conference for the ACM Special Interest Group in Graphics (SIGGRAPH) in the USA is the world's premier forum and focus of the computer graphics industry. The purpose of this evaluation of the AvatarBT was to expose the technology to the industry and the public, gather useful data on the likely reaction of the public to the exhibit in the Dome, and identify some of the problems that might need to be overcome. This occurred in the middle of the booth system development cycle and it consequently allowed time for adjustments to be made accordingly.

Data was gathered through a survey of 189 subjects after people had experienced the attraction — 73% of these subjects were male and 27% female. The breakdown of subjects by profession is shown in Fig 16.24 and indicates that most were technical or creative people — like most of the subjects they will have come from a computer graphics community and therefore have a predisposition to avatar technology.

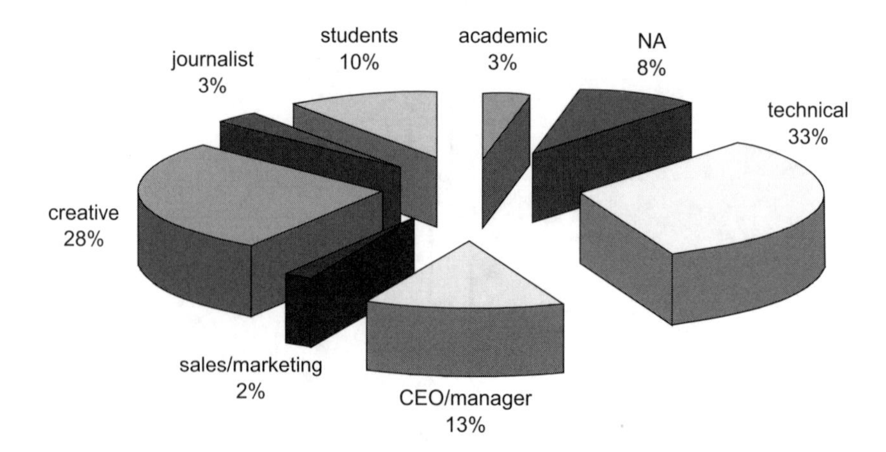

Fig 16.24 Breakdown of users by profession.

16.6.1.2 TalkZone Evaluation

At the time the TalkZone study was done the exhibit had been in operation for seven months and 155 000 people had been scanned. After the press coverage and advertising, the system had settled down into a routine. The survey used at SIGGRAPH '99(see Fig 16.25(a)) had been revised for use in the TalkZone (see Fig 16.25 (b)). The survey was carried out on 104 TalkZone visitors during June 2000 and was done immediately after they had been scanned and seen their avatar in the ET animation player.

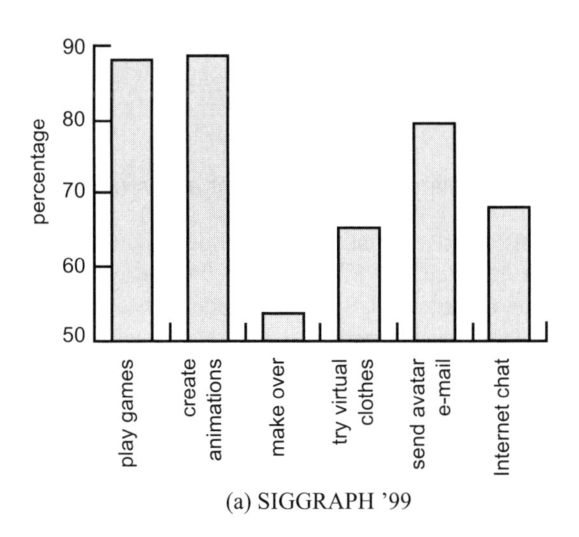

(a) SIGGRAPH '99

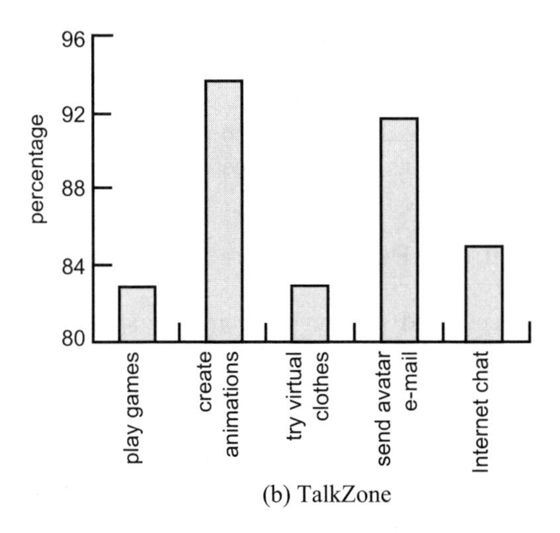

(b) TalkZone

Fig 16.25 Survey of applications.

16.6.2 Results

16.6.2.1 Personal Avatar Applications that Appeal to People

At both SIGGRAPH '99 and the TalkZone, users were asked what applications appealed to them for use with their avatar. The results were surprising in that all applications seem to be popular. In both surveys the most popular answer was that people wanted to create animations of themselves, with 88% of people at SIGGRAPH '99 asking for this and 93% at the TalkZone. At SIGGRAPH '99, 87% of people said they would like to use it to play computer games, whereas in TalkZone only 83% of people wanted this. This might be to do with the fact that the TalkZone was exposed to people from many backgrounds while SIGGRAPH has a large representation from the gaming community.

16.6.2.2 How Much Would People Pay for a Personal Avatar

Price is often the differentiating factor that will determine whether a new piece of technology will be adopted. At SIGGRAPH '99 the average cost people were willing to pay for an avatar was US$15, the mode was between US$5 and US$10. The people in the TalkZone typically fell into two categories, 30% willing to pay between £4 and £5, and 18% willing to pay between £1 and £2 (see Fig 16.26(a) and (b)).

16.6.2.3 What People Thought About their Avatar Quality

When people were asked about the quality of the avatars, this typically referred to the number of blemishes it had. At SIGGRAPH '99 it was quite evident that there was some work to do to improve the quality, with 16% of people being dissatisfied. It was refreshing to see that only 12% were unhappy with their avatar at the TalkZone. More satisfying was that a higher proportion of people at the TalkZone thought their personal avatar was very good (see Fig 16.27).

16.7 The Future of Personalised Virtual Humans

This chapter has concentrated on the creation of graphically authentic virtual humans. This technology is already being refined and commercialised. However, as Ballin [36] points out, believable virtual humans (and virtual characters in general) are the summation of two key components — visual realism and behaviour. Therefore it should come as no surprise that current research by the authors is now focusing equally on extracting the demeanour of the subjects that want personal avatars, as well as on making the visual features of the avatar more faithful. This will make sure that attributes such as the subject's gait, body language, and

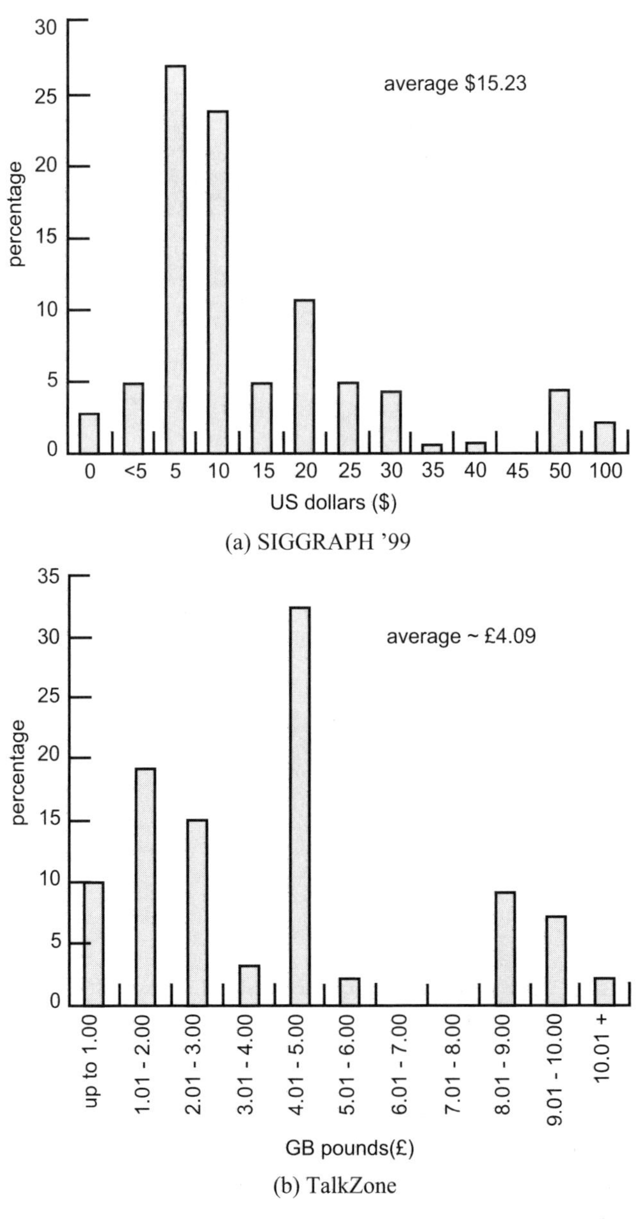

(a) SIGGRAPH '99

(b) TalkZone

Fig 16.26 How much would you pay?

individual mannerisms are captured and expressed in their respective personal avatar.

Personal avatars will not just appear as service differentiators, but become intelligent virtual human interfaces that are part of the everyday fabric of computer

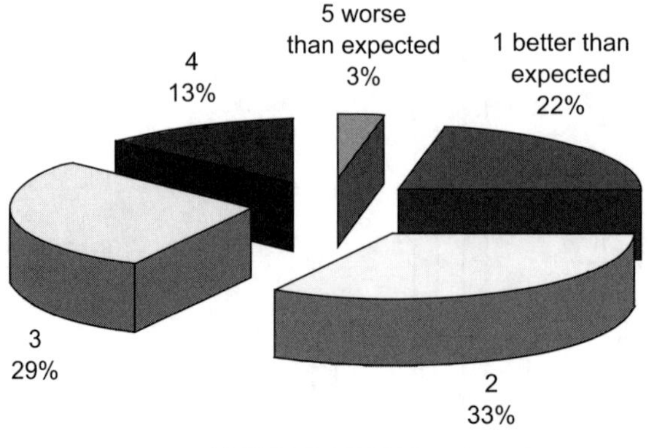

(a) SIGGRAPH '99

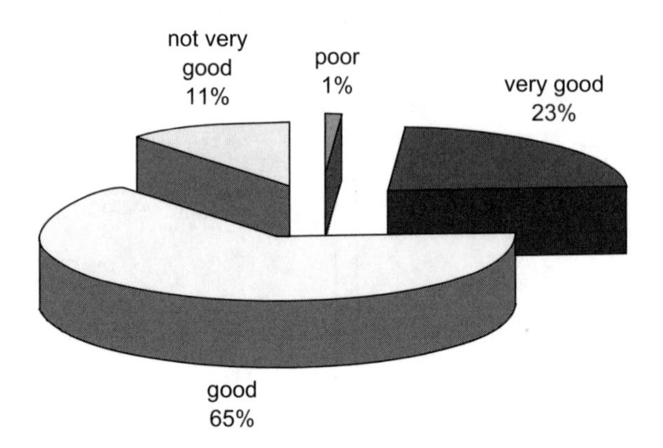

(b) TalkZone

Fig 16.27 Avatar quality.

systems [37, 38]. This will not just be in personal computers, but on all digital convergent devices, including personal digital assistants, mobile telephones, games consoles, and set-top boxes. Customer relationship management is a growth industry and has never been more active [39]. Humans are social animals by their nature, and with the new generation of computer users seeing an electronic channel as their principal interface with a business, industry is looking towards avatar technology to help. These do not have to be full-bodied characters — Fig 16.28 shows a celebrity 3-D talking head that was created by BTexact Technologies using one of their new tools. These on-line personalities are not only important for communication but also for branding, e.g. people immediately associate the virtual marketing character of Connie with AOL.

The other key growth area will be personal avatar experiences like the TalkZone. Eateries, hotels, theme parks, and cinemas are in a strong position to take advantage

Fig 16.28 Celebrity avatars.

of personal avatars. Some sectors will want applications, others booths, others would like booths for short-term use, in support of which BTexact Technologies has developed, in conjunction with its partners, a mobile avatar booth that can be flat packed and hired out. Another experience is that the entertainment industry has recently applied virtual reality to motion and interactive rides; this is an emerging market for life-like characters [36, 40], including personal avatars. Known in the trade as location-based entertainment [41, 42], it refers to motion rides such as 'The Volcano Mind Ride' and large multiplayer video games that are bound together due to size or cost of the equipment. Gaming has already been shown to be popular with the AvatarBT personal avatars, and, with on-line gaming currently being one of the key drivers for broadband, personal avatars will give it an edge. It is technologically not inconceivable that people could also carry their personal avatar around with them on a smartcard that could be swiped in machines and PCs, such as an arcade game.

Two real-world examples of personal avatars that have been successfully deployed have been the focus of this chapter. The case studies highlight the two main fields where we shall see personal avatars, one being 'interfaces' and the other being 'experiences'. Predicting the future is a dangerous game, and can leave people with egg on their face. However, the authors of this chapter have every confidence, from their market evaluation and preliminary sales, that the personal avatar will become a commercially significant ubiquitous technology, making for an exciting future.

References

1 Stephenson, N.: '*Snow Crash*', Bantam Books (1992).

2 Morningstar, C. and Farmer, F. R.: '*The Lessons of Lucasfilm's Habitat*', in Benedikt, M. (Ed): '*Cyberspace: First Steps*', MIT Press, Cambridge, Massachusetts, pp 273-302 (1991).

3 Chat — http://communities.microsoft.com/home/mschat.asp

4 Smith, M., Farnham, S. and Drucker, S.: '*The Social Life of Small Graphical Chats*', in ACM SIG CHI (2000).

5 Blaxxun Interactive — http://www.blaxxun.com

6 Activeworlds — http://www.activeworlds.com

7 Talk Zone — http://www.talkzone.co.uk

8 Ballin, D., Lawson, M., Crampton, S., Child, T. and Hilton, A.: '*Practical, reliable, repeatable: scanning avatars in the Millennium Dome*', Virtual Human Modelling, Scanning, Paris (2001).

9 AvatarBT — http://innovate.bt.com/avatar/

10 Hilton, A., Beresford, D., Gentils, D., Smith, R. J., Sun, W. and Illingworth, J.: '*Whole-body modelling of people from multi-view images to populate virtual worlds*', Visual Computer, International Journal of Computer Graphics, **16**(7), pp 411-436 (2000).

11 Dekker, L., Khan, S., West, E., Buxton, B. and Treleaven, P.: '*Models for understanding the 3D human body form*', IEEE Workshop on Model-based 3D Image Analysis (1998).

12 Humanoid Animation Working Group — http://www.h-anim.org

13 Hilton, A., Beresford, D., Gentils, T., Smith, R. and Sun, W.: '*Virtual people: capturing human models to populate virtual worlds*', IEEE International Conference on Computer Animation, pp 174-185 (2001).

14 TombRaider — http://www.tombraider.com

15 Televirtual — http://www.televirtual.co.uk

16 Carter, R.: '*Mapping the Mind*', Weidenfeld and Nicolson (1998).

17 Cahn, J. E.: '*The generation of affect in synthesized speech*', Journal of the American Voice I/O Society, **8**, pp 1-19 (July 1990).

18 '*L&H RealSpeak*^{TM} *— The New Standard of Text-to-Speech*', BT Forum Whitepaper.

19 Jones, C. M. and Dlay, S. S.: '*The Face of the Future: An Animated Virtual Agent with Life-like Articulation*', in Ballin, D. (Ed): '*Intelligent Virtual Agents*', Proc of IVA'99 (1999).

20 Microsoft Agent — http://www.microsoft.com/msagent/

21 Bates, J.: '*The role of emotion in believable agents*', Communications of the ACM, **37**(7), pp 122-125 (1994).

22 Aylett, R. S. and Cavazza, M.: '*Intelligent virtual environments — a state-of-the-art Report*', Eurographics, Manchester, pp 87-109 (2001).

23 Sanders, G. A. and Scholtz, J.: '*Measure and evaluation of embodied conversational characters*', in Cassell, J. et al (Eds): '*Embodied conversational agents*', MIT Press, Cambridge, Massachusetts (2000).

24 Reeves, B. and Nass, C.: '*The media equation: how people treat computers, television and new media like real people and places*', Cambridge University Press, Cambridge, UK (1996).

25 Thorisson, K.: '*Communicative humanoids, a computational model of psychological dialogue skills*', PhD dissertation, Massachusetts Institute of Technology (1996).

26 Cassell, J. et al: '*Embodied Conversational Agents*', MIT Press, Cambridge, Massachusetts (2000).

27 Laird, J. E. and van Lent, M. A.: '*Human-level AI's killer application interactive computer games*', AI Magazine, American Association of Artificial Intelligence (2001).

28 Cassell, J. and Jenkins, H. (Eds): '*From Barbie to Mortal Kombat: Gender and Computer Games*', MIT Press (1998).

29 Digital Gaming Whitepaper: '*The new face of gaming*', Datamonitor (August 2000).

30 '*New media technology report*', Guardian Unlimited (July 2001).

31 Davies, J.: '*Game On*', Insight Interactive (2001) — http://www.bt.com/insight-interactive/

32 Allbeck, J. M. and Badler, N. I.: '*Consistent communication with control*', in Pelachand, C. and Poggi, I. (Eds): '*NI Workshop on Multimodal Communication and Context in Embodied Agents*', Fifth International Conference on Autonomous Agents, ACM Press (2001).

33 VRML Consortium — http://www.vrml.org/

34 Jewell, M.: '*BT Forum Whitepaper*', (November 1999).

35 Lawson, M.: '*Entering Cyberspace at the Millennium Dome*', University College of London, MSc Thesis (2000).

36 Ballin, D., Aylett, R. and Delgado, C.: '*Towards the development of life-like autonomous characters for interactive media*', BCS Conference on Intelligent Agents for Mobile and Virtual Media, National Museum of Film and Photography (April 2001).

37 Thalmann, D.: '*The foundations to build a Virtual Human Society*', in Aylett, R., de Antonio, A. and Ballin, D. (Eds): '*Intelligent Virtual Agents*', Springer (2001).

38 Aylett, R., de Antonio, A. and Ballin, D. (Eds): '*Intelligent virtual agents*', Proc of Third Workshop on Intelligent Virtual Agents, Springer Lecture Notes in Science (2001).

39 Millard, N. J.: '*Best practice 2001: 21st century customer contact*', Internal BT report (May 2001).

40 '*Time for Virtual Teletubbies: The development of Interactive and Autonomous Children's Television Characters*', Workshop on Interactive Robotics and Entertainment (WIRE2000), Carnegie-Mellon University, pp 109-116 (April 2000).

41 Kerlow, I. V.: '*The Art of 3-D Computer Animation and Imaging*', John Wiley (2000).

42 Bates, J.: '*Virtual reality, art and entertainment*', Presence: The Journal of Tele-operators and Virtual Environments, **1**(1), pp 133-138 (1992).

17

HYPE AND REALITY IN THE FUTURE HOME

D Patel and I D Pearson

17.1 Introduction

We have been bombarded with hype for decades on how the future home will look. Every few years, industry giants and think tanks build 'homes of the future' to showcase their contributions to the development of the technology, hoping they can cash in on future market opportunities or demonstrate their strategic thinking to competitors.

Some of the hype is based on real products and services that people will actually want once they become sufficiently cheap, easy to use and reliable. But much of it is just hype and will never happen. We cannot really be sure about what everyday people will want to use, but we certainly can spot a lot of the things that people will never want.

This chapter aims to separate hype from reality and convey a picture of a future home that is technologically plausible and fits in with what we know about human nature, i.e. ideas that might actually appear in the market-place.

The chapter is split into two parts. The first takes a brief look at the main driving forces behind past and present. We suggest a few reasons why some developments have not taken off or why they may fail in the future. We will then outline our view of how things might need to be in the future home, beyond current technology-oriented thinking, highlighting that the future home will have to integrate technology in a human-centred, affective manner. The second part (section 17.11 onwards) documents a spectrum of emergent technologies and the services they will enable.

17.2 Engineering Fashions — Are You Being Served?

Technologists and futurologists mostly tout similar visions of the future home. Since these predictions feed off one another, they increasingly look and feel the same. The uniform vision at the moment seems to be that the future home will be intelligent, connected and wireless. These visions change in line with popular social and technological fashions. Therefore, if society at large wants to do away with going into an office everyday, our homes will adapt so that we can work from home. If the engineers have figured out how to make things wireless, then everything will be wireless. It is easy to see where our ideas came from because much of our thinking is incremental rather than radically new. The danger of course is that such simple extrapolations often miss major effects from disruptive forces. Technological trends certainly have some momentum, but lateral forces can create rapid changes of direction.

The visions of the 1960s were stimulated by the rapid technological advances accompanying the space race. Every family with a newly acquired car, television and vacuum cleaner were in awe of technology and what it could do. This was the era of the 'labour-saving' device that would revolutionise every homemaker's life. Visions were of electromechanical advances.

In the early 1990s, BT's vision included a software butler, with the appearance and at least some of the functionality of a human servant without any of the social guilt. People would interact with them in everyday language and they would carry out their tasks flawlessly. Such a butler could find and sort all the information we needed, automate boring administration tasks and even make sure that the bath was ready just how we like it, in time for when we arrived home. Some elements of this have been achieved. AskJeeves.com provides answers to everyday questions, and home networking and control kit is available to the dedicated enthusiast. Honda have developed walking robots that they hope will soon do the 'fetch and carry jobs' around the home. And, of course, BTexact Technologies now has a range of avatar technology that will bring natural language interfaces to a talking head display rather like Holly in Red Dwarf. The vision was accurate in part, but was only just beginning to get the network message.

More recently, as the Web developed, engineering focus moved back to appliances, but this time they were on-line and connected. Smart or Internet appliances, allowing remote monitoring and control applications, have often been predicted to be the killer application for home networks. A smart fridge could automatically order fresh milk if the existing supply has passed its best-before date. There have been more useless gadgets invented in the last few years than in any period of human history — 'just because it's buildable doesn't mean it will sell'.

Today the emphasis is not just on connecting devices, but making them connect cordlessly. Short-range radio technologies such as Bluetooth aim to eliminate wires with the 'unplug and play' mantra. Mobility has captured much of the attention for at least a while, until the next big thing appears. Portable used to mean that it was physically possible for a single person to pick something up, but now of course it usually means that it is not very heavy.

It is clear that engineering is very much driven by the fashion of the moment, and occasionally fashion is stimulated by the latest technology. The entire industry changes direction to pursue the latest gizmo. In doing so, we sometimes lose sight of the fact that the gizmo needs to be used by people — so while features multiply, the devices get harder to use and market success becomes more difficult. These distractions have diverted engineering effort away from making things easier to use, and have thus delayed the realisation of the dream. Evidence for the fashion need comes by way of the fact that some shops exist whose only business is to sell covers for mobile telephones. Keeping up with the latest ring tones also generates significant business.

We suspect that we might just be starting to learn. As the market is swamped by products that no-one buys, the general emphasis on good design and ease of use is increasing. People are becoming more and more tired of the plethora of bells and whistles in software that does its basic job very badly. Producers will either get the message soon or die. A refocusing on simplicity and reliability is long overdue.

17.3 Agents

The main approach for the creation of our future Jeeves has been in the field of artificial intelligence (AI). Each of us would have an army of agents, some of which may be roaming free in cyberspace, looking after our interests. Even though many live systems use them today, they are much less sophisticated than we envisaged ten years ago. Agents on the whole have underachieved, and are typically just specialised rule-based algorithms, optimised to achieve one particular limited task — they have no sophisticated intelligence yet. They still lack the ability to use generalised commonsense intelligence that could make them truly useful. Babies acquire a lot of common sense in their first few years from direct experience and from their parents. That knowledge is not generally available in cyberspace yet, though there are ongoing attempts to transcribe common sense in rule lists, which currently run to over ten million rules! But this cannot be the right approach. Other approaches, that allow robots to learn how the world works for themselves via sensors, show much more promise.

Neural networks aim to achieve machine intelligence via a biologically inspired approach, building lots of little artificial brains. Connecting these brains and agents is a fresh challenge — the semantic Web proposed by Tim Berners-Lee hopes to provide an environment in which distributed intelligence and agents can thrive. In a roundabout way, having taken a diversion with the Internet, we are gradually heading towards what telcos used to call the global superhighway, but without the bandwidth.

Eventually, agents will undoubtedly become a fundamental part of middleware, the fabric of a truly intelligent future home, and as indispensable as Jeeves ever was. But do not hold your breath. It will certainly be at least a decade or more before AI becomes as useful as that which we are used to seeing in science fiction, but AI will make a big contribution to home life in the future (Fig 17.1). There will also be a few robots and also some more ‘free-floating’ intelligence, while sensors in the surroundings will form the basis of the interface to humans and other devices.

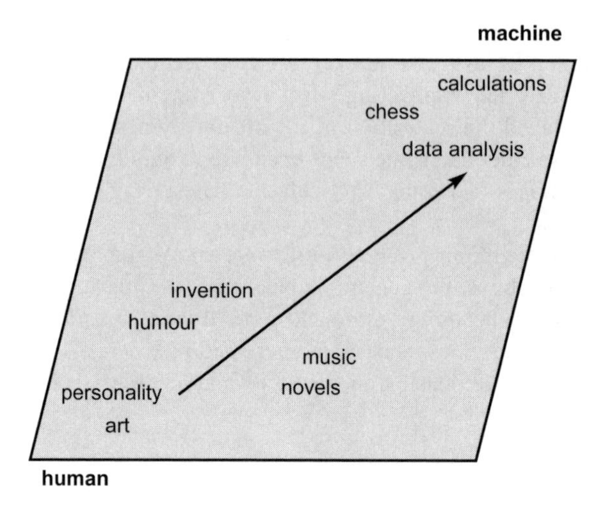

Fig 17.1 AI in the home.

17.4 Does Smart Imitate Life?

Just because a TCP/IP stack can be integrated into almost anything today does not automatically mean that it should. Some of the ongoing convergence and integration is warranted and useful. It often makes sense to integrate communication functionality and computing functionality together in one 'communicator' package; but other efforts are decidedly counter-productive.

A large electronics manufacturer is now adding the ability to browse the Web to all of its camcorders. The camcorder connects wirelessly via a Bluetooth link, which can be used to upload pictures to a computer or connect to the net (via a Bluetooth-enabled modem). But we must ask why anyone would really want to browse the Web on a three-inch LCD and whether it is worth compromising ease of use to make room for the facility. What percentage of the Web sites today can render properly on such a small display? By integrating lots of diverse functionality into one package, we add complexity and worsen the interface. The resultant Frankenstein devices are like electronic Swiss army knives — they can do a lot of jobs, but none well.

Other functions that are added may be used for purposes totally different from those for which they were designed. Free calls have been tried as a market ploy, but some users have exploited the facility to act as a baby alarm. Sony's night-shot facility on some of their camcorders was discovered to have some voyeuristic application since it appeared to see partially through clothing.

Apple's near-term solution is their 'digital hub' approach, which places the personal computer at the centre of the range of our camcorders, personal organisers and music players. It acknowledges that there are many things that cannot really be done properly on anything other than a computer, but rejects functionality that does not genuinely add to usefulness. They use the computer's power both to automate, as much as possible, what we need to do, and to keep the menus down to areas where we really need to have a choice. They add the value through good design, and understand that we do not want choice, we want it to do what we want, when we want it! By this approach of keeping it simple and making it attractive, the home computer will quickly move from the study to the living room, where it will be accompanied by other devices that do things well and do them simply, like TVs and light switches. Although I am quite happy to have the management of the TV volume on a remote control with an on-screen display, I dread the day when the light switch runs Windows and a TCP/IP stack, and insists I download an upgrade to the atmosphere program, from the Internet, before it will switch on, because I know it will probably crash and need to be rebooted first. I will go out and buy candles!

Some strategies for smart appliances do not involve humans directly at all. Some smart printers are already available that can be set up to order new toner or request servicing automatically over the Internet without any user intervention. This could certainly make life easier, but we need to have confidence in the software before we let it spend our money! Unless the default for such functions is that they are switched off, people will object to them in case their bank account is spent on a lorry load of new ink cartridges.

17.5 Is Incrementalism the Enemy of Innovation?

Many scenarios take an incremental evolutionary view, based on continuous innovation, described by established metrics such as Moore's Law. It is easy to predict that things will be smaller, cheaper, more powerful, thinner, lighter and generally better. However, we rarely hear about disruptive technologies, discontinuous innovations, things that seem to come out of nowhere and radically change established thinking — for more information on this area, see Cosier and Hughes [1]. The MP3 music format is a good example of a disrupter, and challenges the traditional record companies' business models. However, sometimes we do not even know something is disruptive until afterwards. The interactions that cause the disruption are not always obvious.

The real power of disruptive technologies lies not in the technology itself, but in the social, political and economic drivers. These put the technology in context and give it the 'disruptive' kick to really change the way things are done, often irreversibly. Also, incremental innovations tend to occur within a single discipline, whereas disruptive innovation often occurs at the boundaries of many disciplines. The differences between these kinds of innovation highlight flaws with the metaphors proposed for the future home and provide us with clues as to why they often fail.

17.6 Bits AND Atoms, Not Bits OR Atoms

One reason why we have not yet reached our utopian future home is that instead of taking a broad, multi-disciplinary approach, developers often take narrow, single-discipline, mutually exclusive paths.

In the beginning, the world was analogue and atom centred. There were no digital bits to play with. Innovation in the home centred around new and easier ways of manipulating matter, hence the term 'labour-saving device'. The precursors to today's white goods arrived — twin-tub washing machines, food processors and vacuum cleaners. (Even the word 'machine' is really a mechanical word — we still do not have proper words for electronic devices.) Visions of an automated home paradise flourished around mechanical servants that would cook, clean and do your bidding, and they would look like people. Then the IT revolution heralded the arrival of bits into the home. As chips could be added to almost any appliance, you could now have your toast exactly as brown as you like it. The new millennium brings us the age of 'devices everywhere' with the home playing host to a digital hub where all our gadgets can connect in some way to one another. We could soon be offered toasters that can burn our e-mail into our toast!

Looking back, it appears that innovation concentrates on the area of current engineering fashion — atoms this decade, bits the next. With micro-electromechanical systems (MEMS) coming quickly on to the scene, fashion is swinging back to atoms and electromechanical innovation again. This seesaw approach sees atoms and bits too often treated independently of one another. But for Jeeves to be as useful as the human version, he must have the ability to manipulate atoms and interact in our analogue world of atoms, as well as be smart in the digital world. This third sector is often overlooked but is just as important as the other two. Surely the next decade of research and development will take a more holistic view. We need to look at what happens when bits and atoms meet, intertwine and intermingle with human minds.

17.7 Well Beings

Just because something is technologically possible, does not mean that people will want to do it or that they will want it. A lot of technology fails because it ignores social, economic and market factors. Since people are the purpose of every home, we need to look at approaches that make their lives easier rather than bombard them with new technology. If tomorrow's technology is anything like today's, then it is likely to make people more miserable.

We need to look towards a future where atoms, bits and minds interact pleasantly and productively (Fig 17.2). Today we have a very difficult relationship with our machines. One of the reasons is obvious. The lack of integration between our devices means we have several remote controls in our living rooms already — there

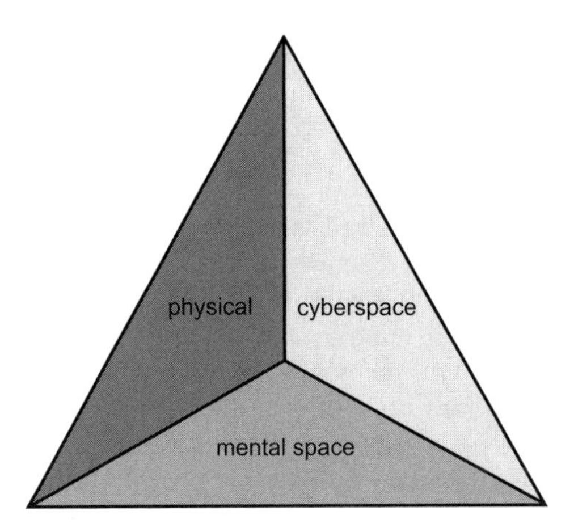

Fig 17.2 Our three-way world.

is a limit to the number of remote controls we can cope with. We will need less rather than more, and they need to be simpler. One of the solutions will be to reduce the number of human interfaces. When there is a high quality of interaction between systems, the human interface disappears from the primary focus. Much like the interface between a talented artist and the pencil, the focus is on the act of creative expression instead. We need to be able to treat the home as a single system, with one interface, or to get as close to that ideal as possible. The era of the extreme interface is here; but extreme should not mean unnatural.

17.8 'Psyence' — Fact or Fiction?

Chips with a given level of capability continue to shrink in size, while those of a fixed size continue to grow in functionality and power. Progress in the materials technology for chip manufacture recently enabled storage and processing to be included on the same substrate. MEMS can also now be integrated with memory and processing capability. With just a little more progress, we can see that energy storage or collection can soon be integrated with telecommunications, storage, processing, sensing and activation capability, enabling a bewildering range of potential devices, many of which can be very small. As devices drop towards and through the millimetre level, they quickly disappear from our conscious visual image of our environment. The technology becomes invisible. This will lead to a new class of technology known as 'digital dust'. Some of the mechanical tricks that were last used decades ago may well re-surface during the next few years, but on a very small scale.

Today's home has lots of technology in view, often very conspicuous. Our future homes could include a vast array of invisible background technology as well as the conspicuous black boxes that we have come to expect. It is hard to say which of the range of applications will be successful, but the enormity of the range of potential applications means that at least some of them will succeed and our homes will become at least a little smarter.

The future home will thus have a psyche of its own. It will be infested by activators and software entities as well as black boxes, hanging around from today's era. These will interact and self-organise in the main, with a few island and archipelago systems dotted in this cyber sea. Even the space itself will have digital co-ordinates. Digital air will bring a range of useful capabilities to our homes. Hopefully, if we get it right, the home's psyche will be complementary to the personalities and sensibilities of those who dwell within it, in tune with our own thinking. If we get it wrong, it will feel as if it is possessed by a poltergeist, but more supercomputer than supernatural. We must get it right — Big Brother will not be popular!

Even if things did work really well and just like magic, it could be extremely disorienting for the human. Mankind has enjoyed the feeling of control for a long time. Anything that challenges that position may be met with resistance.

Human beings are visually oriented, hands-on and very curious by nature. We may want to keep tangible interfaces to the real world, and to retain at least the impression that we know what is going on. Instead of things getting thinner, faster, better, etc, which seems inevitable, there could be occasional human pressure to go in the other direction. Humans like to have at least the option to interact with external things via a tangible interface.

Some people may not trust a non-human intelligence that is trying to act as a human, especially those people who have ever used a PC today. They may not be afraid of a 'please fasten your seatbelt' chime or a 'your headlights are on' buzzer, because we can understand how it works and can cope happily when it does not. But when the technology is intelligent enough to make decisions that influence many areas of our lives, such as relationships, creativity, holidays and so on, there may be more rejection. We are not used to accepting technology in these areas and might not welcome it. We will certainly want to know what it is doing. We accept engine management systems optimising our fuel consumption and traction control stopping us from skidding. How will we react to engine management preventing us from exceeding speed limits?

17.9 Kitchen Rage

We can expect some novel problems around the home too. Anyone using computers is well accustomed to the frustration of trying to get software to do what it says on the box. My experience is that there is usually a problem. In the kitchen, when people are wrestling with time-scales and hungry kids, network failures, software crashes and incompatibility problems will be much more irritating and we can expect regular occurrences of kitchen rage, as people finally lose patience with systems that are not co-operating. Designers will have to make kitchen IT much more robust and easier to use than anything that currently exists in the computer industry, if it is to be successful in the market-place. The same is true to a lesser extent elsewhere in the home too.

17.10 Voices in your Head

Hearing a voice just hanging in the air can be extremely disorienting, unless there is a clear context. If the future home has voices coming out of nowhere, then it will feel spooky. When you hear real humans speak, you can hear them breathing or clearing their throat, or feel their presence before they speak. Voice interfaces will probably need a cuddly interface or an avatar to accompany them, at least some of the time, so that some form of simple relationship can be established. The more natural the interface is, or at least appears, the more likely it is to be widely accepted. Human nature has taken millennia to evolve, and will not change very quickly.

17.11 So What Will Work?

If we take into account these many design principles, and human limitations, we can identify those technologies for the future home which we believe will be successful in the market-place. This section outlines some of the key elements.

17.11.1 Displays

The future home will include various types of display. While a television set today often takes up a large part of our living room, we now expect it to move imminently on to a wall, and blend in as a picture, a fish tank, or window when it is not acting as a TV. If the technology for flat screens continues to improve at current rates, this seems to fall very firmly at the credible end of the projections. But it will not be the only display. There is simply too little real-estate on a single display to cope with all the demands made by a family.

We will have some personal displays too, perhaps based on lightweight goggles. Otherwise, we would all have to compete for the use of the large wall display for all our different services. We may use goggles for playing games, watching movies, shopping, or meeting friends in virtual environments.

Other applications can be better provided for by using video tablets. Tablet-based computing is being pushed heavily by companies such as Microsoft. The format is very appropriate for coffee table use, for applications such as Web browsing, shopping or electronic magazines, and will fit very comfortably into people's life-styles. Other displays are already built into PDAs, computers and communicators.

These will improve, but do not in themselves precipitate any major change to the home.

There are a number of challenges that have not yet been fully solved in the scope of managing and controlling displays. Network protocols already exist today that allow us to separate the display client from the server.

There may well be new kinds of display technology that seem radical today. Displays that could be painted on walls have been suggested. They may use bi-stable materials so that very low power consumption can be achieved. Display pixels in the paint will self-organise to make a display. If this can be successfully developed into a reasonable cost solution, it would appeal to many home owners, and could greatly change the way we decorate our rooms to achieve a highly adaptable ambience. Displays could be anywhere.

Some displays might have a single function, acting as an electronic painting or fish tank or even a star map in a kid's bedroom. Others need to be high-definition and high-refresh rate. The future of the interior designer is going to be increasingly a high-tech job!

Displays can also act as windows to somewhere else. The images could change size and pan in real time according to the position of the user. If you come closer, then the object grows. This sort of facility would work well with in-room positioning and digital air functions.

Displays in children's bedrooms would have a wide range of uses. They could be the standard outlet to watch the Barbie doll soap operas that are happening in the neighbourhood.

17.11.2 Digital Mirror

With the pervasiveness of cheaper high-resolution cameras and future display technology, objects such as the common mirror could take on a new role. The mirror could show what a lady could look like after she has applied the make-up. When she selects the desired image, she can then apply her make-up 'by numbers'.

The mirror could also show her how she would look wearing different outfits from her wardrobe (Fig 17.3). She would even be able to look at snapshots of herself in the past when she used to be able to fit into her jeans!

17.11.3 Robotics

A range of robots already exist for the home, including an assortment of pets, vacuum cleaners and lawn mowers. None are proving irresistible to the average home owner. While electronic pets appeal to a few gadget freaks today, it will be quite some years before they are sophisticated enough to appeal to the mass market. We must remember, though, that we do not employ real pets for their functionality, but because of the sentience, character and even affection that they bring. Successful

Fig 17.3 Digital bathroom mirror.

robot pets will be able to read e-mail, but will also need to have some of the warm characteristics of real pets (Fig 17.4).

Eventually, electronic pets could be an important component in the home human/machine interface, accepting instructions and delivering information in everyday natural language, and will have sophisticated synthetic personalities, but not for a few years. Meanwhile, robotic cleaners and mowers suffer from the more fundamental problems of high prices and low battery life, and still are not intelligent enough to be truly useful, except in niche environments.

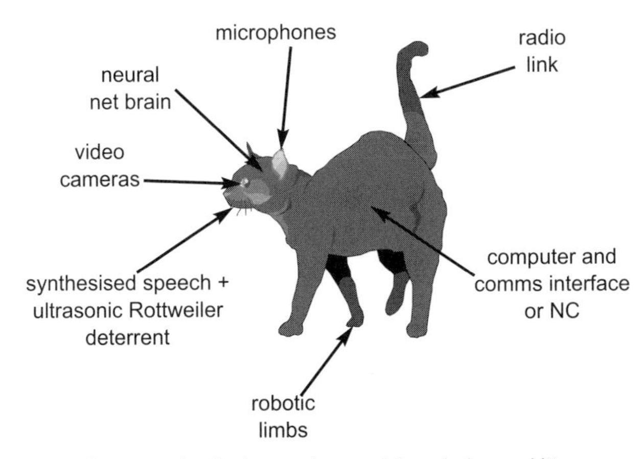

Fig 17.4 Future robot pet interface.

Proposed robots that will kill slugs, pick up bits of paper, or collect the crockery after a meal seem many years away from feasibility as far as everyday home use is concerned. In the short and medium term, their only market will be in corporate demonstrations of homes of the future, just like the Internet fridge and smart waste bin. One day, mobile robotics will be part of home life, but, for today, most of us are content with fixed automata such as dishwashers and washing machines.

Robot pets will have a useful role in caring for people who may need to be monitored constantly in case of falls or illness. The robot might seek out its owner from time to time, and thus could detect if there was a problem and raise an alarm. It could potentially be fitted with a range of sensors that would allow it to convey useful data to a hospital or medical program too, and even monitor some aspects of health while it is being stroked.

17.11.4 Interactive TV

Interactive TV has already appeared in many homes, but it has a long way to go to achieve its full potential of letting people participate in programmes in sophisticated ways (Fig 17.5). As TV and computing capabilities gradually converge, games technology will allow people to interact and compete with other viewers, and be able to explore programme environments more fully. If dinosaurs have been recreated for a mini-series, it should be possible eventually to download their features into a processor to allow viewers to interact with them in full 3-D, rather than just watch them wander around under the director's guidance. The embryonic signs of such interactivity are already with us. With the rapid growth of computer capability, we should not have too long to wait.

This is, of course, good news for education and entertainment. When things are already digital (like the characters in *Toy Story*), making endless variations will be easy. People will be able to train and grow their models and send them back into the

Fig 17.5 Interactive TV.

TV world so that everyone else can see them. Indeed, some Web sites already sell virtual objects and spaces that people can populate and decorate at will.

17.11.5 Tags and Activators

An important part of the chips-everywhere mindset is the simple identifier tag. All this does is to identify an object to a scanner. Identifying a tag on an object allows some networked functionality or data (e.g. price and name) to be associated with that object. While a barcode is read by a laser scanner, many tags are now based on printed inductive loops, where their resonant frequency identifies them from a distance — but they are moving on very quickly from being just simple identifiers. A whole class of capabilities is emerging and we should think instead of enabling activators. When we add an activator to an object, we add intelligence to it and allow it to be recognised, since we can now link it to a record in a database, a Web page, or a logic function, and offer useful services and facilities. We add the capability to communicate with the object, or at least the activator, so that we can track and manage the objects around us. The tag itself is best thought of as the tip of a very large cyberspace iceberg (Fig 17.6).

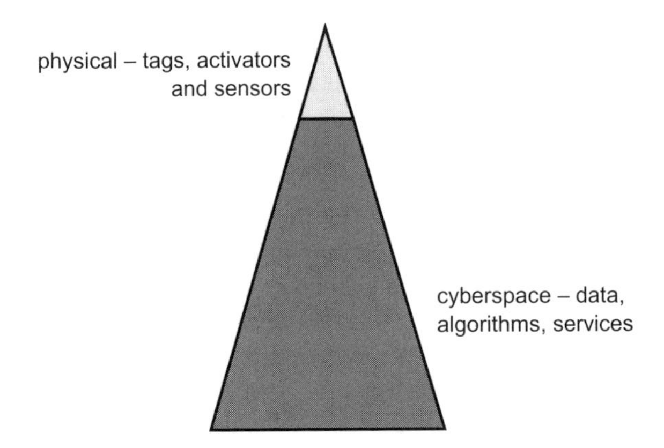

Fig 17.6 Physical size has little relationship to functionality.

This is a natural development of the post-atomic digital confetti and digital dust ideas. We will have to be careful in the design to account for whether these things can be ingested by humans or not. It might even be the case that some activators are partially organic, and so light that they could be breathed in with the air, or ingested on our food, so design will not be trivial. But, once we can mark specks of dust in some way to effectively give them computing power, the possibilities are endless. The mechanics of the architecture and how they network are major areas that will have to be designed. This is a new conceptual field and not much thought has been given to it yet. Most sensor-networking assumptions so far have been based on TCP/IP, but there might be a much better way of solving the problems they face. Maybe answers will come from a more biological approach — this approach is already being used in the field of computer viruses. We are just starting a new research area to look at the whole area of information technology based on ultra-simple approaches.

An example of the potential will be familiar to all of us who grew up with brothers and sisters, or have children of our own. Imagine being able to stick an activator on everything you own, and then putting a scanner on your bedroom door. If a sibling removes an object, the time and culprit can easily be recorded, together with a track of where it went and where it currently resides. More advanced tags can deactivate the object if it is outside its allocated zone. These activators will only cost a few pence each at most, and some could even be recycled from future cornflakes boxes. Objects could be named by the owner and a child could locate it simply by calling for it, even though it may be buried under a pile of laundry in the corner. Simple functionality additions such as these can add genuine quality of life at little cost to children who need to police their property. Even for parents, being able to locate something instantly will often save hours of fruitless searching.

The identifier tags are just one species of activator. Some will provide a positioning field, others will act as sentries. A few might act as bugs! Some will provide storage capability, some will provide help files and instruction manuals. Network activators will set up symbiotic communications networks around the home (Fig 17.7). Processors will provide a distributed processing capability to allow object management to reach all corners of a room. Others will gather sensory data, making sure that the environment remains comfortable and safe, while monitoring any abuse of an object, such as dropping it on hard surfaces. Such data might be needed for insurance claims. Parents could use such devices to enforce audio levels from a hi-fi. The range of potential applications is vast, but we can be almost certain that most will be invented by ordinary customers, who will see uses that we engineers have never imagined. We can be fairly sure that this technology will be ubiquitous, but not precisely what it will be used for.

Our home will thus become aware of its contents. The electronic pet will know what is in the home and where it is, and can save us many hours of searching when we forget where we put something. It might even help us to put our socks in pairs.

17.12 Networks

It is certain that the future home will need an internal network to link the many appliances together. There is already a growing market in this area and lots of hype about future prospects. The Bluetooth concept is now well known, but the chips are taking much longer than projected to come down to the cost where they can become ubiquitous. We do not yet have fully automated homes with every appliance linked

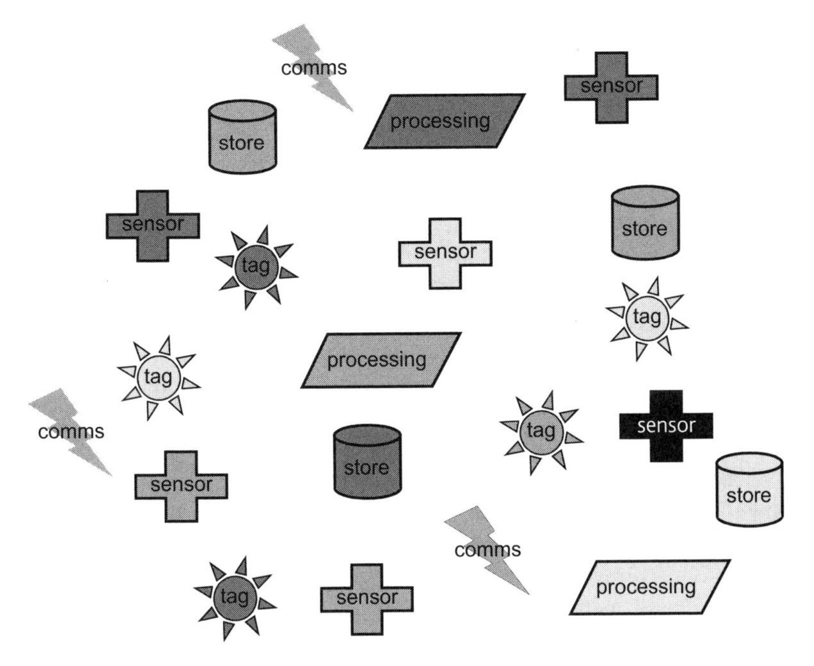

Fig 17.7 Activator-rich home environment.

into the network via Bluetooth. Now of course, we also have strong competition in the same concept space from wireless LAN technologies such as IEEE802.11b — and a, and now g! It will probably take much longer for any of these to take off when the market-place is so fiercely contended. We will also have problems when we upgrade appliances, if we are locked into a dying standard. BT's own passive picocell technology had promised to bring radio into the home on the end of a fibre, that would allow us to extend the network right through to each appliance, but this particular dream seems to be hovering on the edge, awaiting the arrival of the market. At some point it could be a disrupter. Passive picocell was invented several years ago within BTexact Technologies. An optical signal on the fibre is converted to radio at the end. Some of the optical signal is reflected back to the exchange and is modulated by the incoming radio signal to provide a return path.

Any of these technologies could dominate the future home. As far as individual device manufacturers are concerned, cost, data rates, range and power consumption will all influence the decision over compatibility. However, software radio offers the future capability to access any of these networks, since the waveforms and protocols can be synthesised and executed largely in the software domain before finally putting a signal on to an antenna. A device can listen to what networks are available and then subscribe to one of them.

Another key network capability is in setting up symbiotic (*ad hoc* or parasitic) networks. Some devices, such as Cybiko, already search for the presence of compatible terminals and automatically set up communications with those that they find. It is expected that this technology will develop rapidly in the next few years to allow people's communicators to set up symbiotic networks wherever there is a reasonable density of terminals. That means that most built-up areas, shopping areas and transport termini will have such networks.

Radio networks could be linked to a future fibre-to-the-home network by means of passive picocell, but it looks like DSL variants will be the mid-term solution for main network connection for most homes (ADSL in at least the very short term). In-house networking will almost certainly be radio based, probably an IEEE802.11 variant, though Bluetooth will probably take some of the market too.

Sadly, scepticism about potential markets has proved to be self-fulfilling. If we do not believe there is a market for high bandwidths, and therefore do not provide them, we can guarantee that the services that need them will not appear, the lack of demand will then be considered proven, and the sceptics will claim they have made the correct decisions. Such market scepticism has held back our industry badly in the last decade, resulting in the Web becoming a poor shadow of its original vision. A DSL broadband network will provide the bandwidth needed for many applications, but some services will always work better on higher bandwidths, so we will see an ongoing trade-off between market risk and network capability.

Instead of one home network that does everything, there may be many. There could be an overlapping hierarchy of networks, based on functionality, needs, and security.

Since we have generally been bandwidth constrained, much engineering emphasis to date has been on saving bandwidth and sweating the bandwidth we already have. Most applications in the future home will have a very variable bandwidth demand, and this lies strongly in the network's favour, since bandwidths can be adjusted according to instantaneous need and availability. We believe strongly that services should be configured to be as transparent as possible to the device at which they are aimed. Services will be available in different ways, depending on the need, the type of terminal and the available bandwidth. But when we are accessing them on high-capability networks and terminals, we should be able to grab the bandwidth that is needed to give them the appropriate quality.

17.13 Storage-Based Networks

Storage-based networks have been described in some depth elsewhere [2]. Summarising the key principle, networks would make use of large quantities of local storage distributed around the environment to minimise the amount of data that needs to be transmitted, and the distance over which it must be transmitted. A home store would collect information from various network and hard-storage-based sources on a continuous basis, using adaptive personal filters to ensure that most of the information a resident would want would probably be stored locally, and therefore network access in real time would seldom be needed (Fig 17.8). Even when residents leave their home, they would rarely need to download data from mobile networks either, since most of the information they need for the day ahead would be uploaded on to their portable.

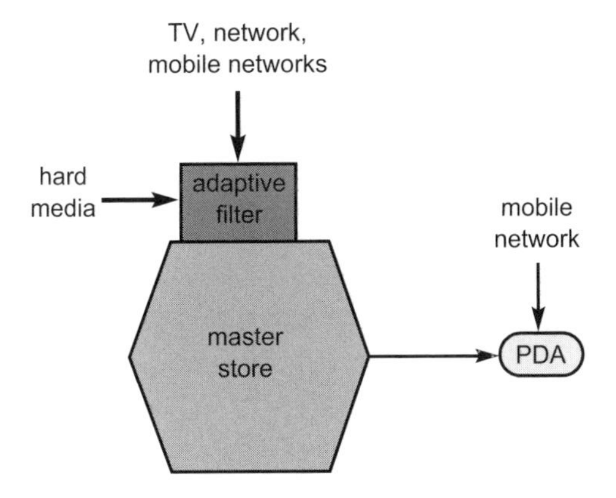

Fig 17.8 Storage-based networking in the home.

Storage-based networks will only work at their best if we have high capacity in the local area for transmission. We need to be able to pipe several channels of DVD-quality video around the home, otherwise we would need a multiplicity of stores.

17.14 Summary

Only products that take good account of basic human needs and nature will be successful in the market-place. Many existing visions of the future home are based too much on technology capability and take too little account of human nature and everyday life-styles, and therefore are doomed to failure. Many of the products hyped for the future home will never achieve significant market success. However, another flaw in many visions is that they assume incremental and evolutionary progress in technology, whereas history has shown repeatedly that markets are changed mostly by disruptive technologies. These cannot be seen so easily by simply applying extrapolation devices such as Moore's Law. They require more lateral thinking, and therefore such insights tend to happen by accident rather than by planning.

Our future homes will be smarter than today's, and much of the intelligence will come from almost invisible systems. Myriads of almost invisible activators will infest our homes and make self-organised communications networks between themselves to enhance many areas of our lives. But we must be careful to develop control systems that allow humans to easily control what is going on, or they will not be accepted. Machines will be welcome in our homes, but only if they know their place.

References

1 Cosier, G. and Hughes, P. M. (Eds): '*Disruption*', Special issue, BT Technol J, **19**(4) (October 2001).

2 Pearson, I. D.: '*What's next*', BT Technol J, **19**(4), pp 98-106 (October 2001).

ACRONYMNS

3G	third generation (wireless)
3GPP	3G Partnership Program
ACK	acknowledgement
ADSL	asymmetric digital subscriber line
AI	artificial intelligence
ANSI	American National Standards Institute
AO	always-on
AP	access point
API	application programming interface
APON	ATM passive optical network
ARP	address resolution protocol
ASP	application service providers
ATM	asynchronous transfer mode
ATSC	Advanced Television Systems Committee
ATVEF	Advanced Television Enhancement Forum
BABT	British Approvals Board for Telecommunications
BAL	Broadband Applications Laboratory
BHPS	British Household Panel Survey
BIOS	basic input output system
BMRB	British Market Research Bureau
BSI	British Standards Institute
BSS	basic service set
BWA	broadband wireless access
B-WLL	broadband wireless local loop

CA	conditional access
CAPI	computer-assisted personal interviewing
CAR	committed access rate
Cat3	Category 3 (wiring)
Cat5	Category 5 (wiring)
CATI	computer-assisted telephone interviewing
CBR	constant bit rate
CC/PP	composite capabilities/preference profile
CCTV	closed circuit TV
CCU	customer connection unit
CDMA	code division multiple access
CDR	call detail record
CEM	customer experience management
CES	consumer electronics show
CI	communication interface
CORBA	Common Object Request Broker Architecture
CPCM	content protection and copy management
CPE	customer premises equipment
CPU	central processing unit
CRC	cyclic redundancy check
CRID	content reference identifier
CRM	customer relationship management
CSCW	computer supported co-operative work
CSMA/CA	carrier sense multiple access/collision avoidance
CSMA/CD	carrier sense multiple access/collision detection
CSP	content service provider
CSS	cascading style sheet
CTS	clear to send
DAVIC	Digital Audio-Visual Council
DBPSK	differential binary phase shift keying

DCOM	Distributed Component Object Model
DECT	Digital Enhanced Cordless Telecommunications
DES	data encryption standard
DHCP	dynamic host configuration protocol
DIX	DEC/Intel/Xerox
DMZ	demilitarised zone
DNG	delivery network gateway
DNS	domain name server
DOCSIS	Data over Cable Service Interface Specification
DOM	document object model
DQPSK	differential quadrature phase shift keying
DRM	digital rights management
DSL	digital subscriber line
DSLAM	digital subscriber line access multiplexer
DSM-CC	digital storage media — command and control
DSSS	direct sequence spread spectrum
DTD	document type definition
DTE	data terminal equipment
DTV	digital TV
DUS	device-unifying service
DVB	digital video broadcast(ing)
DVB-CP	DVB copy protection
DVB-IPI	DVB Internet protocol infrastructure
DVB-RCS	DVB return channel via satellite
DVB-WIN	DVB wireless in-home network
DVD	digital versatile disk
EAS	electronic article surveillance
EB1	Enhanced Broadcasting
ECMA	European Computer Manufacturers' Association
ECML	electronic commerce modelling language

EMC	electromagnetic compatibility
EMOTE	expressive motion engine
ENUM	enumerator
EPD	early packet discard
EPG	electronic programme guide
ESS	extended service set
ETSI	European Telecommunications Standards Institute
EURESCOM	European Institute for Research and Strategic Studies in Telecommunications
FACS	facial action coding system
FAI	fast-track applications infrastructure
FDM	frequency division multiplexing
FDQAM	frequency-diversity quadrature amplitude modulation
FEC	forward error correction
FFFT	forward fast Fourier transform
FHSS	frequency hopping spread spectrum
FITL	fibre in the local loop
FOMA	freedom of multimedia access
FPD	functional processing and decoding
FSAN	Full Service Access Network
FS-VDSL	full service VDSL
ftp	file transfer protocol
GAP	general access profile
G.DMT	G-series discrete multi-tone (ITU-T)
GFSK	Gaussian frequency shift keying
GHA	Glasgow Housing Association
GO	gateway operator
GRE	generic routing encapsulation
GUI	graphic user interface
HACM	hierarchical adaptive curve matching
H-Anim	Humanoid Animation Working Group

HAVi	Home Audio Video Interoperability
HDD	hard disk drive
HDTV	high definition TV
HFC	hybrid fibre coax
HiperLAN	high performance local area network
HMD	head mounted display
HNCD	home network connecting device
HNED	home network end device
HNS	home network segment
Home API	home application programming interface
HomeRF	home radio frequency
H-PNA	Home Phoneline Networking Alliance
HRFWG	HomeRF working group
HTML	hypertext markup language
HTTP	hypertext transfer protocol
HVAC	heating, ventilation and air conditioning
IA1	Internet Access
IB1	Interactive Broadcasting
ICS	Internet connection sharing
ICT	information and communications technology
IDC	insulation displacement connection
IDD	international direct dial
iDVD	interactive DVD
IEC	International Electrotechnical Commission
IEEE	Institute of Electrical and Electronics Engineers
IETF	Internet Engineering Task Force
IFFT	inverse fast Fourier transform
IM	instant messaging
IPC	international private circuit
IPI	Internet protocol infrastructure

IPsec	Internet protocol security
IPX	internetwork packet exchange
IrDA	Infra-red Data Association
ISDN	integrated services digital network
ISER	Institute for Social and Economic Research
ISM	industrial, scientific and medical
ISO	International Standards Organisation
ISP	Internet service provider
IST	Information Society Technologies (EU project)
IT	information technology
ITU-T	International Telecommunication Union — Telecom Sector
IXP	Internet exchange point
jar	Java archive
JIT	just in time
JPEG	Joint Photographic Experts Group
L2TP	layer 2 tunnelling protocol
LAN	local area network
LAPQ	limited automatic repeat request
LBS	location-based service
LCD	liquid crystal display
LDAP	lightweight directory access protocol
LEP	light emitting polymer
LLC	logical link control
LLU	local loop unbundling
LMA	Laban Movement Analysis
LMDS	local multipoint distribution system
MAC	media access control
MEMS	micro-electromechanical system
MHP	Multimedia Home Platform
MIS	management information system

MIT	Massachusetts Institute of Technology
MoU	memorandum of understanding
MP3	MPEG layer 3 (audio)
MPEG	Moving Picture Experts Group
MSDW	Morgan Stanley Dean Witter
NAT	network address translation
NAV	network allocation vector
NetBEUI	network BIOS extended user interface
NIC	network interface card
NTE	network terminating equipment
OCAP	OpenCable Application Platform
OFDM	orthogonal FDM
OLT	optical line termination
ONT	optical network termination
ONU	optical network unit
OS	operating system
OSGi	Open Services Gateway Initiative
OSI	open system interconnection
OSS	operational support systems
OTG	On-The-Go
P2P	peer-to-peer
PAN	personal area network
PBX	private branch exchange
PC	personal computer
PCMCIA	Personal Computer Memory Card International Association
PDA	personal digital assistant
PDR	personal digital recorder
PEP	performance enhancing proxy
PHY	physical layer (device)
PNG	portable network graphics

PON	passive optical network
POTS	plain old telephone service
PPD	partial packet discard
PPoE	point-to-point over Ethernet
PPTP	point-to-point tunnelling protocol
PSTN	public switched telephone network
PVC	permanent virtual circuit
PVR	personal video recorder
QMUL	Queen Mary University of London
QoS	quality of service
QPSK	quadrature phase shift keying
RAS	remote access server
RFI	radio frequency interference
RFID	radio frequency identification
RG	residential gateway
ROBO	robust orthogonal (FDM)
ROI	return on investment
RTP	real time protocol
RTS	request to send
SDSL	symmetric DSL
SG	services gateway
SGA	standard generic avatar
SGA2	standard generic avatar (second generation)
SIG	special interest group
SIGGRAPH	Special Interest Group in Graphics
SIM	subscriber identity module
SIN	Supplier's Information Note
SLM	Salutation manager
SME	small to medium enterprises
SMS	short message service

SNG	satellite news gathering
SOAP	simple object access protocol
SOHO	small office home office
SP	service provider
SPX	sequenced packet exchange
SSL	secure sockets layer
STB	set-top box
STG	Software-To-Go
SUNA	scenario-based user needs analysis
SVC	switched virtual circuit
SW	software
SWAP	shared wireless access protocol
TCP/IP	transmission control protocol/Internet protocol
TDMA	time division multiple access
ToS	type of service
TTS	text-to-speech
TVAF	TV-Anytime Forum
TVRO	TV receive-only
UBR	unspecified bit rate
UDDI	Universal Description, Discovery and Integration
UDP	user datagram protocol
UMTS	Universal Mobile Telecommunications System
UPnP	Universal Plug and Play
URL	uniform resource locator
USB	universal serial bus
UTP	unshielded twisted pair
VDC	volts direct current
VDSL	very-high-bit-rate DSL
VIRTUE	Virtual Team User Environment
VoD	video-on-demand

VoDSL	voice over digital subscriber line
VoHPNA	voice over HPNA
VoIP	voice over IP
VPN	virtual private network
VR	virtual reality
VRML	Virtual Reality Modelling Language
VSAT	very small aperture terminal
VToA	voice and telephone over ATM
VTP	VDSL termination processing
VTPD	VTP and decoding
VTU-R	VDSL transceiver unit remote terminal
W3C	World Wide Web Consortium
WAP	wireless application protocol
WCDMA	wideband CDMA
WDM	wavelength division multiplexing
WEP	Wired Equivalent Privacy
WIM	wireless identity module
WIMP	windows, icon, mouse, pointer
WLL	wireless local loop
WMF	Wireless Multimedia Forum
WRAP	Web-ready appliances protocol
WSDL	Web services description language
WWW	World Wide Web
XML	extensible markup language

INDEX